POLLEN

BIOTECHNOLOGY

POLLEN

BIOTECHNOLOGY

GENE EXPRESSION AND

ALLERGEN CHARACTERIZATION

EDITED BY

SHYAM S. MOHAPATRA
Department of Immunology
University of Manitoba, Winnipeg, Canada

AND

R. BRUCE KNOX
Department of Botany
University of Melbourne, Melbourne, Australia

CHAPMAN & HALL

New York • Albany • Bonn • Boston • Cincinnati • Detroit • London • Madrid • Melbourne
Mexico City • Pacific Grove • Paris • San Francisco • Singapore • Tokyo • Toronto • Washington

For more information, contact:

Chapman & Hall
115 Fifth Avenue
New York, NY 10003

Chapman & Hall
2-6 Boundary Row
London SE1 8HN
England

Thomas Nelson Australia
102 Dodds Street
South Melbourne, 3205
Victoria, Austrailia

Chapman & Hall GmbH
Postfach 100 263
D-69442 Weinheim
Germany

Nelson Canada
1120 Birchmount Road
Scarborough, Ontario
Canada M1K 5G4

International Thomson Publishing Asia
221 Henderson Road #05-10
Henderson Building
Singapore 0315

International Thomson Editores
Campos Eliseos 385, Piso 7
Col. Polanco
11560 Mexico D. F.
Mexico

International Thomson Publishing - Japan
Hirakawacho-cho Kyowa Building, 3F
1-2-1 Hirakawacho-cho
Chiyoda-ku, 102 Tokyo
Japan

1 2 3 4 5 6 7 8 9 10 XXX 01 00 99 98 97 96 95

Library of Congress Cataloging-in-Publication Data

Pollen biotechnology : gene expression & allergen characterization /
    editors, Shyam S. Mohapatra & R. Bruce Knox.
        p.   cm.
    Includes bibliographical references and index.
    ISBN 0-412-03521-9
    1. Allergens--Biotechnology.   I. Mohapatra, Shyam S., 1955-
II. Knox, R. Bruce, 1938-
TP248. 65.A39P65   1995   94-42327
616.2'02--dc20   CIP

British Library Cataloguing in Publication Data available

To order for this or any other Chapman & Hall book, please contact **International Thomson Publishing, 7625 Empire Drive, Florence, KY 41042.** Phone: (606) 525-6600 or 1-800-842-3636. Fax: (606) 525-7778. e-mail: order@chaphall.com.

For a complete listing of Chapman & Hall's titles, send your request to
Chapman & Hall, Dept. BC, 115 Fifth Avenue, New York, NY 10003.

Contents

Preface

The last decade has seen tremendous progress in our knowledge of the pollen development and gene expression on one hand and the characterization of pollen specific proteins on the other. In compiling the chapters for this volume, we have pragmatically categorized these basic developments in pollen molecular biology and biotechnology into two sections based on their applications in agriculture and implications in medicine.

Pollen developmental biology and gene expression: applications in agriculture. Pollen development is an extremely complex process encompassing a series of biochemical, physiological and genetic events. At the basic level, sporophytically expressed genes may expound our knowledge of unique processes of cellular differentiation which ultimately give rise to a full-fledged organism. At the applied level, the studies on the pollen and male sporophyte-specific gene expression, and of promoters and transcription factors of relevant genes have the potential to manipulate the fertility in certain cash crops leading to agricultural biotechnology.

The first chapter provides a comprehensive review of biology of male gametophyte development in plants, from the perspective of whole plant, specifically in Barley. The next three chapters present the latest advances on the anther and pollen specific gene expression in model plant systems such as, maize representing monocot, and in sunflower, brassica, and arabidopsis, representing typically the dicot plants. Arabidopsis is particularly important since it has been dubbed as "drosophila" of the plant kingdom in terms of the developmental biology. The last chapter presents an elegant review of the recent advances in *in vitro* studies relating to pollen cultures and its implications for emerging pollen and agricultural biotechnology.

Cloning, expression and structure-activity relationships of pollen allergen proteins: implications in biomedicine. A number of pollen proteins, referred to as allergens, are specifically important in biomedicine. Allergens induce the culprit IgE antibodies, which are responsible for the development of allergies and asthma. The last five years have seen spectacular advances in the field of allergen characterization. The application of recombinant DNA techniques to the characterization of allergens has recently led to rapid advances in (a) molecular cloning of several major allergens, (b) determination of their primary structures, (c) delineation of the B- and T-cell epitopes of some of these allergens, (d) characterization of cross-reactivities that exist among aeroallergens, and (e) determination of biological functions of these proteins. To-date about thirty proteins that constitute the major allergens of diverse plant pollens have been cloned, this number has tripled since 1992 and 3 to 5 new allergenic proteins are being added to this list every year. In addition to the cloned allergens, a number of other allergens have been reported, presumably the cloning of these allergens is currently in progress. The recombinant allergens have been synthesized mostly by *Escherichia coli* but also by yeast and baculovirus systems.

Recombinant allergens are expected to be useful for the allergen injection immunotherapy (AIIT), which has been the "holy grail" of clinical allergology since its introduction in early 1900s. AIIT has been practiced to date as the one and the only immunologically specific therapy against allergic rhinitis and asthma. AIIT involves repeated injections of the appropriate allergens over prolonged periods. It is to be noted that AIIT as currently practiced, in addition to being cumbersome, expensive and associated with the risk of fatal side reactions including systemic anaphylaxis, leads often to no improvement of the patient's condition. This lack of predictable efficacy is due to the fact that the available commercial preparations consist of heterogeneous (crude) extracts of the corresponding pollen(s) and may contain the relevant allergens in only minute and irreproducible amounts leading up to 100x variation in allergenic potency. Also the extracts often contain constituents unrelated to the few allergens, to which the patient is actually allergic; clearly the former components may lead to the induction of IgE Abs with new specificities (3). The main drawbacks that presently limit the efficacy and use of AIIT are (i) most importantly, the lack of sufficient amounts of pure and standardized preparations of allergens, and (ii) lack of knowledge of the precise structural relationships among the allergens.

The advent of the recombinant allergens has also played a significant role in furthering our knowledge of the structure-function relationships of allergens and understanding of the allergen-induced IgE-mediated allergies. The B- and T-cell epitopes have been defined only for a few allergens. Whereas some of the B-cell epitopes appear to be linear others appear to be conformational in nature. Moreover, the knowledge of the primary structures, and B and T cell epitopes have led to the elucidation of cross-reactivities among different allergens. Sequence homologies among various allergens and their epitopes appear to be a principal reason for extensive cross-reactivities observed among allergens. In addition, recently a method for the analysis of structural motifs within and among diverse groups of aeroallergens was suggested as an approach to identify the major IgE-binding motifs on allergens. Knowledge of T cell epitopes is considered very important since this may lead to devising peptide based immunotherapy.

The chapters presented in this section review and discuss specific advances in a number of important pollen allergens and their cloning, structure-function relationships and the therapeutic implications.

We hope that a combination of the molecular aspects of pollen genes and proteins from agricultural and biomedical perspectives will be useful to the academics, researchers and students alike and will cross fertilize the future research in these fields.

Shyam S. Mohapatra, Ph.D.
Associate Professor
Departments of Immunology
University of Manitoba
Winnipeg, Canada

Bruce Knox, Ph.D.
Professor
Department of Botany
University of Melbourne
Melbourne, Australia

The Editors

Shyam S. Mohapatra, Ph.D., currently holds the position of Associate Professor in the Department of Immunology at the University of Manitoba, in Winnipeg, Canada.

Dr. Shyam Mohapatra, a native of Cuttack (India), obtained his M.Sc. degree in G.B. Pant University of Agriculture and Technology and then undertook postgraduate studies on microbial genetics at the Australian National University in Canberra, Australia, where he completed his Ph.D. in 1983. He was then awarded the Alexander von Humboldt Research Fellowship for his post-doctoral studies (1984–85) at the Institute of Genetics in Bielefeld, Germany. In 1985, he moved to Canada to take up the position of a Research Associate at McGill University, Montreal.

In 1988, he was appointed as Assistant Professor in the Department of Immunology, University of Manitoba, then headed by Professor Alec Sehon. He built a strong research group on molecular biology and immunology of allergens with the collaborative support of the MRC Group for Allergy Research. He became Associate Professor in 1992. He is a strong believer of the application of recombinant allergens in diagnosis and therapy of allergic disorders.

Dr. Mohapatra was awarded the 1992 Pharmacia Allergy Research Foundation International Award for his pioneering research on recombinant allergens. He is actively involved in organization of research conferences, symposia and workshops and has delivered several plenary lectures at international congresses and symposia. He holds the membership in many professional societies including American Association of Immunologists, American Academy of Allergy and Clinical Immunology and International Aerobiology Association. He works as a external referee for a number of national and international granting agencies and also as a peer-reviewer for several journals including J. Immunology, J. Clinical Investigation, Int. Archives of Allergy and Immunology, and Allergy. He has received research support from the Canadian Medical Research Council, a number of competitive grants and biotechnology firms. He has published over 65 full-length papers. He is a dedicated teacher and scientist, who believes in innovative research and technology transfer.

Presently his research interests focus on the immunomodulation of allergic responses with allergen vaccines and the identification and molecular analysis of genetic predisposition factors involved in hyper-IgE and airways responsiveness in a murine model.

R. Bruce Knox, PhD, DSc, FAA is Professor of Botany at the School of Botany, University of Melbourne. He is a distinguished researcher and Fellow of the Australian Academy of Science. He specializes in the reproductive biology of flowering plants, particularly at the cellular, developmental and molecular levels. He has contributed to our knowledge of how pollen causes asthma and hay fever. With his research group, he pioneered the isolation of genes encoding allergens from grass pollen, and the new knowledge gained has been used by a US pharmaceutical company in the design of a new generation of vaccines against hay fever.

He has written more than 200 scientific papers and is an editor and author of several scientific books, including the recent BIOLOGY, which he co-edited as a textbook for first year biology students.

Contributors

James D. Astwood
Monsanto Company
700 Chesterfield Village Parkway North
St. Louis, MO 63198
314-537-6356

Rachel Baltz
Institut de Biologie Moleculaire des
    Plantes (C.N.R.S.)
Université Louis Pasteur
12, Rue de General Zimmer
67084 Strasbourg, Cedex, France

Julian F. Bond
ImmuLogic Pharmaceutical Corporation
One Kendall Square, Building 600
Cambridge, MA 02139

Catherine M. Counsell
ImmuLogic Pharmaceutical Corporation
One Kendall Square, Building 600
Cambridge, MA 02139

C. De Santo
International Institute of Genetics and
    Biophysics
National Research Council
Via Marconi, 12
80125 Napoli, Italy

Claire Domon
Institut de Biologie Moleculaire des
    Plantes (C.N.R.S.)
Université Louis Pasteur
12, Rue de General Zimmer
67084 Strasbourg, Cedex, France

Michael Duchêne
Institute of General and Experimental
    Pathology
AKH, University of Vienna
Währinger Gürtel 18–20
A-1090 Vienna, Austria

Jean-Luc Evrard
Institut de Biologie Moleculaire des
    Plantes (C.N.R.S.)
Université Louis Pasteur
12, Rue de General Zimmer
67084 Strasbourg, Cedex, France

Fatima Ferreira
Institute of General and Experimental
    Pathology
AKH, University of Vienna
Währinger Gürtel 18–20
A-1090 Vienna, Austria

Irwin J. Griffith
ImmuLogic Pharmaceutical Corporation
One Kendall Square, Building 600
Cambridge, MA 02139

Monika Grote
Institute of Medical Physics
University of Munster
Münster, Germany

Douglas Hamilton
Department of Biological Sciences
State University of New York at Albany
Albany, NY 12222

Erwin Heberle-Bors
Institute of Microbiology and Genetics
Vienna Biocenter
University of Vienna
Dr. Borgasse 9
A-1030 Vienna, Austria

Robert D. Hill
Department of Plant Science
University of Manitoba
Winnipeg, MB.R3T 2N2, Canada

Dietrich Kraft
Institute of General and Experimental
  Pathology
AKH, University of Vienna
Währinger Gürtel 18–20
A-1090 Vienna, Austria

David G. Marsh
Johns Hopkins Asthma and Allergy
  Center
301 Bayview Blvd.
Baltimore, MD 21224

William J. Metzler
Bristol-Myers Squibb Pharmaceutical
  Research Institute
Princeton, NJ

Jay P. Morgenstern
ImmuLogic Pharmaceutical Corporation
One Kendall Square, Building 600
Cambridge, MA 02139

Egil Olsen
Department of Immunology
University of Manitoba
730 William Ave.
Winnipeg, MB, R3E 0W3 Canada

E. K. Ong
School of Botany
University of Melbourne
Parkville
Victoria 3052, Australia

Thorunn Rafnar
Johns Hopkins Asthma and Allergy
  Center
301 Bayview Blvd.
Baltimore, MD 21224

Bruce L. Rogers
ImmuLogic Pharmaceutical Corporation
One Kendall Square, Building 600
Cambridge, MA 02139

Sergio Romagnani
Division of Clinical Immunology and
  Allergy
University of Florence
Istituto di Clinica Medica 3
Viale Morgagni
85-Firenze 50134, Italy

Anna Ruffilli
International Institute of Genetics and
  Biophysics
National Research Council
Via Marconi, 12
80125 Napoli, Italy

Otto Scheiner
Institute of General and Experimental
  Pathology
AKH, University of Vienna
Währinger Gürtel 18–20
A-1090 Vienna, Austria

Anna Scotto d'Abusco
International Institute of Genetics and
    Biophysics
National Research Council
Via Marconi, 12
80125 Napoli, Italy

Mohan B. Singh
School of Botany
University of Melbourne
Parkville
Victoria 3052, Australia

Penelope M. Smith
School of Botany
University of Melbourne
Parkville
Victoria 3052, Australia

André Steinmetz
Institut de Biologie Moleculaire des
    Plantes (C.N.R.S.)
Université Louis Pasteur
12, Rue de General Zimmer
67084 Strasbourg, Cedex, France

Eva Stöger
Institute of Microbiology and Genetics
Vienna Biocenter
University of Vienna
Dr. Bohrgasse 9
A-1030 Vienna, Austria

Ines Swoboda
Institute of Microbiology and Genetics
University of Vienna
Dr. Bohrgasse 9
A-1030 Vienna, Austria

Alisher Touraev
Institute of Microbiology and Genetics
Vienna Biocenter
University of Vienna
Dr. Bohrgasse 9
A-1030 Vienna, Austria

Rudolf Valenta
Institute of General and Experimental
    Pathology
AKH, University of Vienna
Währinger Gürtel 18–20
A-1090 Vienna, Austria

Oscar Vicente
Institute of Microbiology and Genetics
Vienna Biocenter
University of Vienna
Dr. Bohrgasse 9
A-1030 Vienna, Austria

Susanne Vrtala
Institute of General and Experimental
    Pathology
AKH, University of Vienna
Währinger Gürtel 18–20
A-1090 Vienna, Austria

Huiling Xu
School of Botany
University of Melbourne
Parkville
Victoria 3052, Australia

Ming Yang
Department of Immunology
University of Manitoba
Winnipeg
Manitoba R3EQOW3, Canada

Viktor Zarsky
Institute of Microbiology and Genetics
Vienna Biocenter
University of Vienna
Dr. Bohrgasse 9
A-1030 Vienna, Austria

Lei Zhang
National Research Council
Institute of Biology, Rm 3148
100 Sussex Drive
Ottawa, ONT. K1A OR6, Canada

PART I
Gene Expression

1

Molecular Biology of Male Gamete Development in Plants—An Overview*

James D. Astwood and Robert D. Hill

Introduction

This chapter describes the biology of pollen development emphasizing (where possible) biochemical, physiological, and genetic events as they are known to occur in cereals generally and barley (*Hordeum vulgare*) specifically. Inasmuch as it is ultimately impossible to separate these phenomena, attention to their discrete consequences will be instructive. Pollen mother cells undergo reductional divisions and further divide twice to become mature tricellular haploid gametes in a process that can last several weeks and even years in some gymnosperms. Throughout this program, enzyme activity, translation, transcription, and DNA replication all proceed in a differential but scheduled manner. That schedule is fatally tied to environmental conditions, hormonal conditions, and expression of specific genes in the reproductive phase plant.

As agriculturalists seek to control fertility in important cash crops such as cereals, a renewed interest in pollen and male sporophyte-specific gene expression has evolved. Thus, several approaches to identifying the cast of genetic characters necessary for successful reproduction have been attempted. Initially, sporophytically expressed genes were examined in the hope that pollen-specific variants could explain the unique processes of cellular differentiation that give rise to the three-celled free-living haploid pollen organism. Recently, pollen and anther expressed (pollen-specific and anther-specific) genes have been identified by techniques such as differential screening. Finally, the role of a class of pollen-specific proteins known as pollen allergens, proteins for which considerable

*This chapter is based on J. D. Astwood's Developmental and Molecular Characterization of Pollen Allergen Cognates in Barley, Ph.D. thesis. Copyright © 1993 by the University of Manitoba; by permission.

physicochemical data are available, is being examined in the context of the developmental program. These approaches have extended our understanding to the point where it may now be possible to envisage and implement direct genetic engineering of the fertilization process.

Microsporogenesis and Pollen Development

General Biology (Anatomy)

A general discussion of the development of the inflorescence in *gramineae* is beyond the scope of this review. However, the following conceptual landmarks of developmental morphology will be useful in defining biochemical and genetic events as they occur. In the first instance, a distinction is made between vegetative and floral development of the angiosperm plant. Floral development is said to have been initiated when manifest signs of floral evocation (the irreversible commitment of the plant meristems to organ development) are observable. Environmental stimuli such as photoperiod, vernalization, temperature, hormone balance, nutritional status, and age are all known effectors of floral evocation (Bernier, 1988). Once vegetative meristems are committed to floral meristem development, floral morphogenesis proceeds by differentiating male (stamens) and female (pistils) organs; as well as supporting structures—petals and sepals—that resemble leaves. Organs undergo a final transition by differentiating tissue types and cells that support gamete production (Gasser, 1991).

The *gramineae* inflorescence is composed of spikelets carried on a raceme, each containing one (e.g., barley) or more (e.g., wheat) flowers or florets attached to the spike-shaped panicle by a rachilla. The floret itself is subtended by two bracts (leaflike structures) known as the lemma and the palea (Williams, 1966). In barley, there is one whorl of three stamens surrounding a monocarpelate pistil with two styles (Cass and Jensen, 1970). Each stamen is composed of an elongated anther having two locules (giving the appearance of four lobes) and a filament to which the anther is attached. "Six-row" cultivars of barley have three fully developed spikelets arising from a common branch-point of the spike, whereas "two-row" cultivars develop only one mature spikelet (Briggs, 1978), apparently aborting the growth of two branches (sterile glumes often occur).

Within the cereal anther, *microsporogenesis* is said to have begun when pollen mother cells (PMCs) arise from archesporium cells via five to 10 rounds of mitosis (Bennett et al., 1973). This premeiotic phase of cell division continues until all pollen mother cells (also known as microsporocytes) become arrested at the GI phase of the somatic cell cycle and contain 2C DNA. Thus, all PMCs become synchronized for microsporogenesis by meiosis. While PMCs are synchronizing, nuclei of the nutritive tapetal cell layer (found on the inside wall of the locules) undergo synchronous DNA synthesis, increasing their DNA content from 2C to 4C. Also, a callose (1,3-β-glucan) cell wall on the surface of the now separated PMCs becomes apparent. Upon synchronization, all PMCs enter a phase of DNA synthesis (S1 phase) increasing their DNA content to 4C.

Pollen tetrads are formed by PMC meiosis, reducing them to haploid (1N) cells containing 1C DNA. Strangely, during PMC meiosis the nuclei of the tapetal cells undergo one round of mitosis, producing a binucleate tapetal cell layer (most eukaryotic cells have only one nucleus*). *Pollen development* is said to have begun when callose-bound tetrads begin to lose the callose wall and become spherical microspores. After a time microspores become vacuolated, so that the nucleus appears pressed against the pollen cell wall, and are referred to as "unicellular" stage pollen. The germinal pore is usually visible on the surface of unicellular pollen. Unicellular pollen undergo two rounds of mitosis. The two cells produced during pollen grain mitosis I initially have a similar appearance, but soon differentiate into the vegetative cell and the generative cell. The nucleus of the vegetative cell appears larger, but is more diffusely stained by aceto-carmine than the generative cell nucleus. Two-celled pollen are referred to as bicellular (older literature refers to these grains as "binucleate" because pollen appeared coenocytic). In all *gramineae* species, the generative cells undergo a second round of mitosis, yielding a tricellular pollen grain. The two generative cells eventually become elongated and are subsequently referred to as the sperm cells. Throughout pollen mitosis, starch granules become abundant and readily visible, eventually swelling the pollen grain and obscuring the view of the sperm cells (Bennett et al., 1973).

In barley it is impossible to distinguish the appearance of the two sperm cells (Cass, 1973). However, in many species sperm dimorphism is observable and is thought to reflect the differing functional roles of the sperm cells within the so-called male germ unit (reviewed by Mogensen, 1992). It has been suggested that heterospermy is due to the fact that one sperm cell fertilizes the ovum (egg) and the other fertilizes one polar nucleus (Russell et al., 1990). Recent studies of sperm cell mobility have emphasized the notion that the sperm cells act cooperatively during pollen germination (Heslop-Harrison et al., 1988). Perhaps heterospermy is due to the expression of different genes by each sperm cell.

Biochemistry—Transcription, Translation and Storage Reserves

Early biochemical studies of microsporogenesis and pollen development were concerned mainly with bulk accumulation of RNA, protein, and storage reserves such as starch. To assess "gene activity," ribosomal RNA (rRNA), transfer RNA (tRNA) and messenger RNA (mRNA) synthesis were measured for each definable stage of development. ^{32}P pulse-labeling of total RNA in lily (*Lilium longiflorum*)

---

*Exceptions in the plant kingdom include multinucleate green algae (Siphonales), some fungi (Phycomycetes), and the coenocytic endosperm of angiosperms.

showed that most rRNA was synthesized after DNA synthesis by PMCs but prior to pollen mitosis (Steffensen, 1966). Transcription of RNA classes has been further defined in *Tradescantia paludosa* (Mascarenhas, 1975) to reveal that 4S tRNAs as well as the 5S, 16S, and 25S rRNAs are synthesized prior to mitosis, with RNA synthesis declining sharply thereafter. Furthermore, the use of transcription inhibitors such as actinomycin D have shown that no new RNA synthesis occurs during initial phases of pollen germination, and only resumes after tube elongation is well underway (Mascarenhas, 1975). Apparently, pollen contains all of the translational machinery necessary for germination at maturity and relies on that apparatus to initiate tube elongation.

However, messenger RNA transcription does not stop at pollen grain mitosis. Tupy (1982) showed by hybridization with tritiated poly-uridylic acid that poly(A$^+$)RNA accumulated through pollen development up to maturity. In tobacco, the proportion of poly(A$^+$)RNA increased ninefold to 2.7% of total RNA, which is reported to be similar to that for stored mRNA in seed. By examining RNA sedimentation profiles at different stages, Tupy (1982) further showed that the average message size increased from 700 to 2,100 nucleotides. By exhaustively cataloguing ^{35}S labeled proteins prepared by in vitro translation of mRNA extracted from staged developing pollen, Schrauwen et al. (1990) was able to define three groups of mRNA; those present at a constant level throughout, those increasing quantitatively until anthesis, and those expressed transiently. Transiently expressed mRNA was found before mitosis, after mitosis, and occasionally overlapping more than one stage. It appears that transcribed message could be performing stage-specific tasks and that the accumulated mRNA primes pollen for germination. For example, the 69 kDa pollen glycoprotein mRNA accumulates during development, but has been shown to be translated only during germination (Storchova et al., 1994).

Changes in mRNA populations are reflected in subsequent patterns of protein synthesis as revealed by SDS-PAGE of extracts from different developmental stages (Zarsky et al., 1985). Total protein also seemed to increase up to maturity, declining somewhat during maturation when starch accumulation was evident. Vergne and Dumas (1988) also showed that stage-specific proteins from developing wheat pollen (*Triticum aestivum*) could be discriminated by isoelectric focusing and by SDS-PAGE. Clearly distinguishable proteins were evident prior to pollen mitosis I and pollen mitosis II.

Storage reserve accumulation begins during PMC differentiation and is found in all species (Baker and Baker, 1979). Starch is found primarily in the vegetative cell plastids, which also increase in number, whereas lipid bodies form in the cytoplasm (Wetzel and Jensen, 1992). Dictyosome vesicles (containing polysaccharide) are also evident and are first used in the formation of the intine cell wall; they later increase in number and remain in the cytoplasm as storage reserves (Wetzel and Jensen, 1992).

Gametophyte vs. Sporophyte Genes

Two questions arise out of the knowledge that transcription and translation are occurring during male gametophyte development: how many genes are expressed by the haploid genome, and are they unique to the male gametophyte? Estimates show that pollen expresses fewer genes than the sporophyte, but they are expressed more abundantly and many of them are gametophyte specific.

To estimate the diversity (number) of genes expressed in pollen, Willing and Mascarenhas (1984) have analyzed the hybridization kinetics of radiolabeled *Tradescantia* pollen cDNA with mRNA populations from pollen and shoots. In both *Tradescantia* and *Zea mays* (Willing et al., 1988), three abundant classes of mRNA were observed in the hybridization Cot curve. Approximately 15% of the mRNA was present in low copy number, but represented more than 17,000 discrete sequences, while 245 discrete sequences accounted for another 35% of the mRNA population (Table 1-1).

The total number of genes expressed by pollen estimated by this technique is approximately two-thirds of the number expressed in shoots (taken as representative of the sporophyte). However, the fraction of abundant, semiabundant, and rare messages, and the number of unique sequences, is roughly equivalent for shoots and pollen. For all three classes, the copy number per cell is 100-fold greater in the pollen grain than in the shoot. Mascarenhas (1989) pointed out that there may be a requirement for protein synthesis during the final stages of maturation and/or germination and early pollen tube elongation, and this requirement is reflected in the message copy number. One caveat is that only mRNAs from mature pollen were examined, so generalizations about all haploid gene expression (i.e., during other stages of development) may not be valid.

The extent to which the genes expressed by pollen are unique has been assessed by comparing isozymes expressed in shoots and pollen. In tomato (*Lycopersicon esculentum*) isozymes for nine dimeric enzymes representing 28 genes were examined (Tanksley et al., 1981). Two conclusions were drawn from the tomato data. First, haploid gene expression occurs without influence from the sporophyte. This was supported by the finding that for plants heterozygous for single locus genes, only one isozyme is found in the pollen (heterodimers would form if sporophytic gene expression contributed to the population of proteins). Second,

Table 1-1 mRNA Population in Maize

mRNA Class	Copies/Cell	Number (diversity) of Unique Sequences	Fraction of Total mRNA Population
Abundant	32,000	245	35%
Semi-Abundant	1,700	6,260	50%
Rare	195	17,250	15%

the haploid genome expresses unique genes as evidenced by the discovery of new pollen-specific isozymes (e.g., esterase-A). Subsequent studies in maize (*Zea mays*) (Gorla et al., 1986) and barley (Pedersen et al., 1987) confirmed that new isozymes are present in pollen, and further showed that approximately 60% of all isozymes are expressed in both the sporophyte and the gametophyte, 30% are expressed only by the sporophyte, and only 10% are pollen specific. The total expression found in the gametophyte (10% plus 60%) by isozyme analysis is in good agreement with estimates of total gene expression found by Cot hybridization analysis (66%) described previously.

Environmental and Hormonal Factors

Pollen production can influence the reproductive success of plants and ultimate grain yield of cereal species. This has generated a wide variety of agronomic studies on the physiological ontogeny of pollen. Studies have shown that *gramineae* pollen development is especially sensitive to environmental conditions that can disrupt pollen development. Male sterility in cereals is induced by water stress (Saini and Aspinall, 1981), heat stress (Saini and Aspinall, 1982a), modified photoperiod (Batch and Morgan, 1974), micronutrient deficiency (Graham, 1975), unusual plant hormone balances (Morgan, 1980), and even air pollution (Wolters and Martens, 1987). The exact basis of these environmental effects remains elusive, but some or all are likely to result in modified gene expression. Several lines of evidence support this general hypothesis. Levels of endogenous growth regulators alone can influence the developmental program as shown for abscisic acid (Morgan, 1980; Saini and Aspinall, 1982b). Naturally occurring "white pollen" mutants (Coe et al., 1981) and genetically engineered chalcone synthase mutants (Taylor and Jorgensen, 1992) lack flavonoids. These mutants have a functionally male sterile phenotype in which otherwise apparently normal mature pollen fail to germinate unless exogenous flavonoids are present.

Water stress in cereals appears to affect only male fertility, not female fertility (Bingham 1966). Why this should be so is unclear. Saini and Aspinall (1981) showed that water stress sensitivity correlated with postmeiotic events (during microspore release from tetrads). Morgan (1980) showed that exogenous application of ABA, a stress-related plant growth regulator (Zeevaart and Creelman, 1988), alone could induce male sterility in wheat and that ABA mimicked water stress effects. Heat stress too (Saini and Aspinall, 1982a) caused abnormal development. The critical period in which the most damage to the program occurs, in all three cases, is near completion of meiosis. When the stress is removed, recovery does not occur. Typically, pollen development proceeds but is characterized by incomplete pollen grain mitosis I and rarely does pollen complete mitosis II. Cytologically, mature pollen is devoid of starch granules and viability is nil. The physiological and molecular basis of this effect is not known, but apparently something irreversible happens to stressed unicellular pollen.

Temperature and Heat Shock Proteins

Temperatures five or more degrees above optimal growing temperatures arrest synthesis of most proteins and mRNAs, except for a small group of proteins that are induced: the heat-shock proteins (HSP). HSP in plants (Vierling, 1991) have structural and sequence similarities with HSP found in other eukaryotes and some prokaryotes and appear to be ubiquitous. Otherwise lethal doses of high temperature are survivable if pretreatment with sublethal temperatures occurs, giving rise to acquired thermotolerance. Conditions that lead to thermotolerance are the same conditions that induce HSP, which explains why it is widely believed that HSP are the agents of thermotolerance (Veirling, 1991).

It was observed that heat stress can inhibit pollen germination, and that mature germinating *Tradescantia* pollen does not seem to synthesize HSP when challenged (Mascarenhas and Alschuler, 1983). Subsequent studies (Schrauwen et al., 1986) of two other species, lily and petunia (*Petunia hybrida*), confirmed this observation. Paradoxically, maize plants, when pretreated during pollen development with sublethal elevated temperatures, develop thermotolerant mature pollen capable of germination apparently in the absence of synthesis of HSP (Xiao and Mascarenhas, 1985). In an effort to sort out this phenomenon. Frova et al., (1989) initiated an elegant series of experiments in which pollen taken at unicellular, bicellular, tricellular and mature stages of development were assayed for the capacity to synthesis HSP. ^{35}S-methionine was incorporated into newly synthesized pollen proteins by exising anthers from immature maize tassels and incubating them at the desired temperature in a liquid medium. Pollen was separated from the sporophytic anther tissue by agitation and filtration through cheesecloth. With this method it was discovered that as pollen matures its capacity to express HSP declines and is eliminated by maturity. Unicellular, premitosis microspores were found to have the greatest capacity for HSP synthesis, both in terms of quantity of individual proteins and in number of individual proteins. Unicellular microspores synthesized the typical 102, 84, 72 and 18 kDa HSP in addition to three other previously undescribed HSP, whereas tricellular pollen produces two "faint" polypeptides of 72 and 64 kDa only. No other tissue in plants displays incompetence for HSP synthesis except pretorpedo stage somatic embryos, which are also thermosensitive (Pitto et al., 1983). Whether the *in vitro* anther culture method adequately simulates conditions *in planta* is uncertain. Nevertheless, the fact remains that germinating pollen does not seem to synthesize HSP, which is consistent with a hypothesis that during development, a genetic switch is thrown, turning off HSP synthesis and conferring thermosusceptibility on the haploid organism.

Chalcone Synthase and Flavonoids in Development

Flavonoids have been implicated as a key component of the pollen germination apparatus. Chalcone synthase (CHS) deficient maize mutants (Coe et al., 1981), referred to as "white pollen" (*whp*) because of a white pollen color, lack flavonoids. White pollen mutants appear male sterile by failing to produce functional pollen tubes. Yellow coloured flavonoid compounds can account for 2–4% of the dry weight of pollen at maturity (Zerback et al., 1989). Chalcone synthase is the key regulatory enzyme in the flavonoid biosynthesis pathway which gives rise to flavonols, isoflavonoids, flavanones, and anthocyanins (Heller and Forkmann, 1988). Also, flavonoids are responsible for floral and fruit pigments in many plant species, including petal color in petunia and snapdragon (*Antirrhinum majus*), testa color in maize, and aleurone color in barley (Harborne, 1986). Other roles for flavonoid compounds include pathogen defenses (phytoalexins) (Dixon, 1986), *Rhizobium* nodulation signals in legumes (Redmond et al., 1986) and protection from UV light (Chappel and Hahlbrock, 1984). Interestingly, flavonoids from pollen can be inducers of the Ti plasmid *vir* region (virulence genes) of *Agrobacterium tumafaciens* (Zerback et al., 1989). Perhaps flavonoid-packed pollen can serve as suitable targets for *Agrobacterium*-mediated gene transfer in recalcitrant cereal crops.*

In pollen the requirement for flavonols in germination and the lack thereof in CHS mutants causes conditional male fertility (CMF) (Taylor and Jorgenson, 1992). White pollen mutants in maize are fully germinable *in vitro* if flavonols are present. Biochemical complementation of this sort implicates flavonols as an important class of regulatory molecules (plant growth regulators). CHS cosuppression mutants (Napoli et al., 1990) and antisense-RNA CHS constructs (van der Meer et al., 1992) in petunia also behave as CMF plants, and can be rescued by addition of flavonols. Mo et al., (1992) have shown that the active flavonol found in mature pollen of petunia and maize is kaempferol. Kaempferol is also found in small quantities in wild-type stigmas of petunia, which explains why petunia stigmas have the capacity to rescue pollen germination (stigma fertility is unaffected by flavonols).

Since CHS is implicated in the developmental program of anthers, Pollack et al., (1993) examined the relationship between CHS levels and flavonol accumulation. CHS protein appears after pollen mitosis I in developing petunia anthers

---

*Many cereal and other monocotyledonous species are resistant to *Agrobacterium*-mediated transformation. One theory (Potrykus, 1991) suggests that this recalcitrance to transformation is due to the physiological state of the plant. Ag-transformation requires wounding in compatible species (so that the bacteria can get inside the plant). He suggests that only the "appropriate" wound-response physiology will allow successful integration of the Ti plasmid. Chalcone synthase induction is considered a classic wound-response mechanism, which therefore implicates flavonoids as contributors to this "appropriate" physiological state.

(i.e., bicellular stage), reaches a maximum of 0.2% of the total anther protein, and rapidly declines as anthers mature. CHS activity is not present in mature pollen. Presumably flavonols are produced only by the anther and are somehow transferred or absorbed by developing pollen. However, developing pollen contains no free flavonols; instead they occur as flavonol glycosides—mainly kaempferol 3-glucosyl-galactoside. The carbohydrate moiety of this compound is unusual: a β_{1-2} linked glucosyl-galactoside. During imbibition and germination the aglycone form of kaempferol-glycoside may be released by a specific β-glycosidase. It will be interesting to see if mutants deficient in this glycosidase can be found in which germination is prevented by permanently sequestering flavonols in the carbohydrate bound form.

In any case, kaempferol is the first example of a pollen-specific, and apparently ubiquitous (present in maize and petunia), growth regulator directly involved in the developmental program. Its importance to the understanding of basic mechanisms of pollen gene regulation, especially during germination (Mascarenhas, 1993), and practical applications, such as transformation of recalcitrant crops, can not be overstated. Already, this system is being used to enhance seed set (Vogt et al., 1994).

Genetic Factors

It has already been seen that some genetic factors can affect the pollen development program (i.e., *whp,* CHS). In maize, there are at least a dozen genes, all of which have been mapped to specific chromosomes, and in some cases regions of chromosomes, which affect pollen development or function (*Maize Genetics Cooperation* Newsletter **65,** 1991). The phenotype in most cases is male sterility and sometimes female sterility. Presumably, female gametogenesis shares many genetic components with male gametogenesis, especially those which regulate meiosis. *Ameiotic (am 1)* is one example of a gene that adversely affects meiosis by disrupting the normal chromosomal movements choreographed by the cytoskeleton (Staiger and Cande, 1992).

Zhao and Weber (1989) examined the effects of maize nullisomic (chromosome deletion) lines on male fertility. With this approach, some chromosome arms were found to be important to the developmental program (i.e., their absence disrupted pollen development), whereas others had no effect. Many genes are required to make a viable mature pollen. Perhaps these genes cluster on specific chromosomes, and genes that have the misfortune of falling outside the male gametogenesis regions are repressed, for example genes responsible for heat shock proteins. Unfortunately, there are relatively few pollen-specific genes identified for any given species, so that mapping them would not yet be informative. The test of this hypothesis will come when many pollen genes are known and many genes of all types are mapped in one species.

Programmed Genes

If it is true that the haploid genome expresses its own unique set of genes, then it is important to identify them and to define roles for them, in an attempt to understand why they should be necessary to the gametophyte life cycle. The genesis of the gamete itself is a unique phenomenon, so that genes directly involved in meiosis play a crucial role. But what distinguishes pollen-ness from egg-ness? Which genes invoke and control the developmental process, and when do they turn on? Finally, which genes control if, when, and where a pollen will germinate? Some of these questions are beginning to be answered as developmentally regulated pollen genes and their promoters have been characterized. As recently as 1987, the first male-gametophyte specific genes were identified (Stinson et al., 1987) and characterized (Hanson et al., 1989). Since then dozens of developmentally programmed genes have been examined. Meanwhile, the answers to other questions remain variously unresolved; what are the functions of these genes, and what regulates the genes and processes associated with self-incompatibility and recognition during germination? Three tactics used to identify programmed genes have been successful components of the overall strategy to understand and control microsporogenesis and pollen development. They are outlined below.

Approach 1: Analysis of Enzymes in Development

Early attempts to dissect the developmental program in pollen looked at the temporal expression of enzymes that could easily be assayed (see Table 1-2).

The first enzyme to be examined for stage-specific expression was alcohol dehydrogenase (ADH) in maize (Stinson and Mascarenhas, 1985). ADH was a good choice for study at the time: ADH was one of the best-studied enzymes, its genetic locus was known, it had been cloned and sequenced (Dennis et al., 1984), and it was known to be under control of the haploid genome. ADH occurs at two unlinked loci, *adh-1* and *adh-2;* and *adh-1* occurs in many allelic isozymic forms. Microspores were harvested at the tetrad, unicellular, bicellular, tricellular, and mature stages of development and quantitatively assayed for ADH activity. ADH was not found in microsporocytes, PMCs, and postmeiotic tetrads.

Table 1-2 Isozyme Analysis in Developing Pollen

Isozyme	Stage(s) of Expression	References
Alcohol dehydrogenase (ADH-1)	Postmeiosis to maturity	Stinson et al., 1985
β-galactosidase (β-GAL)	Postmeiosis, decline at maturity	Singh et al., 1985
Catalase-1 (CAT-1)	Postmitosis I to maturity	Frova, 1990
Catalase-3 (CAT-3)	Tetrad stage to mitosis I	Frova, 1990
Glutamate-oxalacetate-transaminase (GOT-1)	Postmitosis I to maturity	Frova, 1990

However, once the callose wall of the tetrads began to degrade, ADH became detectable, increasing progressively to maturity. From this protein data, it was concluded that transcription of *adh-1* from the haploid genome must occur very soon after meiosis is completed. Northern analysis of *adh-1* transcripts in developing maize pollen has never been reported, although in petunia (Gregerson et al., 1991), *adh-1* transcripts are found only in the later stages of development. Unfortunately, these authors do not state precisely what stage the pollen has progressed to when transcription occurs. Since petunia pollen grains are bicellular at maturity, and pollen grain mitosis occurs much closer to maturity, onset of *adh-1* transcription in petunia and maize may occur at equivalent functional times—just before mitosis I. In addition, no *adh-1* mRNA is found in mature pollen of either species.

A variety of other enzyme isozymes has been examined. The results are summarized in Table 1-2. Two classes of gene activity are apparent from this analysis: genes expressed "early" (ADH-1, β-GAL, CAT-3) and genes expressed "late" (CAT-1, GOT). Whether this fact reveals anything about the function of these enzymes remains to be seen.

Approach 2: Differentially Expressed cDNAs

Differential screening has frequently been used to gain insights into the development of the independent male gametophyte. This tactic assumes that knowing where and when specific genes are expressed will lead to understanding how developmental processes are regulated. The essence of the tactic is to clone the genes (usually cDNAs) that are expressed at a given point in development (or in a specific tissue) by eliminating all the genes in a library that are also expressed in a reference time point (or tissue).

Typically, pollen-specific genes have been identified by differentially screening pollen clone libraries with leaf mRNA or cDNA. This method has been used successfully, yielding some of the first identified pollen-specific genes, notably Zmc13 from maize (Stinson et al., 1987), LAT52, LAT56, and LAT59 from tomato (Twell et al., 1989) and P2 from *Oenothera organensis* (Brown and Crouch, 1990). A key attribute of these genes is that their expression is maximal in pollen grains. However, in situ hybridization studies of the LAT series of genes in tomato (Ursin et al., 1989) revealed that some (see Table 1-3) begin to be expressed as early as tetrad formation (postmeiosis). One curiosity is that in mature anthers, expression of LAT genes also occurs in the endothecium (which marks maturity in the anther). Apparently, the LAT genes are expressed by sporophytic and gametophytic tissues.

In an effort to identify pollen-specific, early expressed genes, Albani et al. (1990) isolated microspores from *Brassica napus* immature anthers (tetrad stage). Differential screening of a genomic DNA library was performed using ^{32}P-labeled poly (A[+]) RNA extracted from the microspores and ^{32}P-labeled leaf, stem, and silique poly (A^+) RNA. Three clones were identified by this technique. All show expression exclusively in developing microspores, and all decline at maturity (see Table 1-3). Only one, Bp10, was found to be expressed in mature pollen.

To identify genes that might be expressed, even transiently, throughout the developmental program, Scott et al. (1991) prepared two cDNA libraries from immature *Brassica napus* anthers: one called the sporogenesis (S) library because immature anthers contained pre- and postmeiosis phase microspores (PMCs and tetrads); and a second called microspore development (MD) because immature anthers contained mitotically active pollen (uni-, bi-, and tricellular pollen). This approach led to the identification of clones expressed at various times throughout the entire developmental period (see Table 1-3). In situ hybridization experiments showed that all of the clones selected from the S library were in fact expressed only in tapetal cells of anthers (clones A3, A6, and A8). In contrast, all the clones from the MD library were microspore specific, some being expressed earlier and others later (E2, F2S, and I3). Prior to tetrad formation, transcription in the tapetum appears to dominate transcription in PMCs. Once microspores are formed, transcription in developing pollen dominates. Smith et al., (1990) also prepared a tetrad stage cDNA library in tomato and were able to define several tapetum specific genes. Other tapetum, anther, and microspore specific genes have been identified and are listed in Table 1-3. In all, markers for the entire spectrum of microspore development, pollen development and germination have been observed. Table 1-3 categorizes developmental markers by time of onset of expression (usually assayed by northern blots).

Approach 3: Characterization of Allergens

The widespread occurence of pollen allergies has created an interest in the cloning and characterization of pollen-expressed genes in the biomedical community (Scheiner, 1992). Pollen allergen proteins have been isolated, purified, and in some cases cloned. Table 1-4 lists the pollen allergens which have been cloned and characterized, and referenced by genetic database accession number (GenBank, EMBL, PIR, and SwissProt). Immunologists think of allergens as being "major" if the majority (more than 50%) of allergic patients have IgE antibodies which react with the specific protein, and "minor" if fewer than 50% react. Pollen allergens are further subdivided into "groups", mostly on the basis of order of identification. As a result, a rational nomenclature for allergens has been developed (Marsh et al., 1988; King et al., 1994).

Although many allergens have been characterized from a physicochemical point of view, their role in the biology of pollen has been difficult to consider until recently. However, in the majority of cases, even for pollen allergens that have been cloned, the spatial and temporal regulation of the protein and mRNA, and its gene, remain unknown.

Recently, the major group IX pollen isoallergens from kentucky bluegrass (*Poa pratensis*) (Mohapatra et al., 1990; Silvanovich et al., 1991), major group I and group II allergens from ryegrass (*Lolium perenne*) (Griffith et al., 1991; Ansari et al., 1989) and the major group I and group II allergens from birch (*Betula verrucosa*) (Beiteneder et al., 1989; Valenta et al., 1991), have been shown to be exclusively expressed in pollen. In addition, many allergen clones from different species share regions of sequence similarity, particularly in domains thought to be highly immunogenic (Silvanovich et al., 1991). Sequence similarity has been exploited to identify allergen cognates in cereals such as maize (*Zea mays*) (Broadwater et al., 1993) and barley (*Hordeum vulgare*) (Astwood et al., 1995). Indeed, pollen allergen homologues have been found in a wide variety of wild and domesticated *gramineae* species (Singh and Knox, 1985; Zhang et al., 1991; Astwood et al., 1995).

Some allergens share sequence similarity with better characterized proteins—*Bet v 2*, for example, is similar to profilin (Valenta et al., 1991). The expression of others, whose function cannot be inferred from sequence similarity, have been characterized in anther development. The expression pattern of the barley pollen allergen homologue *Hor v 9* is one (Astwood, 1993).

Molecular Control: Promoters and Transcription Factors

To understand developmental regulation of gene expression it has been necessary to examine the structure of the genes and their controlling elements, especially pollen and anther specific transcription promoters. Since it is known that a majority of the genes expressed in the male gametophyte are also expressed in the sporophyte whereas others are expressed only in the gametophyte, it might be anticipated that there are at least two types of controlling elements. For genes expressed in both places, tissue-specific promoters may be aligned to the same gene (or two copies of the same ORF may exist under the control of different promoters). Alternatively, only one promoter may exist, but it may be under the control of two (or more) phase-specific transcription factors. For gametophyte-expressed-only genes, unique pollen-specific promoters may exist. There is evidence for both types of genetic control.

What makes a stamen not a pistil? From the point of view of the entire flower, it is already well established that unique sets of transcription factors control the fate of whorl primordia. This control is exerted to cause some primordial whorls to become anthers and others to become pistils, leaves, and petals. This revelation was due to the molecular genetic characterization of floral homeotic mutants—mutants that fail to express some organs or that express them at the inappropriate time or place (for a review, see Coen, 1991). Homeotic mutants, like *apetela, ovulata, deficiens, sepaloidea,* and *super-man* (it is easy to guess the phenotype of these) turn out to be deficient in whorl-specific transcription factors, usually

Table 1-3 Patterns of Pollen and Anther Gene Expression (by time of onset)

Gene	PMC	Tetrad	Uni	Bi	Tri	Pollen	Germ	Reference
Meiosis								
A6[b]	-,+	+	-	-	-	-	-	Scott et al., 1991
A8[b]	+	+	-	-	-	-	-	Scott et al., 1991
A9[b]	+	+	-	-	-	-	-	Scott et al., 1991
A3[b]	-,+	+	+,-	-	-	-	-	Scott et al., 1991
Satap35/44[b]	-,+	+	+,-	-	-	-		Staiger et al., 1994
TOMA92[b]	+	+	+,-	-	n/a	-		Chen et al., 1994
3C12	-,+	-,+	+	+	+,+	+,+		Allen and Lonsdale, 1993
Post Microsporre Meiosis								
TA-29[b]	-	-,+	+,-	-	-	-	-	Koltunow et al., 1990
TomA5B[b]	-	+,+	+	+,-	n/a	-	-	Aguirre and Smith, 1993
Bp4		-,+	+	+	+,-	-		Albani et al., 1990
Bp19		-,+	+	+	+,-	-		Albani et al., 1991
MFS18[e]		+	+	+	+,-	-		Wright et al., 1993
clone 108		-,+	+	+	n/a	+,-		Smith et al., 1990
LAT52		-,+	+	+	n/a	+		Twell et al., 1989
Bp 10		-,+	+	+,+	+,-	+,-		Albani et al., 1992
MFS1		+	+	+	+	+		Wright et al., 1993
APG[d]	-	-,+	+,+	+,-	+,-	+,-	+,-	Roberts et al., 1993
Pollen Mitosis I								
OSC4[d]			+	-	-	-		Tsuchiya et al., 1992
E2	-	-	-,+	+,-	-	-	-	Scott et al., 1991
F2S	-	-	-,+	+,+	+,-	-	-	Scott et al., 1991
OSC6[d]			+,+	+	+,-	-		Tsuchiya et al., 1992
MFS14[b]			+	+	+,-	-		Wright et al., 1993
Npg1	-	-	-,+	+	n/a	+,+		Tebbutt et al., 1994
sta 44.4		-	-,+	+,+	n/a	+,+		Robert et al., 1993
PPE1		-	-,+	+	n/a	+,+	+,+	Mu et al., 1994a
tub 3/tub 4			-,+	+	+,+	+,+	+,+	Rogers et al., 1994

16

Pollen Mitosis II

	PMC	Tetrad	Uni	Bi	Tri	Pollen	Germ	Reference
I3	—				+,−		—	Scott et al., 1991
TUA1	—		—,+	—,+	n/a	+,+		Carpenter et al., 1992
LAT59	—		—,+	—,+	n/a	+		Ursin et al., 1989
Bcp1[a]	—		—,+	+	+	+,+		Theerakulpisut et al., 1991
Tp44/70			—,+	+	n/a	+	+,−	Stinson et al., 1987
LAT56			—,+	—,+	n/a	+	+	Wing et al., 1989
NPT			—,+	—,+	n/a	+	+,+	Weterings et al., 1992
P2	—		—,+	—,+	+	+	+	Brown and Crouch, 1990
69 kDa Prot.	—		—,+	+	+	+	+	Storchova et al., 1994
PRK-1			—,+	—,+	n/a	+,+	+,+	Mu et al., 1994b
Pollen Maturation								
Zm30	—		—	—	—,+	+		Stinson et al., 1987
PG1	—		—	—	—,+	+		Niogret et al., 1991
SF2/18[c]	—		—	—	—,+	+,−		Evrard et al., 1991
LMP131A				+	+	+		Kim et al., 1993
tac 25						+,+	+,+	Thangavelu et al., 1993
SF3						+	+	Blatz et al., 1992a
Zmc13	—		—	—	—	+	+	Hanson et al., 1989

a: Expressed in anther.
b: Expressed in anther (tapetum).
c: Expressed in anther (epidermis).
d: Expressed in both anthers and pollen.
e: Expressed in anthers, pollen and glumes.

Developmental Stages:
PMC = Pollen mother cells
Tetrad = Pollen tetrads
Uni = Unicellular pollen
Bi = Bicellular pollen
Tri = Tricellular pollen
Pollen = Mature pollen
Germ = Germinating pollen

Notes:
n/a = not applicable
− = not expressed
+ = expressed
−,+ = expressed and increasing
+,+ = expressed maximally
+,− = expressed but declining
blank = not tested

Table 1-4 Characterized Pollen Allergens (Protein and DNA Sequences*)

Species	Common Name	Allergen	Synonym(s)	Accession
Agrostis alba	bent grass	*Agr a* 1	group I	E37396
Alnus glutinosa	alder tree	*Aln g* 1	*Bet v* 1 homologue	S50892
Ambrosia artemisiifloia	ragweed (short)	*Amb a* 1.1	antigen E	A39099
		Amb a 1.2	antigen E isoform	B39099
		Amb a 1.3	"	C39099
		Amb a 1.4		D53240
		Amb a 2	antigen K	E53240
		Amb a 3	Ra3	P00304
		Amb a 5	Ra5	A03371
Ambrosia trifida	ragweed (tall)	*Amb t* 5	Ra5 homologue	S39336
Ambrosia psilostachya	ragweed	*Amb p* 5 (A2)	Ra5 homologue	L24465
		Amb p 5 (A3)	*Amb p* 5 isoform	L24466
		Amb p 5 (B1)	"	L24467
		Amb p 5 (B2)	"	L24468
		Amb p 5 (B3)	"	L24469
Anthoxanthum odoratum	sweet vernal grass	*Ant o* 1	group I	G37396
Artemisia vulgaris	mugwort	*Art v* 2	glycoprotein allergen	A38624
Betula verrucosa	birch tree	*Bet v* 1	pathogenesis related (PR)	S05376
		Bet v 1(c)	*Bet v* 1 isoform	X77265
		Bet v 1(d)	"	X77266
		Bet v 1(e)	"	X77267
		Bet v 1(f)	"	X77268
		Bet v 1(g)	"	X77269
		Bet v 1(h)	"	X77270
		Bet v 1(i)	"	X77274
		Bet v 1(j)	"	X77271
		Bet v 1(k)	"	X77272
		Bet v 1(l)	"	X77273
		Bet v 1(m)	"	X81972
		Bet v 1(n)	"	X82028
		Bet v 2	profilin	B45786
		Bet v 3	calmodulin-like	X79267
		Bet v 4	calmodulin-like	n.a.[1]

Species	Common name	Allergen	Description	Accession
Carpinus betulus	hornbeam tree	*Car b* 1(a)	*Bet v* 1 homologue	X66932
		Car b 1(b)	"	X66918
		Car b 1(2)	"	X66933
Castanea sativa	European chestnut	*Cas s* 1	*Bet v* 1 homologue	PC2001
Corylus avellana	hazel tree	*Cor a* 1.5	*Bet v* 1 homologue	S30053
		Cor a 1.6	"	S30054
		Cor a 1.11	"	S30055
		Cor a 1.16	"	S30056
Cryptomeria japonica	Japanese cedar	*Cry j* 1-A		D26544
		Cry j 1-B		D26545
		Cry j 2		D29772
Cynodon dactylon	Bermuda grass	*Cyn d* 1	BG60a	A61226
Dactylis glomerata	orchard grass	*Dac g* 2	Ag DGI	S45354
		Dac g 3		A60359
Festuca elator	reed fescue	*Fes e* 1-A		C37396
		Fes e 2-B		D37396
Glycine max	soybean	cim 1	cytokinin-inducible	U03860
Holocus lanatus	meadow velvet	*Hol l* 1	30K allergen	Z27084
Hordeum vulgare	barley	*Hor v* 9	group IX homologue	U06640
Lolium perenne	ryegrass	*Lol p* 1.	group I	M57474
		Lol p 1b	"	M59163
		Lol p 2-A	group II	A34291
		Lol p 2	"	A48595
		Lol p 3	group III	A33422
		Lol p 4	group IV	A60737
		Lol p 9	group IX	L13083
		Lol p 30K	30K group V allergen	S38290
		Lol p 34K	34K group V allergen	S38289
		Lol p 50K	50K allergen	S38288
Olea europea	olive tree	*Ole e* 1		S36872
		Ole e 1.4		X76397
Oryza sativa	rice	*Ory s* 1	*Lol p* I homologue	n.a.[2]
Parietaria judaica	parietaria	*Par j* 1	allergen I	X77414
Parietaria officinalis	parietaria	*Par o* 1		A53252

continued

19

Table 1-4 Continued.

Species	Common Name	Allergen	Synonym(s)	Accession
Phleum pratense	Timothy grass	*Phl p* 1	group I	X78813
		Phl p 1	"	Z27090
		Phl p 2	group II	X75925
		Phl p 5a	group V, group IX	X70942
		Phl p 5b	group V	Z27083
		Phl p 6		Z27082
		Phl p 32K	group V-like	S38294
		Phl p 38K	group V-like	S38293
		Phl p 11	group X/profilin	P35079
Poa pratensis	Kentucky bluegrass	*Poa p* 1	group I	F37396
		Poa p 9	(KBG31) group IX isoform	M38342
		Poa p 9	(KBG41) group IX isoform	M38343
		Poa p 9	(KBG60) group IX isoform	M38344
Quercus alba	oak tree	*Que a* 1	*Bet v* 1 homologue	D53288
Secale cereale	cultivated rye	*Sec c* 30K	30K group V allergen	S38292
Sorgum haplense	Johnson grass	*Sor h* 1	*Lol p* 1 homologue	n.a.[3]
Triticum aestivum	bread wheat	*Tri a* 2.1	*Bet v* 2 homologue/profilin	S72384
		Tri a 2.2	*Tri a* 2 isoform	S72374
		Tri a 2.3	"	S72375
Zea mays	maize	*Zea m* 1	*Lol p* I homologue	JC1524

*Sequences retrieved from GenEMBL/GenPept ver. 87, Swissprot ver. 30, PIR ver. 41.

n.a.[1] = not available in the genetic databases but reported in Engel et al., (1995).
n.a.[2] = not available in the genetic databases but reported in Xu et al., (1995).
n.a.[3] = not available in the genetic databases but reported in Avjioglu et al., (1993).

of the leucine zipper type (Yanofsky et al., 1990; Sommer et al., 1990), which implies that they act in dimeric, cooperative ways. Most of the homeotic genes contain a putative DNA binding domain, the MADS-box, found in other known transcription factors [such as serum response factor (SRF) which regulates *c-fos*, a human proto-oncogene, and sarcomeric C-binding factor (CBF) regulating α-tubulin in human muscle] (Schwarz-Sommer et al., 1992). Based on this, genetic models have been proposed in which as few as three dimeric transcription factors can determine the fate of all four organ whorls (Schwarz-Sommer et al., 1990). It turns out that floral evocation and phyllotaxis, which occur prior to the establishment of whorl identity, are also controlled by specific homeotic genes that have been likened to *Drosophila melanogaster* embryo segmentation genes which are also expressed early and transiently (Coen et al., 1990; Weigel et al., 1993).

What differentiates pollen within the anther? As noted earlier, chalcone synthase and chalcone flavanone isomerase are key regulatory enzymes in flavonoid biosynthesis. These enzymes are encoded by gene families. The chalcone flavanone isomerase family has two members: CHI-A and CHI-B. The CHI-B gene is expressed exclusively in immature anthers (van Tunen et al., 1988). An odd observation from initial characterizations by northern analysis was that a third, anomalously large CHI transcript was found in mature pollen. Structural analysis of the CHI genes showed that *chi-A* actually has two promoters, P_{A1} and P_{A2}, and that P_{A2} produced the 0.4 kb longer transcript. The *chi-B* gene has only one promoter, P_B. While all three promoters share two consensus regions, it was possible to define an anther-specific region, called the "anther-box" in P_B. (van Tunen et al., 1989). The chalcone synthase gene family members *chs-A* and *chs-J* also maintain the anther-box in their promoters (van der Meer et al., 1990). It turns out that one other anther-only expressed gene contains the same anther box, namely Bp19 of *Brassica napus* (Albani et al., 1991). Two seemingly unrelated tapetum-specific promoters have been identified, but not fully defined, for TA-29 (Block and Debrouwer, 1993) and A9 (Paul et al., 1992).

In addition to promoter P_{A2}, several other pollen-specific promoters have been identified for genes that are expressed in both the sporophyte and the gametophyte. Notable, *adh-1*, seems to have a cis element in the TATA box region which causes a 5′ elongated transcript to be expressed in pollen only, much the same way P_{A2} does (Kloeckener-Gruissem et al., 1992). Using transgenic osmotin (OSM) promoter: :GUS constructs, Kononowicz et al., (1992) found that OSM promoter driven GUS expression is found in mature pollen, especially at dehiscence when the pollen grains are undergoing desiccation. Osmotin is a salt-stress, water-stress, and ABA inducible gene (Singh et al., 1989). One "housekeeping" gene that also shows expression of one of its family members in a pollen-specific manner is α-tubulin I (TUA1). TUA1 transcripts accumulate in postmitotic pollen and germinating pollen tubes (Carpenter et al., 1992). Microtubules are not found in mature pollen, but are intimately involved in the germinating pollen tube and in sperm cell mobility during fertilization (Peirson and Cresti, 1992). Other pollen-specific gene promoters are known, including those for LAT52 (Twell, 1992) and Zm13 (Hamilton et al., 1992).

If there are pollen-specific promoters, might there also be pollen-specific transcription factors? One pollen gene, SF3 (Table 1-3), turns out to be a zinc finger protein that contains the LIM motif common to many developmental regulators, such as rat insulin gene enhancer binding protein (ISL-2) Drosophila APTEROUS protein, and mammalian onco-proteins (TTG) (Blatz et al., 1992b).

From a regulatory point of view, pollen seem to have promoters in various combinations and even unique transcription factors. Moreover, these elements are different than those driving anther-specific genes. How these components come together to build the mature, functioning, pollen grain will take some time to sort out; but it is comforting to note that the same molecular genetic toolkit is used in the development of many other organisms such as *Drosophila* and humans.

Function of Anther and Pollen Genes

So far, little direct evidence of the actual function of pollen specific genes is available. While the function of known enzymes such as ADH, β-GAL, and CHS are clear, the precise role of these enzymes in pollen is open to speculation. As for pollen-only expressed genes such as P2, their function can only be inferred on the basis of sequence similarity with other well-characterized proteins and from the knowledge of their tissue specificity determined by localization of the protein and/or transcript. In general, it might be expected that early expressed anther-specific genes are important in the development and maintenance of anther tissue while pollen are developing within, whereas genes expressed transiently at specific points in the developmental program perform unique functions required for that stage. At maturity, we already know that many transcripts are poised for translation at germination, so many of these probably perform germination-specific functions such as tube growth and pollen-pistil recognition. However, to date, no clear evidence for any of these processes is available for any gene, so that only educated guesses are possible. These hypotheses are outlined in Table 1-5 and discussed below.

Many pollen proteins appear to be pectinases. Pectin is found in the cell walls of all plants. Degradation of pectin results from the action of pectinases including polygalacturonase (PG), pectin lyase (PL) and pectin esterase (PE).

The P2 gene family members (Brown and Crouch, 1990) are very similar (54% identity) to polygalacturonase found in fruits. Polygalacturonase in fruits is responsible for digestion of the middle lamellar cell walls (composed mainly of homo- and polygalacturonans), causing the entire fruit to soften (ripen). P2 protein is found in mature pollen and germinating pollen tubes, which suggests

Table 1-5 Inferred Function of Anther and Pollen Genes Based on Sequence Similarity

Name(s)	Similarity	Tissue Localization
P2, PG1, sta44.4, 3c12	fruit polygalacturonase	pollen and pollen tubes
LAT56, Zmc58, Npg1	fungal pectate lyases	anthers, pollen, and pollen tubes
Bp19, PPE1	pectin esterase	immature pollen
Bp10; NPT303	ascorbate oxidase	pollen and pollen tubes
E2	phospholipid transferase	tapetum and unicellular pollen
I3	elosin	immature pollen
A9; TomA5B; *fil* 1	α-amylase inhibitor	tapetum
LAT 52; Zmc13	kunitz trypsin inhibitor	immature pollen and tubes
SF3	transcription factor	mature pollen
Bet v 1	disease resistance (Pr)	mature pollen and leaves
Bet v 2	profilin	mature pollen
Bet v 3, *Bet v* 4	calmodulin	pollen and pollen tubes
LMP13A	cofilin, destrin	mature pollen
tac 25	actin	mature pollen
TAU 1	α-tubulin	mature pollen
tub3, tub4	β-tubulin	immature and mature pollen
PCP	self-recognition of stigmas	mature pollen

that it may digest cell wall pectins of the pistils during germination. Alternatively, softening of the pollen cell walls may be required for germination and/or tube elongation. A similar polygalacturonase (PG) gene family has also been found in maize pollen (Niogret et al., 1991; Allen and Lonsdale, 1993).

In characterizing the tomato anther genes LAT56 and LAT59, Wing et al. (1989), discovered that the deduced proteins show some similarity to pectate lyase (PL) isozymes PLe and PLb. The homologous regions of LAT56, LAT59, Ple, and Plb coincide with two regions that are thought to be important to the function of the protein and that are conserved in seven other pectate lyases of the fungal plant pathogens *Erwinia carotovora* and *Erwinia chrysanthemi*. The maize pollen clone Zmc58 (Mascarenhas, 1990) also shows similarity to PL. Pectate lyases, like PGs, could be required in the germination process.

The early expressed gene, Bp19 (Albani et al., 1991) is similar to pectin esterases found in tomato fruit and *Erwinia*. Pectin esterases remove methoxyl groups from pectin, which is a requirement for subsequent PG and PL activity. Bp19 expression occurs earlier in the developmental program than the PG and PL genes discussed previously. Experiments have not determined whether the protein persists into latter stages by incorporation into the cell wall or is active early and is degraded. In either case, a de-esterified pectin substrate required by pollen PGs and PLs might be available for cell wall reorganization during germination.

Pollen proteins are involved in lipid metabolism. Pollen cell walls have a unique lipid constituent, known as sporopollenin (for review, see Wiermann and

Gubatz, 1992), which is laid down soon after tetrads are formed. The tapetum-specific and unicellular pollen-specific gene E2 (Foster et al., 1992) shows high (50–60%) similarity to phospholipid transfer proteins (PLTPs), including all conserved structural features. PLTPs are capable of transferring lipids between membranes in vitro, however their function in vivo may be limited to the extracellular milieux because of the presence of signal peptide leader sequences that may target them for secretion. Thus, PLTPs may be responsible for cuticle formation. In this context, a homologous function for PLTPs in sporopollenin assembly may be considered. I3 appears to encode a pollen oleosin (Roberts et al., 1993) which may be important to oil body formation in mature pollen.

Two genes, one expressed maximally in bicellular microspores (Bp10, Albani et al., 1992) and the other expressed during microsporogenesis and germination (NTP303, Wetering et al., 1992), have sequence similarity to the metabolic copper-containing enzyme ascorbate oxidase (AAO). However, neither gene conserves important copper-binding histidine residues, which makes them unlikely to have bona fide AAO activity. Instead, it could be argued that AAO and these pollen genes have a common ancestry.

Pollen proteins may be storage proteins. Both LAT52 (Twell et al., 1989) and Zmc13 (Hanson et al., 1989) have sequence similarity to the kunitz trypsin inhibiter (Kti) family of seed storage proteins. It is clear from antisense experiments that LAT52 is essential for normal pollen development and function (Muschietti et al., 1994). Another tapetum-specific gene, A9 (Paul et al., 1992), expressed only at the tetrad stage, appears to be a member of the seed-storage protein α-amylase/trypsin inhibitor superfamily* (Kreis et al., 1985). Since the A9 protein shares certain structural feature with the α-amylase/trypsin inhibitor subfamily, Paul et al., (1992) suggest that A9 protects the anther and pollen from insect or microbial damage. This role would be consistent with the fact that antisense experiments show A9 is not required for pollen function. A9 would seem to be needed under unknown condition (Turgut et al., 1994). Likewise the stamen-specific genes TomA5B (Aguirre and Smith, 1993) and *fil* 1 (Nacken et al., 1991) contain the multiple cysteine motif of this superfamily. Expression of TomA5B is maximal in postmeiotic tetrads and is localized to the tapetal layer. The presence of a putative signal peptide in TomA5B may result in the deposition of this protein on the exine of maturing pollen (Aguirre and Smith, 1993).

---

*Members of the superfamily include seed storage proteins such as prolamines from cereals and 2S globulins from dicots, and α-amylase and trypsin inhibitors from cereals. Three important regions (A,B,C) are common to all family members, and one (B) is especially conserved and contains multiple Cys-Cys motifs. Members that have long distances between the A and B regions are usually seed storage proteins, whereas those with short distances are usually α-amylase/trypsin inhibitors. The seed storage members also have unusual amino acid content—either high in methionine (sulphur-rich) or glutamine (nitrogen-rich). The function of the inhibitors is elusive but may involve protection from animal or insect attack.

However, the proteins for A9 and TomA5B have not been traced directly, so that their persistence late in development is unknown. Alternatively, these proteins may function as amino acid reserves required later in development or during pollen germination.

Pollen allergens may be disease resistance genes. The *bet v* 1 pollen allergen has a very strong sequence homology (70%) with a pea (*Pisum sativum*) disease resistance (I49) gene (Brieteneder et al., 1989). I49 is one member of a small gene family that is induced when peas are challenged by *Fusarium solani* or *Pseudomonas spp.* (Fristensky et al., 1988). Low levels of Bet *v* 1 mRNA has been detected in leaves, which Brieteneder et al., (1989) construe to mean that it is involved in general disease defense. In addition, the fact of such high homology between proteins from birch trees (*Betulaceae*) and peas (*Leguminosae*) suggests that *Bet v* 1 has a significant, albeit unknown role.

Allergens and other pollen proteins may be cytoskeletal regulators. Profilin (for a review, see Pollard and Cooper, 1986) is a small (15 kDa) protein found in platelet cells of vertebrates and sperm cells of invertebrates. The importance of profilin is that it regulates the polymerization of actin. Profilactin is formed when profilin binds actin monomers, effectively stopping the growth of microfilaments and is reversible by high pH or by binding of phosphatidylinositol 4,5 bisphosphate found in membranes. During the fertilization process of invertebrates, the acrosomal reaction of sperm is invoked by a sudden polymerization of actin microfilaments. Platelet formation in mammals is similar in this respect.

In birch pollen, a 14.7 kDa allergen, called *Bet v* 2, shows significant sequence similarity to slime mold (*Myxomycetes*), yeast (*Saccharomyces cerevisiae*), and human profilins (Valenta et al., 1991). *Bet v* 2 was also found in a wide variety of other plant species, including *Gramineae*. Direct evidence that *Bet v* 2 acts as a regulator of actin organization has been obtained by microinjection of recombinant *Bet v* 2 into the stamen hair cells of *Tradescantia* (Staiger et al., 1994). Immediate and irreversible changes in cellular organization and arrested cytoplasmic streaming were observed. Regulation of profilin may prove to be a key switch controlling pollen germination in plants. The role of actin in germination is well documented (Heslop-Harrison and Heslop-Harrison, 1989; Heslop-Harrison et al., 1988).

Kim et al., (1993) have recently cloned pollen genes from *Brassica napus* and *Lilium* which encode proteins with significant sequence similarity to animal actin depolymerization factors (ADFs): cofilin and destrin. Although it is presently unclear if these genes have a unique and important role in anthers or pollen relative to other tissues, it is worth noting that ADFs regulate the equilibrium between polymerized and depolymerized actin filaments. It is possible that ADFs control pollen cell movements, especially those which occur during germination and mobilization of sperm cells. Pollen also express unique actin (Thangavelu et al., 1993) and tubulin (Carpenter et al., 1992; Rogers et al., 1993) gene family members relative to other tissues. Although the molecular mechanisms regulating pollen germination and tube growth are presently poorly understood (Mascarenhas, 1993), control of cytoskeletal elements appears to be central.

Pollen express signal transduction molecules. The ultimate goal of pollen development is to allow germination and fertilization to take place. Evidence is mounting that this process is mediated by pollen-specific signal transduction proteins. Bet v 3 and Bet v 4 have extensive homology with calcium binding proteins (calmodulins) and have demonstrated calcium binding activities (Seiberler et al., 1994). PRK-1, a pollen germination expressed protein, has homology with receptor tyrosine kinases (Mu et al., 1994b).

Do pollen "coating-borne peptides" act in cell communication? Self-incompatibility (SI) is a common form of genetic control over fertilization. Large numbers of alleles exist, many of which have been cloned and characterized, and all appear to be expressed solely in pistils. One way in which S-alleles, S-linked glycoproteins (SLG), S-linked receptors (SLR), and S-linked kinases (SLK) may operate is by a phosphorylation cascade involving protein kinases and transmembrane protein kinase receptors (for a recent review, see Nasrallah and Nasrallah, 1993). Other S-alleles have homology with RNases, and are thought to function by preferentially degrading pollen germination-specific RNAs during incompatible reactions (Haring et al., 1990) and have been reviewed in Newbigin et al., (1993). In either case, no protein counterparts to the SI genes have been found in pollen, even though pollen phenotype determines compatibility.

In an analysis of the S-allele-Promoter: :GUS construct expression in transgenic Brassica, GUS activity was found in tapetal cells when pollen were tetrads, as well as in mature stigmas (where S-allele expression is expected) (Toriyama et al., 1991). Brassica displays "sporophytic" self-incompatibility, which means that the genotype of the diploid sporophytic tissue (anthers) determines the genetic phenotype of the pollen. The determinants of phenotype are laid down on the pollen grains, presumably at the tetrad stage as indicated by the GUS experiment. Even so, no actual S-allele proteins (SLGs) have been detected in anthers or pollen. Perhaps the same promoter drives a different protein in the tapetum, one that defines the male SI determinant (Thorsness et al., 1991).

In an effort to identify pollen compounds that might interact with S-alleles (which are heavily glycosylated), Doughty et al., (1993), discovered a 7 kDa polypeptide resident on the surface of mature pollen (so-called coating-borne peptides) which bind SLGs. The interaction product (IP) formed when the 7 kDA peptide binds the SLG shows a protein isoelectric point increase of 2.0 pH units. Regulation of the germination process might be thought of in terms of the physiological consequences of the interaction product. IPs might be expected to operate through an SLK kinase activity on the pollen or stigmatic membranes. Perhaps the 7 kDa polypeptides are fragments of another type of coat-borne protein, the pollen allergens.

It is intriguing to note that some pollen allergens share these properties: They are glycosylated and are found on the surface of the pollen grain (for a review see Howlett and Knox, 1982), they have high isoelectric points, and some have low (approximately 4–10 kDa) molecular weights. Glycosylation does not seem to be important to allergenicity of these antigens (Nilsen et al., 1991b). In addition, a large number of high isoelectric point proteins become apparent late in barley pollen development (Astwood and Hill, 1993). The physicochemical properties, the time of expression, and the extracellular location of this class of allergenic proteins also implies a role in late reproductive events such as cell communication and recognition of self.

Genetically Engineered Fertility Control

As key regulatory elements of pollen- and anther-specific genes have come to be identified, new opportunities to genetically engineer fertility have arisen. Control of fertility, usually in the form of male sterility, is a convenient way to produce hybrid seed. One way to create male sterility is to transform plants with a construct containing a toxic gene driven by a pollen or anther specific promoter*. To do this, the anther tapetum-specific TA-29 gene promoter (Koltunow et al., 1990) of tobacco was built into a construct containing a bacterial RNase, called barnase. Tobacco plants transformed with the TA-29: :barnase construct proved to be male sterile—barnase degraded all mRNA produced by the tapetum (Mariani et al., 1990). Generating male sterility has been done before: Many chemicals and physiological conditions induce male sterility. However, barnase was different because there is also an inhibitor of barnase, called barstar. The barstar RNase inhibitor could be engineered into a similar construct, again using the TA-29 promoter, and when transformed back into the transgenic Barnase male sterile tobacco, the barstar construct completely restored fertility. When the TA-29: :bar-star construct was transformed into a wild-type strain, the wild-type/barstar transgenic plant could be used to restore fertility when crossed with the barnase male sterile line (Mariani et al., 1992). The result of all this sophisticated work was the development of genetically engineered male sterility complete with genetically engineered male fertility restorer lines. This approach has been extended to *Brassica napus* (Block and Debrouwer, 1993) and a variety of other crops including maize (R. Goldberg, unpublished). Other male-sterile lines have been developed by similar methods (van der Meer et al., 1992; Napoli et al., 1990), but none has achieved the elegance of the Barnase/Barstar system.

---

*In fact, genetic ablation of reproductive tissues in transgenic plants has been used primarily as a tool to dissect promoter specificities, typically using the diphtheria toxin A gene (DT-A) to knockout pistils or anthers (Thorsness et al., 1991).

New Opportunities

As the physicochemical and developmental data concerning pollen expressed genes accumulate, it should be possible to exploit potential synergies between techniques and approaches used by the biomedical and botanical scientific communities. We predict that the accumulated knowledge of pollen allergens and of differentially regulated anther and pollen genes will lead to exciting new insights into both the natural history of pollen allergy and the fundamental molecular machinery of plant reproduction.

References

Aguirre, P. J. and A. G. Smith, 1993. Molecular characterization of a gene encoding a cysteine-rich protein preferentially expressed in anthers of *Lycopersicon esculatum*. *Plant Mol. B.* 23:477–487.

Albani, D., L. S. Robert, P. A. Donaldson, I. Altosaar, P. G. Arnison and S. F. Fabijanski, 1990. Characterization of a pollen-specific gene family from *Brassica napus* which is activated during early microspore development. *Plant Mol. B.* 15:605–622.

Albani, D., I. Altosaar, P. G. Arnison and S. F. Fabijanski, 1991. A gene showing sequence similarity to pectin esterase is specifically expressed in developing pollen of *Brassica napus*. Sequences in its 5' flanking region are conserved in other pollen-specific promoters. *Plant Mol. B.* 16:501–513.

Albani, D., R. Sardana, L. S. Robert, I. Altosaar, P. G. Arnison and S. F. Fabijanski, 1992. A *Brassica napus* gene family which shows sequence similarity to ascorbate oxidase is expressed in developing pollen. Molecular characterization and analysis of promoter activity in transgenic tobacco plants. *Plant J.* 2:331–342.

Allen, R. L. and D. M. Lonsdale, 1993. Molecular characterization of one of the maize polygalacturonase gene family members which are expressed during late pollen development. *Plant J.* 3:261–271.

Ansari, A. A., P. Shenbagamurthi and D. G. Marsh, 1989. Complete amino acid sequence of a *Lolium perenne* (perennial rye grass) pollen allergen, *Lol p* II. *J. of Biol. Chem.* 264:11181–11185.

Astwood, J. D. 1993. Developmental and molecular characterization of allergen cognates in barley. Ph.D. diss., The University of Manitoba.

Astwood, J. D., S. S. Mohapatra, H. Ni and R. D. Hill, 1995. Pollen allergen homologues in barley and other crop species. *Clin. Exp. Allergy* 25:66–72.

Avijoglu, A., M. Singh and R. B. Knox, 1993. Sequence analysis of *Sor h* I, the group I allergen of Johnson grass pollen and its comparison to *Lol p* I. *J. of Allergy and Clin. Immunol.* 91:340.

Baker, H. G. and I. Baker, 1979. Starch in angiosperm pollen grains and its evolutionary significance. *Am. J. Bot.* 66:591–600.

Batch, J. J. and D. G. Morgan, 1974. Male sterility induced in barley by photoperiod. *Nature* 250:165–167.

Bennett, M. D., M. K. Rao, J. B. Smith and M. W. Bayliss. 1973. Cell development in the anther, ovule, and the young seed of *Triticum aestivum* L. var. Chinese Spring. *Proc. Royal Soc. London* B 266:39–81.

Bernier, G. 1988. The control of floral evocation and morphogenesis. *Ann. Rev. Plant Physiol. Mol. Biol.* 39:175–219.

Bingham, J. 1966. Varietal responses in wheat to water supply in the field, and male sterility caused by a period of drought in a glasshouse experiment. *Ann. Applied Biol.* 57:365–377.

Blatz, R., C. Domon, D.T.N. Pillay and A. Steinmetz, 1992a. Characterization of a pollen-specific sunflower cDNA encoding a zinc finger protein. *Plant J.* 2:713–721.

Blatz, R., J-L. Evrard, C. Domon and A. Steinmetz, 1992b. A LIM motif is present in a pollen-specific protein. *Plant Cell* 4:1465–1466.

Block, M. D. and D. Debrouwer, 1993. Engineered fertility control in transgenic *Brassica napus* L.: Histochemical analysis of anther development. *Planta* 189:218–225.

Breiteneder, H., K. Pettenburger, A. Bito, R. Valenta, D. Kraft, H. Rumpold, O. Scheiner and M. Breitenbach, 1989. The gene coding for the major birch allergen *Bet v* I, is highly homologous to a pea disease resistance response gene. *EMBO J.* 8:1935–1938.

Briggs, D. E. 1978. *Barley*. Chapman and Hall, London.

Broadwater, A. H., A. L. Rubinstein, C. H. Chay, D. G. Klapper and P. A. Bedinger, 1993. *Zea m* I, the maize homolog of the allergen-encoding *Lol p* I gene of rye grass. Gene 131:227–230.

Brown, S. M. and M. L. Crouch, 1990. Characterization of a gene family abundantly expressed in *Oenothera organensis* pollen that shows sequence similarity to polygalac-turonase. *Plant Cell* 2:263–274.

Carpenter, J. L., S. E. Ploense, P. Snustad and C. D. Silflow, 1992. Preferential expression of an α-tubulin gene of Arabidopsis in pollen. *Plant Cell* 4:557–571.

Cass, D. D. 1973. An ultrastructural and Nomarski-interference study of the sperms of barley. *Can. J. Bot.* 51:601–605.

Cass, D. D. and W. A. Jensen, 1970. Fertilization in barley. *American Journal of Botany* 57:62–70.

Chappel, J. and K. Hahlbrock, 1984. Transcription of plant defence genes in response to UV light or fungal elicitor. *Nature* 311:76–78.

Chen, R., P. J. Aguirre and A. G. Smith, 1994. Characterization of an anther- and tapetum-specific gene encoding a glycine-rich protein from tomato. *J. Plant Physiol.* 143:651–658.

Coe, E. H., S. M. McCormick and S. A. Modena, 1981. White pollen in maize. *J. Heredity* 72:318–320.

Coen, E. S. 1991. The role of homeotic genes in flower development and evolution. *Ann. Rev. Plant Physiol. Mol. Biol.* 42:241–279.

Coen, E. S., J. M. Romero, S. Doyle, R. Elliot, G. Murphy and R. Carpenter, 1990.

Floricaula: A homeotic gene required for flower development in Antirrhinum majus. *Cell* 63:1311–1322.

Dennis, E. S., W. L. Gerlacj, A. J. Pryor, J. L. Bennetzen, A. Inglis, D. Llewellyn, M. M. Sachs, R. J. Ferl and W. J. Peacock, 1984. Molecular analysis of the alcohol dehydrogenase (*adh 1*) gene of maize. *Nucleic Acids Res.* 12:3983–3999.

Dixon, R. A. 1986. The phytoalexin response, elicitation, signalling and control of host gene expression. *Biol. Rev.* 61:239–291.

Doughty, J., F. Hedderson, A. McCubbin and H. Dickinson, 1993. Interaction between a coat-borne peptide of the *Brassica* pollen grain and stigmatic S (self-incompatibility)-locus-specific glycoproteins. *Proc. Nat. Acad. Sci. USA* 90:467–471.

Engle, E., B. Stegbuchner, B. Kramer, C. Ebner, M. Breitenbach, K. Richter and F. Ferreira, 1995. cDNA cloning and characterization of a birch pollen allergen *Bet v* 4, homologous to EF-hand calcium binding proteins. (unpublished).

Evrard, J-L., C. Jako, S-G. Agnes, J-H. Weil and M. Kunz, 1991. Anther-specific, developmentally regulated expression of genes encoding a new class of proline-rich proteins in sunflower. *Plant Mol. Biol.* 16:271–281.

Foster, G. D., S. W. Robinson, R. P. Blundell, M. R. Roberts, R. Hodge, J. Draper and R. J. Scott, 1992. A *Brassica napus* mRNA encoding a protein homologous to phospholipid transfer proteins, is expressed specifically in the tapetum and developing microspores. *Plant Sci.* 84:187–192.

Fristensky, B., D. Horovitz, and L. A. Hadwiger, 1988. cDNA sequence for pea disease resistance genes. *Plant Mol. Biol.* 11:713–715.

Frova, C. 1990. Analysis of gene expression in microspores, pollen, and silks of *Zea mays* L. *Sexual Plant Reprod.* 3:200–206.

Frova, C., G. Taramino and G. Binelli, 1989. Heat-shock proteins during pollen development in maize. *Develop. Genetics* 10:324–332.

Gasser, C. 1991. Molecular studies on the differentiation of floral organs. *Annu. Rev. Plant Physiol. Plant Mol. Biol.* 42:621–649.

Graham, R. D. 1975. Male sterility in wheat plants deficient in copper. *Nature* 254:514–515.

Gregerson, R., M. McLean, M. Beld, A.G.M. Gerats and J. Stommer, 1991. Structure, expression, chromosomal locations, and product of the gene encoding ADH1 in Petunia. *Plant Mol. Biol.* 17:37–48.

Griffith, I. J., P. M. Smith, J. Pollock, P. Theerakulpisut, A. Avjioglu, S. Davies, T. Hough, M. B. Singh, R. J. Simpson, L. D. Ward and R. B. Knox, 1991. Cloning and sequencing of *Lol p* I, the major allergenic protein of rye-grass pollen. *FEBS Letters* 279:210–215.

Gorla, M. S., C. Frova, G. Binelli and E. Ottaviano, 1986. The extent of gametophytic-sporophytic gene expression in maize. *Theoret. and Appl. Genet.* 72:42–47.

Hamilton, D. A., M. Roy, J. Rueda, R. K. Sindhu, J. Sanford and J. P. Mascarenhas, 1992. Dissection of a pollen-specific promoter from maize by transient transformation assay. *Plant Mol. Biol.* 18:211–218.

Hanson, D. D., D. A. Hamilton, J. L. Travis, D. M. Bashe and J. P. Mascarenhas, 1989. Characterization of a pollen-specific cDNA clone from *Zea mays* and its expression. *Plant Cell* 1:173–179.

Harborne, J. B. 1986. Nature, distribution and function of plant flavonoids. In *Plant Flavonoids in Biology and Medicine: Biochemical, Pharmacological, and Structure-Activity Relationships,* (eds.) V. Cody, E. Middleton and J. B. Harborne, pp. 25–42. A. R. Liss, New York.

Haring, V., J. E. Gray, B. A. McClure, M. A. Anderson and A. E. Clarke, 1990. Self-incompatibility: a self-recognition system in plants. *Science* 250:937–941.

Heller, W. and G. Forkman, 1988. Biosynthesis. In *The Flavonoids,* (ed.) J. B. Harborne, pp. 399–425. Chapman and Hall, London.

Heslop-Harrison, J. and Y. Heslop-Harrison, 1989. Conformation and movement of the vegetative nucleus of the angiosperm pollen tube: association with the actin cytoskeleton. *Cell Sci.* 93:299–308.

Heslop-Harrison, J., Y. Heslop-Harrison, M. Cresti, A. Tiezzi and A. Moscatelli, 1988. Cytoskeletal elements, cell shaping and movement in the angiosperm pollen tube. *Cell Sci.* 91:49–60.

Howlett, B. J. and R. B. Knox, 1982. Allergic Interactions. *Encyclopedea of Plant Physiology* 17:655–673.

Jarolim, E., H. Rumpold, A. T. Endler, H. Ebner, M. Breitenbach, O. Scheiner and D. Kraft, 1989. IgE and IgG antibodies of patients with allergy to birch pollen as tools to define the allergen profile of *Betula verrucosa*. *Allergy* 44:385–395.

Kim, S-R., Y. Kim and G. An, 1993. Molecular cloning and characterization of anther-preferential cDNA encoding a putative actin-depolymerizing factor. *Plant Mol. Biol.* 21:39–45.

King, T. P., D. Hoffman, H. Lowenstein, D. G. Marsh, T.A.E. Platts-Mills, and W. Thomas, 1994. Allergen nomenclature. *Int. Archives of Allergy and Immunol.* 105:224–233.

Kloeckener-Gruissem, B., J. M. Vogel and M. Freeling, 1992. The TATA box promoter region of maize *Adh1* affects its organ-specific expression. *EMBO J.* 11:157–166.

Koltunow, A. M., J. Truettner, K. H. Cox, M. Wallroth and R. B. Goldberg, 1990. Different temporal and spatial gene expression patterns during anther development. *Plant Cell* 2:1201–1224.

Kononowicz, A. K., D. E. Nelson, N. K. Singh, P. M. Hasegawa and R. A. Bressan, 1992. Regulation of the osmotin gene promoter. *Plant Cell* 4:513–524.

Kreis, M., B. G. Forde, S. Rahman, B. J. Miflin, and P. R. Shrewry, 1985. Molecular evolution of the seed storage proteins in barley, wheat and rye. *J. of Mol. Biol.* 183:449–502.

Mariani, C., M. DeBeuckeleer, J. Truettner, J. Leemans and R. B. Goldberg, 1990. Induction of male sterility in plants by a chimeric ribonuclease gene. *Nature* 347:737–741.

Mariani, C., V. Gossele, M. DeBeuckeleer, M. DeBlock, R. B. Goldberg, W. DeZGreed and J. Leemans, 1992. A chimeric RNase inhibitor gene restores fertility to male sterile plants. *Nature* 357:384–387.

Marsh, D. G., L. Goodfriend, T. P. King, H. Lowenstein and T.A.E. Platts-Mills, 1988. Allergen nomenclature. *Internat. Arch. Allergy Appl. Immunol.* 85:194–200.

Mascarenhas, J. P. 1975. The biochemistry of angiosperm pollen development. *Bot. Rev.* 41:259–314.

Mascarenhas, J. P. 1989. The male gametophyte of flowering plants. *Plant Cell* 1:657–664.

Mascarenhas, J. P. 1990. Gene activity during pollen development. *Ann. Rev. Plant Physiol. Plant Mol. Biol.* 41:317–338.

Mascarenhas, J. P. 1993. Molecular mechanisms of pollen tube growth and differentiation. *Plant Cell* 5:1303–1314.

Mascarenhas, J. P. and M. Alschuler, 1983. The response of pollen to high temperatures and its potential applications. In *Pollen: Biology and Implications for Plant Breeding,* (eds.) D. L. Mulcahy and E. Ottaviano, pp. 3–8. Elsevier, New York.

Mo, Y., C. Nagel, and L. P. Taylor, 1992. Biochemical complementation of chalcone synthase mutants defines a role for flavonols in functional pollen. *Proc. Nat. Acad. Sci.* (USA) 89:7213–7217.

Mogensen, H. L. 1992. The male germ unit: concept, composition and significance. *Internat. Rev. Cytol.* 140:72–126.

Mohapatra, S. S., R. Hill, J. Astwood, A.K.M. Ekramoddoullah, E. Olsen, A. Silvano-vich, T. Hatton, F. T. Kisil and A. C. Sehon, 1990. Isolation and characterization of a cDNA clone encoding an IgE-binding protein from kentucky bluegrass (*Poa pratensis*) pollen. *Internat. Arch. Allergy Appl. Immunol.* 91:362–368.

Morgan, J. M. 1980. Possible role of abscisic acid in reducing seed set in water-stressed wheat plants. *Nature* 285:655–657.

Mu, J-H., J. P. Stains and T. Kao, 1994a. Characterization of a pollen-expressed gene encoding a putative pectin esterase of *Petunia inflata*. *Plant Mol. Biol.* 25:539–544.

Mu, J-H., H-S. Lee and T. Kao, 1994b. Characterization of a pollen-expressed receptor-like kinase gene of *Petunia inflata* and the activity of its encoded kinase. *Plant Cell* 6:709–721.

Muschietti, J., L. Dircks, G. Vancanneyt and S. McCormick, 1994. LAT52 protein is essential for tomato pollen development: pollen expressing antisense LAT52 RNA hydrates and germinates abnormally and cannot achieve fertilization. *Plant J.* 6:321–338.

Nacken, W.K.F., P. Huijer, J. Beltran, H. Saedler and H. Sommer, 1991. Molecular characterization of two stamen-specific genes, *tap* 1 and *fil* 1, that are expressed in wild type, but not in the deficiens mutant of *Antirrhinum majus*. *Mol. Gen. Genet.* 229:129–136.

Napoli, C., C. Lemieux and R. Jorgensen, 1990. Introduction of a chimeric chalcone synthase gene into petunia results in reversible co-suppression of homologous genes in trans. *Plant Cell* 2:279–289.

Nasrallah, J. B. and M. E. Nasrallah, 1993. Pollen-stigma signalling in the sporophytic self-incompatibility response. *Plant Cell* 5:1325–1335.

Newbigin, E., M. A. Anderson and A. E. Clarke, 1993. Gametophytic self-incompatibility systems. *Plant Cell* 5:1315–1324.

Nilsen, B. M., K. Sletten, B. S. Paulsen, M. O'Neill and H. van Halbeek, 1991. Structural analysis of the glycoprotein allergen *Art v* II from the pollen of mugwort (*Artemesia vulgaris* L.) *J. Biol. Chem.* 266:2660–2668.

Niogret, M-F., M. Dubold, P. Mandaron and R. Mache, 1991. Characterization of pollen polygalacturonase encoded by several cDNA clones in maize. *Plant Mol. Biol.* 17:1155–1164.

Paul, W., R. Hodge, S. Smartt, J. Draper and R. Scott, 1992. The isolation and characterization of the tapetum-specific *Arabidopsis thaliana* A9 gene. *Plant Mol. Biol.* 19:611–632.

Pedersen, S., V. Simonsen and V. Loeschcke, 1987. Overlap of gametophytic and sporophytic gene expression in barley. *Theor. Appl. Genet.* 75:200–206.

Peirson, E. S. and M. Cresti, 1992. Cytoskeleton and cytoplasmic organization of pollen and pollen tubes. *Internat. Rev. Cytol.* 140:72–126.

Pitto, L., F. LoSchiavo, G. Guiliano and M. Terzi, 1983. Analysis of heat-shock protein pattern during somatic embryogenesis of carrot. *Plant Mol. Biol.* 2:231–237.

Pollard, T. D. and J. A. Cooper, 1986. Actin and actin-binding proteins. A critical evaluation of mechanisms and functions. *Ann. Rev. Biochem.* 55:987–1035.

Pollak, P. E., T. Vogt, Y. Mo and L. P. Taylor, 1993. Chalcone synthase and flavonol accumulation in stigmas and anthers of *Petunia hybrida*. *Plant Physiol.* 102:925–932.

Potrykus, I. 1991. Gene transfer to plants: assessment of published approaches and results. *Ann. Rev. Plant Physiol. Plant Mol. Biol.* 42:205–225.

Redmond, J. W., M. Baley, M. A. Djordjevic, R. W. Innes, P. L. Kuempel and B. G. Rolfe, 1986. Flavones induce expression of the nodulation genes in Rhizobium. *Nature* 323:632–635.

Roberts, M. R., R. Hodge, J.H.E. Ross, A. Sorensen, D. J. Murphy, J. Draper and R. Scott, 1993. Characterization of a new class of oleosins suggests a male gametophyte-specific lipid storage pathway. *Plant J.* 3:629–636.

Robert, L. S., S. Allard, J. L. Gerster, L. Cass and J. Simmonds, 1993. Isolation and characterization of a polygalacturonase gene highly expressed in *Brassica napus* pollen. *Plant Mol. Biol.* 23:1273–1278.

Rogers, H. J., A. J. Greenland and P. J. Hussey, 1993. Four members of the maize β-tubulin gene family are expressed in the male gametophyte. *Plant J.* 4:875–882.

Russell, S. D., M. Cresti and C. Dumas, 1990. Recent progress on sperm characterization in flowering plants. *Physiologia Plantarum* 80:669–676.

Saini, H. S. and D. Aspinall, 1981. Effect of water deficit on sporogenesis in wheat (*Triticum aesitivum* L.) *Ann. Botany* 48:623–633.

Saini, H. S. and D. Aspinall, 1982a. Abnormal sporogenesis in wheat (*Triticum aesitivum* L.) induced by short periods of high temperature. *Ann. Botany* 49:835–846.

Saini, H. S. and D. Aspinall, 1982b. Sterility in wheat induced by water defecit or high temperature: possible mediation by ABA. *Austral. J. Plant Physiol.* 9:529–537.

Scheiner, O. 1992. Recombinant allergens: biological, immunological, and practical aspects. *Internat. Arch. Allergy Appl. Immunol.* 98:93–96.

Schrauwen, J.A.M., W.H. Reijnen, H.C.G.M. DeLeeuw and M.M.A. van Herpen, 1986. Response of pollen to heat stress. *Acta Bot. Neer.* 35:321–327.

Schrauwen, J.A.M., P.F.M. de Groot, M.M.A. van Herpen, T. van der Lee, W. H. Reynen, K.A.P. Weterings and G. J. Wullems, 1990. Stage-related expression of mRNAs during pollen development in lily and tobacco. *Planta* 182:298–304.

Schwarz-Sommer, Z., P. Huijser, W. Nacken, H. Sadler and H. Sommer, 1990. Genetic control of flower development by homeotic genes in *Antirrhinum majus*. *Science* 250:931–936.

Schwarz-Sommer, Z., I. Hue, P. Huijser, P. J. Flor, R. Hansen, F. Tetens, W-E. Lonning, H. Saedler and H. Sommer, 1992. Characterization of the *Antirrhinum* floral homeotic MADS-box gene *deficiens:* evidence for DNA binding and autoregulation of its persistent expression throughout flower development. *EMBO J.* 11:251–263.

Scott, R., E. Dagless, R. Hodge, W. Paul, I. Soufleri and J. Draper, 1991. Patterns of gene expression in developing anthers of *Brassica napus*. *Plant Mol. Biol.* 17:195–207.

Seiberler, S., O. Scheiner, D. Kraft, D. Lonsdale and R. Valenta, 1994. Characterization of the birch pollen allergen, *Bet v* III, representing a novel class of Ca^{2+} binding proteins; specific expression in mature pollen and dependence of patients' IgE binding on protein-bound Ca^{2+}. *EMBO J.* 13:3481–3486.

Silvanovich, A., J. Astwood, L. Zhang, E. Olsen, F. Kisil, A. Sehon, S. Mohapatra and R. Hill, 1991. Nucleotide sequence analysis of three cDNAs coding for *Poa p* IX isoallergens of kentucky bluegrass pollen. *J. Biol. Chem.* 266:1204–1210.

Singh, M. B. and R. B. Knox, 1985. Grass pollen allergens: antigenic relationships detected using monoclonal antibodies and dot blotting immunoassay. *Internat. Arch. Allergy Appl. Immunol.* 78:300–304.

Singh, M. B., P. M. O'Neill and R. B. Knox, 1985. Initiation of post-meiotic β-galactosidase synthesis during microsporogenesis in oilseed rape. *Plant Physiol.* 77:225–228.

Singh, N. K., D. E. Nelson, D. Kuhn, P. M. Hasegawa and R. A. Bressan, 1989. Molecular cloning of Osmotin and regulation of its expression by ABA and adaptation to low water potential. *Plant Physiol.* 90:1096–1101.

Smith, A. G., C. S. Gasser, K. A. Budelier and R. T. Fraley, 1990. Identification and characterization of stamen- and tapetum-specific genes from tomato. *Mol. Gen. Genet.* 222:9–16.

Sommer, H., J-P. Beltran, P. Huijser, H. Pape, W-E. Lonning, H. Saedler and Z. Schwarz-Sommer, 1990. *Deficiens*, a homeotic gene involved in the control of flower morphogenesis in *Antirrhinum majus:* the protein shows homology to transcription factors. *EMBO J.* 9:605–613.

Staiger, C. J. and W. Z. Cande, 1992. *Ameiotic*, a gene that controls meiotic chromosome and cytoskeletal behavior in maize. *Devel. Biol.* 154:226–230.

Staiger, D., S. Kappeler, M. Muller and K. Apel, 1994. The proteins encoded by two tapetum-specific transcripts, Satap35 and Satap44, from *Sinapis alba* L. are localized in the exine cell wall layer of developing microspores. *Planta* 192:221–231.

Staiger, C. J., M. Yuan, R. Valenta, P. J. Shaw, R. M. Warn and C. W. Lloyd, 1994. Microinjected profilin affects cytoplasmic streaming in plant cells by rapidly depolymerizing actin microfilaments. *Current Biol.* 4:215–219.

Steffensen, D. M. 1966. Synthesis of ribosomal RNA during growth and division of *Lilium*. *Exp. Cell Res.* 44:1–12.

Stinson, J. and J. P. Mascarenhas, 1985. Onset of alcohol dehydrogenase synthesis during microsporogenesis in maize. *Plant Physiol.* 77:222–224.

Stinson, J., A. J. Eisenberg, R. P. Willing, M. E. Pe, D. D. Hanson and J. P. Mascarenhas, 1987. Genes expressed in the male gametophyte of flowering plants and their isolation. *Plant Physiol.* 83:442–447.

Storchova, H., V. Capkova and J. Topy, 1994. A *Nicotiana tabacum* mRNA encoding a 69-kDa glycoprotein occurring abundantly in pollen tubes is transcribed but not translated during pollen development in anthers. *Planta* 192:441–445.

Tanksley, S. D., D. Zamir and C. M. Rick, 1981. Evidence for extensive overlap of sporophytic and gametophytic gene expression in *Lycopersicon esculentum*. *Science* 213:453–445.

Taylor, L. P. and R. Jorgensen, 1992. Conditional male fertility in chalcone synthase-deficient petunia. *J. Heredity* 83:11–17.

Tebbutt, S. J., H. J. Rogers and D. M. Lonsdale, 1994. Characterization of a tobacco gene encoding a pollen-specific polygalacturonase. *Plant Mol. Biol.* 25:283–297.

Thangavelu, M., D. Belostsky, M. W. Bevan, R. B. Flavell, H. J. Rogers and D. M. Lonsdale, 1993. Partial characterization of the *Nicotiana tabacum* actin gene family: evidence for pollen-specific expression of one of the gene family members. *Mol. Gen. Genet.* 240:290–295.

Theerakulpisut, P., H. Xu, M. B. Singh, J. M. Pettit and R. B. Knox, 1991. Isolation and developmental expression of Bcp1, an anther-specific cDNA clone expressed in *Brassica campestris*. *Plant Cell* 3:1073–1084.

Thorsness, M. K., M. K. Kandasamy, M. E. Nasrallah and J. B. Nasrallah, 1991. A *Brassica* S-locus gene promoter targets toxic gene expression and cell death to the pistil and pollen of transgenic *Nicotiana*. *Devel. Biol.* 143:173–184.

Toriyama, K., M. L. Thorsness, J. B. Nasrallah and M. E. Nasrallah, 1991. A *Brassica* S. locus gene promoter directs sporophytic expression in the anther tapetum of transgenic *Arabidopsis*. *Developmental Biology* 143:427–431.

Tsuchiya, T., K. Toriyama, M. E. Nasrallah and S. Ejiri, 1992. Isolation of genes abundantly expressed in rice anthers at the microspore stage. *Plant Mol. Biol.* 20:1189–1193.

Tupy, J. 1982. Alterations of polyadenylated RNA during pollen maturation and germination. *Biol. Plantarum* 24:331–340.

Turgut, K., T. Barsby, M. Craze, J. Freeman, R. Hodge, W. Paul and R. Scott, 1994.

The highly expressed tapetum-specific A9 gene is not required for male fertility in *Brassica napus*. *Plant Mol. Biol.* 24:97–104.

Twell, D. 1992. Use of nuclear-targeted β-glucuronidase fusion protein to demonstrate vegetative cell-specific gene expression in developing pollen. *Plant J.* 2:887–892.

Twell, D., R. Wing, J. Yamaguchi and S. McCormick, 1989. Isolation and expression of an anther-specific gene from tomato. *Mol. Gen. Genet.* 217:240–245.

Ursin, V. M., J. Yamaguchi and S. McCormick, 1989. Gametophytic and sporophytic expression of anther-specific genes in developing tomato anthers. *Plant Cell* 1:727–736.

Valenta, R., M. Duchene, K. Pettenburger, C. Sillaber, P. Valent, M. Breitenbach, H. Rumpold, D. Kraft and O. Scheiner, 1991. Identification of profilin as a novel pollen allergen; IgE autoreactivity in sensitized individuals. *Science* 253:557–560.

van der Meer, I. M., C. E. Spetlt, J.N.M. Mol and A. R. Stuitje, 1990. Promoter analysis of the chalcone synthase (*chsA*) gene of *Petunia hybrida*: a 67 bp promoter region directs flower-specific expression. *Plant Mol. Biol.* 15:95–109.

van der Meer, I. M., M. E. Stam, A. J. van Tunen, J.N.M. Mol and A. R. Stuitje, 1992. Antisense inhibition of flavonoid biosynthesis in petunia anthers results in male sterility. *Plant Cell* 4:253–262.

van Ree, R., V. Voitenko, W. A. van Leeuwen and R. C. Aalberse, 1992. Profilin is a cross-reactive allergen in pollen and vegetable foods. *Int. Arch. Allergy Immunol.* 98:97–104.

van Tunen, A. J., S. A. Hartman, L. A. Mur and J.N.M. Mol, 1989. Regulation of chalcone flavone isomerase (CHI) gene expression in *Petunia hybrida*: the use of alternative promoters in corolla, anthers and pollen. *Plant Mol. Biol.* 12:539–551.

van Tunen, A. J., R. E. Kroes, C. E. Spelt, K. R. van der Krol, A. R. Stuitje and J.N.M. Mol, 1988. Cloning of the two chalcone flavanone isomerase genes from *Petunia hybrida*; coordinate, light regulated and differential expression of flavonoid genes. *EMBO J.* 7:1257–1263.

Vergne, P. and C. Dumas, 1988. Isolation of viable wheat male gametophytes of different stages of development and variations in their protein patterns. *Plant Physiol.* 88:969–972.

Vierling, E. 1991. The roles of heat shock proteins in plants. *Ann. Rev. Plant Physiol. Plant Mol. Biol.* 42:579–620.

Vogt, T., P. Pollak, N. Tarlyn and L. Taylor, 1994. Pollination- or wound-induced kaempferol accumulation in petunia stigmas enhances seed production. *Plant Cell* 6:11–23.

Weterings, K., W. Reijnen, R. van Aarssen, A. Korstee, J. Spijkers, M. van Herpen, J. Schrauwen and G. Wullem, 1992. Characterization of a pollen-specific cDNA clone from *Nicotiana tabacum* expressed during microgametogenesis and germination. *Plant Mol. Biol.* 18:1101–1111.

Wetzel, C.L.R. and W. A. Jensen, 1992. Studies of pollen maturation in cotton: the storage reserve accumulation phase. *Sexual Plant Repro.* 5:117–127.

Wiermann, R. and S. Gubatz, 1992. Pollen wall and sporopollenin. *Internat. Rev. Cytol.* 140:35–72.

Williams, R. F. 1966. Development of the inflorescence in *gramineae*, In *The growth of cereals and grasses* (eds.) F. L. Milthorpe and J. D. Ivins, pp. 59–73. Butterworth, Toronto.

Willing, R. P. and J. P. Mascarenhas, 1984. Analysis of the complexity and diversity of mRNAs from pollen and shoots of *Tradescantia. Plant Physiol.* 75:865–868.

Willing, R. P., D. Bashe and J. P. Mascarenhas, 1988. An analysis of the quantity and diversity of messenger RNAs from pollen and shoots of *Zea mays. Theor. Appl. Genet.* 75:751–753.

Wing, R. A., J. Yamaguchi, S. K. Larabell, V. M. Ursin and S. McCormick, 1989. Molecular and genetic characterization of two pollen-expressed genes that have sequence similarity to pectate lyases of the plant pathogen *Erwinia. Plant Mol. Biol.* 14:17–28.

Wolters, J.H.B. and M.J.M. Martens, 1987. Effects of air pollutants on pollen. *Bot. Rev.* 53:372–414.

Wright, S. Y., M-M. Suner, P. J. Bell, M. Vaudin and A. J. Greenland, 1993. Isolation and characterization of male flower cDNAs from maize. *Plant J.* 3:41–49.

Xiao, C-M. and J. P. Mascarenhas, 1985. High temperature-induced thermotolerance in pollen tubes of *Tradescantia* and heat-shock proteins. *Plant Physiol.* 78:887–890.

Xu, H., N. Theerakulpisut, C. Goulding, P. Suphioglu, P. Bhalla and M. B. Singh, 1995. Molecular analysis of *Ory s* 1, a major allergen of rice pollen (unpublished).

Yanofsky, M. F., H. Ma, J. L. Bowman, G. N. Drews, K. A. Feldmann and E. M. Meyerowitz, 1990. The protein encoded by the *Arabidopsis* homeotic gene *agamous* resembles transcription factors. *Nature* 346:35–39.

Zarsky, V., V. Capkova, E. Hrabetova and J. Tupy, 1985. Protein changes during pollen development in *Nicotiana tabacum* L. *Biol. Plant.* 27:438–444.

Zeevaart, J.A.D. and R. A. Creelman, 1988. Metabolism and physiology of abscisic acid. *Ann. Rev. Plant Physiol. Mol. Biol.* 39:439–473.

Zerback, R., M. Bokell, H. Geiger and D. Hess, 1989. A kaempferol 3-glucosylgalactoside and further flavonoids from pollen of *Petunia hybrida. Phytochemistry* 28:897–899.

Zhang, L., F. T. Kisil, A. H. Sehon and S. S. Mohapatra, 1991. Allergenic and antigenic cross-reactivities of group IX grass pollen allergens. *Internat. Arch. Allergy Appl. Immunol.* 96:28–34.

Zhao, Z-Y and D. F. Weber, 1989. Male gametophyte development in monosomics of maize. *Genome* 32:155–164.

2

Anther-Specific Gene Expression in *Brassica* and *Arabidopsis*

Huiling Xu, R. Bruce Knox, and Mohan B. Singh

Introduction

Pollen, the male gametophyte of flowering plants, plays a vital role in sexual reproduction and hence seed set. Pollen grains are the haploid products of the meiotic divisions of the diploid microsporocytes, which reside in the anther (Fig. 2-1). This developmental process requires an intimate interaction between two generations: the diploid sporophyte including the microsporocytes and anther tissues and the gametophyte (Knox, 1984). More than 26 developmental characters of male reproductive development have now been described, although for only a limited range of taxa (Blackmore et al., 1987). The character states for *Brassica campestris* pollen development are presented in Figure 2-2, and Table 2-2 which is based on data collected by Knox (1987).

An ontogenetic process such as pollen development requires the sequential expression of many genes, some of which will be uniquely expressed in anther tissues. It has been estimated that approximately 20,000–24,000 genes are expressed in mature pollen grains, of which only about 10–20% are specific to pollen (Willing and Mascarenhas, 1984; Willing et al., 1988).

Genes essential for pollen development and functions are likely to be expressed exclusively in pollen and anther tissues. With the advent of molecular techniques, it is possible to isolate these genes, to determine the nature of their products and ultimately, to understand their function in pollen development. These kinds of investigations should provide much needed information regarding the molecular processes that control pollen development. A subsequent opportunity that arises is the exploitation of these genes in creating male sterile systems for hybrid seed production. With this in mind, anther-specific genes and proteins have been a recent focus of research around the world (see reviews by Mascarenhas, 1992; Davies et al., 1992; Scott et al., 1991). As a result, an increasing number of

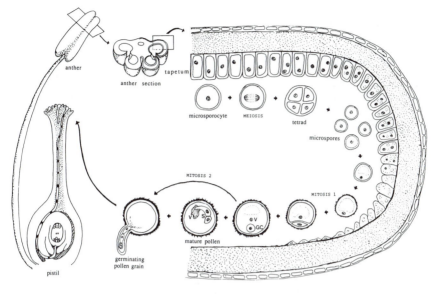

Figure 2-1   Pollen development in higher plant.

anther-specific genes have been isolated from various plants including some of agronomic importance, such as oilseed rape, *B. campestris* and *B. napus,* tobacco, *Nicotiana* spp., and maize, *Zea mays* (see review by Mascarenhas, 1992). Here, we provide an overview of the progress that has been made in the identification and characterization of anther-specific genes in *Brassica,* in particular, and a comparison with studies of other plants. The putative role of these genes in pollen development will be considered.

Identification of Anther-Specific Genes

Genes that are expressed exclusively in the anther or pollen have been isolated using a variety of methods. Differential screening of pollen or anther cDNA

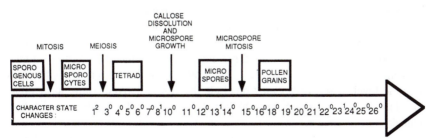

Figure 2-2   Diagram of pollen development in *Brassica* with appropriate character states assigned.

Table 2-1   Summary of Anther-Specific Genes Isolated from Brassica, Arabidopsis, and Other Genera of Family Brassicaceae

Species	Clone Names	Tissue/cell Expression	Putative Function	Minimum Promoter Region	Reference
B. campestris	Bcp1	Tapetum, pollen	Unknown		Theerakulpisut et al., 1991
B. campestris	Bgp1	Tapetum, pollen	Unknown	−168	Xu et al., 1993
B. napus	Bp4	Microspores	Pollen wall protein	−235	Albani et al., 1990
B. napus	Bp10	Microspores	Ascorbate oxidase		Albani et al., 1992
B. napu	Bp19	Microspores	Pectin esterase		Albani et al., 1991
B. napus	E2	Tapetum, microspores	Phospholipid transfer protein		Foster et al., 1992
B. napus	I3	Microspores	Structural protein		Roberts et al., 1991
Arabidopsis thaliana	APG	Tapetum, microspores	Unknown	−530	Roberts et al., 1993
Arabidopsis thaliana	A9	Tapetum	Protease inhibitor, storage protein	−353	Paul et al., 1992

libraries has been the technique most widely used to isolate clones corresponding to anther-specific genes. This method relies on the differential hybridization of two or more probes to individual recombinant clones, in which screening is accomplished by selecting plaques or colonies unique to anther cDNA from cDNA library. Plaque or colony filter lifts are hybridized with a labeled probe made from anther RNA and with a probe made from vegetative tissue. Hybridization signals produced by each clone on the filters are compared. Clones that hybridize to the anther probe but not to the vegetative tissue probes are considered to be putative anther-specific clones. Examples of the successful use of this procedure include the isolation of an anther-specific gene from *B. compestris* (Theerakulpisut et al., 1991) and three microspore-specific genes from *B. napus* (Albani et al., 1990, 1991, 1992). One limitation of this screening approach is that the level of sensitivity may preclude the clones derived from the medium and low abundance transcripts which may fail to give detectable hybridization signals.

To overcome this limitation, a "cold-plaque" screening approach was developed recently in *B. napus* (Hodge et al., 1992). Cold-plaques refer to the recombinant clones that do not give a detectable hybridization signal after hybridization with appropriate probes derived from selected tissues, such as anther or seedling. These cold-plaques are considered to be putative clones derived from low abundance messages and are selected for further analysis. Using this technique, Hodge et al. (1992) have successfully identified 34 anther-specific clones out of 100 cold-plaques from a *B. napus* anther cDNA library. As expected, the majority of them were of low abundance with the peak concentration below 0.02%. The cold-plaque technique is, therefore, particularly useful for isolation of low or medium abundance clones from libraries.

Anther-Specific Genes Display Highly Regulated Patterns of Temporal and Spatial Expression

Table 2-1 summarizes anther-specific genes isolated so far from *Brassica* and *Arabidopsis*. These genes show highly regulated patterns of temporal and spatial expression as determined by RNA blot analysis and in situ hybridization. The majority of the anther-specific genes described to date are found to express in either tapetum or pollen, or in both tissues. Because the tapetum and pollen are highly active and interactive tissues in the anther, it does not come as a surprise that the genes are confined to these tissues.

Tapetum-Specific Genes

The tapetum is the innermost layer of the anther wall and is the tissue in closest contact with the developing microspores (Fig. 2-1). Previous cytochemical and cytological studies revealed that the tapetum is crucial for the successful production of mature, viable pollen. Several functions have been proposed for the tapetum. These functions include production of enzymatic materials for the dissolution of the tetrads (Stieglitz and Stern, 1973), providing nutrition for the developing pollen grains (Pacini, 1990) and production of pollen wall materials (Knox and Heslop-Harrison, 1969). Genes that are expressed exclusively in the tapetum could be responsible for these functions and consequently have attracted much interest.

During early anther development the tapetum is the transcriptionally dominant tissue in the anther. Consequently, the cDNA libraries constructed from immature anthers contain a high proportion of tapetum-specific transcripts. For example, Scott et al. (1991) examined three anther-specific cDNA clones, A3, A8, and A9, isolated from an immature anther cDNA library of *B. Napus* and found that all exhibited tapetum specificity. Furthermore, all tapetum-specific transcripts fall into the high abundance class messages with a peak concentration of 0.04–0.4% (Scott et al., 1991).

The expression of tapetum-specific genes coincides with the differentiation and degeneration of this tissue. For example, the transcripts of the A3 gene from *B. napus* were first detected at the sporogenous cell stage shortly after the differentiation of the tapetum and declined to an undetectable level at the early microspore stage when the tapetum degenerated (Scott et al., 1991).

Pollen-Specific Genes

Degeneration of the tapetum and appearance of the young microspores in anthers leads to a change in the constitution of anther-specific transcripts at the late stage of anther development. The transcriptional activity of the postmeiotic anther becomes dominated by the microspores or pollen.

Two distinct sets of specific transcripts are expressed during pollen development (Stinson et al., 1987). The first, "early," set of genes becomes active soon after meiosis is completed, the genes reach their maximum expression in the microspore and then decrease their activity substantially before pollen maturity. These genes are presumably involved in processes unique to microspore development.

In *Brassica*, eight early genes have been described to date (Albani et al., 1990, 1991, 1992; Scott et al., 1991). Transcripts of these microspore-specific genes are expressed throughout pollen development (Table 2-2). For example, the A6 gene is activated shortly after the completion of meiosis and its activity diminishes substantially prior to microspore release from the tetrad (Scott et al., 1991). In comparison, the I3 gene is expressed from the time of microspore release until the completion of the second pollen mitosis (Roberts et al., 1991). The presence or absence of these microspore-specific genes coincides with the series of unique events that occurs during microspore development, indicating that different gene activity is required for different developmental events.

The second, "late," set of genes becomes active after microspore mitosis and the genes increase their activities during pollen maturation. The transcripts of these late genes accumulate at maximum levels in mature pollen and persist in germinating pollen tubes. These genes are more likely to be involved in the processes of pollen germination and early tube growth. The majority of the pollen-specific clones isolated from mature pollen cDNA libraries appear to correspond to these late genes. Examples of the late genes include the *Bcp1* gene from *B. campestris* (Theerakulpisut et al., 1991), the *Zmc13* gene from *Zea mays* (Hanson et al. 1989), the P2 gene family from *Oenothera* (Brown and Crouch 1990) and three genes, *LAT52, LAT56,* and *LAT59,* from tomato (Twell et al., 1989; Wing et al., 1989).

Genes Expressed in Both Tapetum and Pollen

A unique pattern of gene expression spanning both sporophytic and gametophytic phases of the life cycle has been described for a gene in *Brassica* and *Arabidopsis* (Theerakulpisut et al., 1991; Xu, 1992; Roberts et al., 1991). Using in situ hybridization techniques, Theerakulpisut et al. (1991) identified an anther-specific

Table 2-2   Developmental Characters of Male Reproductive Development

Microspore Formation

1. Type of cytokinesis after meiosis, including whether microspores are of the successive type (1^0), a modified type 1^1) or simultaneous (1^2).
2. Type of cell partitioning during meiosis, either by cell plates (2^0) or constriction furrows (2^1).
3. Tetrads of microspores held within a callose special cell wall, either by common and individual callose walls (3^0) present, common callose wall only (3^1), both walls deleted (3^2), and other variants.
4. Cytomictic channels between microspores in callose special wall, either sealed during meiosis (4^0), absent (4^1), or persisting after meiosis as channels between adjacent tetrads (4^2).
5. Partitioning of microspore walls formed after mieosis separates the microspores from each other (5^0), or wall deposition may be locally delayed (5^1) or deleted (5^2) leaving cytoplasmic channels between the microspores of a tetrad.
6. Configuration of microspores at tetrad stage reflects orientations of the spindles in the second division of meiosis, and five possible configurations occur but are not necessarily mutually exclusive. Tetrads multiplanar, tetrahedral, or decussate (6^0) or uniplanar and tetragonal (6^1); rhomboidal (6^2); T-shaped (6^3); linear (6^4).
7. Deposition of primexine matrix, a thin microfibrillar polysaccharide layer around the microspore surface at late tetrad period (7^0), or microfibrillar surface coat absent (7^1).
8. Deposition of primexine occurs when sporopollenin precursors of the ectexine are assembled on the microspore surface within the matrix (8^0); restricted to small, isolated areas of microspore surface (8^1); or absent (8^2).
9. Ectoapertures defined early in tetrad stage (9^0) or not defined (9^2).
10. Exine differentiates by accumulation of sporopollenin at organized sites (10^0) or sporopollenin accumulates throughout primexine (10^1) or sporopollenin absent (10^2).

Microspore Development

11. Release of microspores from callose special cell wall when microspores immediately separate as monads (11^0) or do not separate as monads until after mitosis (11^1); or remain connected permanently (11^2).
12. Formation of endexine: sporopollenin accumulates on white line centered lamellae (12^0), but sporopollenin can be reduced (12^1), absent (12^2), or form without presence of lamellae (12^3).
13. Formation of endoapertures by localized interruption of endexine deposition: endoapertures not formed and continuous endexine (13^0) or only formed from where endexine deposition is interrupted (13^1) or not formed in absence of endexine (13^2).
14. Vacuolation and microspore growth: vacuolation in free microspores (14^0) or delayed until after mitosis (14^1) or vacuolate stage absent (14^2).

Pollen development

15. Microspore mitosis is asymmetric and produces a large vegetative cell and smaller generative cell: first mitotic division after completion of exine and vacuolation of microspore (15^0); or precocious division at tetrad stage before completion of exine (15^1).
16. Deposition of intine: the pecto-cellulosic intine is the final wall to be deposited around pollen grains and is the only wall of the mature pollen grain invariably present. It may be divided into several zones: intine with at least two distinct layers, one rich in pectins, the other of cellulosic polysaccharides (16^0); or one thick layer with or without elaborate tubules or channels (16^1), or absent except at apertures (16^2).

continued

Table 2-2 Continued.

17. Presence of zwischenkorper, an unusual lens-shaped structure lying above the apertural intine, which plays a part in rehydration and germination: absent (17^0) or present (17^1).

18. Generative cell rounds off within vegetative cell as a cell within a cell, bounded by its own and inner vegetative cell plasma membranes (18^0), or generative cell remains attached to intine (18^1).

19. Generative cell divides to form a pair of sperm cells in pollen tube after germination (19^0) or generative cell divides before tricellular pollen is shed (19^1).

20. Male germ unit formed in which generative cell or pair of sperm cells are linked to vegetative nucleus (20^0) or unit not formed (20^1).

21. Vegetative cell accumulates cytoplasmic storage reserves as starch granules (21^0) or as lipids (21^1).

22. Pollen coat formed from breakdown products of the tapetum that cover exine at pollen maturation (22^0) or absent (22^1).

23. Exine differentiated without substructure (23^0) or columellate (23^1), granular (23^2) or absent (23^3).

24. Apertures of mature pollen are sites at which pollen tubes may emerge during germination, either apertures present (24^0) or absent (24^1).

25. Pollen dehydrated at time of dispersal (25^0) or remaining fully hydrated (25^1).

26. Pollen grains produce one pollen tube during germination (26^0) or multiple pollen tubes (26^1).

Source: Blackmore et al. 1987.

Note: These characters represent sequential landmarks in pollen development that presumably are controlled by different suites of genes. The characters are primarily derived from light and electron microscopic data, including the program of nuclear divisions from meiosis in the microsporocytes to the formation of sperm cells and the processes of wall formation of the complex patterned exine.

cDNA clone, *Bcp1* in *B. campestris*. The *Bcp1* transcript was detected in both tapetal cells and microspores at an early stage of anther development. In the tapetum, the *Bcp1* transcripts decay as the tapetum senses in the later stages of anther development. In the pollen, the *Bcp1* transcripts accumulate to maximum level at the time of pollen maturation and persist throughout pollen germination. The gametophytic activity of the *Bcp1* gene is of the late class. Recently, a homologous gene, *Agp1*, has been isolated from a closely related species, *Arabidopsis thaliana* (Xu, 1992). *Agp1* shows 81% sequence identity with *Bcp1*. Interestingly, the spatial and temporal expression pattern of this gene is identical to that of *Bcp1*, that is, expresses in both haploid pollen and diploid tapetum (Fig. 2-3).

Roberts et al. (1991) identified a sporophytic and gametophytic expressed gene, APG in *B. napus* and *A. thaliana*. The expression pattern of APG was determined in transgenic tobacco by using a reporter gene fusion that consisted of the APG promoter and the β-glucuronidase (*gus*) gene. Analysis of *gus* gene activity in transgenic plants indicates that APG is first active in the tapetum during early microspore development. During mid- and late-microspore development, the gene is expressed in the microspores and in the tapetum, the stomium, and the anther wall. In contrast to *Bcp1*, the gametophytic expression of the

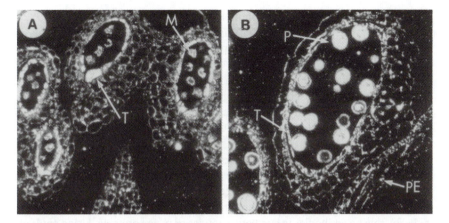

Figure 2-3 In situ hybridization analysis of *Apg 1* transcripts in developing anthers of *Arabidopsis thaliana*. Biotin-labeled antisense *Bcp 1* RNA was used to probe sections containing flowers. Hybridization signals were detected by antibiotin-gold/silver enhanced system and viewed under dark-field microscope. (*A*) *Apg 1* gene expression in immature anthers, high level of expression is observed in the tapetal cells and microspores. (*B*) *Agp 1* expression in mature anthers, very high level of expression was detected in pollen. Tapetum, T; Microspore, M; Pollen, P; Petal, PE. (Xu, 1992).

APG gene is of the early class. Expression reached its maximum around the time of microspore mitosis and declined dramatically during pollen maturation.

Functions of Anther-Specific Genes during Pollen Development

Anther-specific genes are presumably involved in processes that are unique to pollen development such as microsporogenesis, sperm cell formation, pollen germination, and pollen/pistil interactions. To date, relatively little is known about the role of anther-specific genes. Evidence has been obtained in some cases by searching for sequence similarity with known proteins in databases (Table 2-1).

The microspore-specific gene *Bp19* codes for a protein that has striking sequence similarity to pectin esterase from tomato and the pathogenic bacterium *Erwinia chrysanthemi* (Albani et al., 1991). Since pectin is a primary component of the plant cell wall and is also present in the intine wall of the pollen grain, Albani et al. (1991) suggested that *Bp19* coded for a protein associated with the formation of the pollen wall, most likely the intine. This proposed function is further supported by the fact that the timing of *Bp19* expression coincides with the timing of intine deposition in the pollen wall. Similarly, the functional identity of the tapetum/microspore expressed gene, *E2*, has been inferred from its sequence similarity with phospholipid transfer proteins (Foster et al., 1992). Accordingly, the protein encoded by *E2* may involve the transfer of lipids in both the tapetum and developing microspores for the formation of new membranes and storage lipids.

However, sequence similarity may in some cases give ambiguous information. For example, the microspore-specific gene family *Bp10* codes for proteins that showed sequence similarity to ascorbate oxidase (Albani et al., 1992). However, the ascorbate oxidase (AAO) activity centers are not conserved in *Bp10* gene products and no AAO enzyme activity has been detected. This suggests that the *Bp10* gene family is more likely to have an evolutionary relationship with ascorbate oxidase rather than function as the enzyme, as predicted from the sequence similarity analysis (Albani et al., 1992).

Further characterization of protein products of anther-specific genes could be aided by raising antibodies to synthetic peptides or fusion proteins. Using antibodies raised against peptides synthesized according to the predicted amino acid sequence, Theerakulpisut et al. (1991) identified the protein product of *Bcp1* gene from *B. campestris*. The *Bcp1* polypeptide is approximately 12 kD and is not present in other floral or vegetative tissues. In addition, the timing of accumulation of the *Bcp1* polypeptide in pollen more or less coincides with *Bcp1* gene activity (Xu, 1992).

A more direct approach that offers great promise is to use antisense RNA to create a mutant line in which the level of the target gene is selectively reduced or abolished (van der Krol et al., 1988). This is done by introducing a strand of RNA that is complementary to the sequence of the targeted gene to the cell. This results in specific RNA: RNA duplex being formed by base pairing between the target mRNA and antisense RNA. Subsequently, the duplex is rapidly degraded, or mRNA is impaired, or its activity blocked for translation into proteins. If expression of a gene is silenced, a mutant phenotype can be detected and used to assess the function of the gene. Antisense RNA technology has been used successfully to elucidate the biological roles of several plant genes whose functions were previously unknown (Smith et al., 1988, 1990; Hamilton et al., 1989; Visser et al., 1991). For example, by expressing the antisense RNA to the *pTOM13* gene in transgenic tomato plants, Hamilton et al. (1989) demonstrated that *pTOM13* is involved in ethylene metabolism.

The use of antisense RNA technology to ascertain the biological roles of anther-specific genes so far remains largely unexplored. With the availability of anther-specific promoters and genes, it is feasible to apply this technique to investigate the roles of anther-specific genes in pollen development.

The Regulation of Anther-Specific Gene Expression

Anther-specific genes display highly regulated patterns of spatial and temporal expression during pollen development. Extensive analyses have been performed in attempts to identify the promoter regions that control transcriptionally anther-specific gene expression. In most instances, the upstream regions of the anther-specific genes were analyzed by fusing with a reporter gene, β-glucuronidase (*gus*), and subsequently introducing the gene fusions into plants (Jefferson et al., 1986, 1987). Transgenic plants are shown dual expression of *gus* reporter gene and anther-specific genes. The 5' sequences from anther-specific genes analyzed to date, in general, are capable of directing *gus* gene expression in transgenic plants in a pattern identical to that of the endogenous genes (Roberts et al., 1991; Xu et al., 1993). However, in a number of instances the promoters showed a different pattern of expression in transgenic plants from that of the source species. For example, the anther-specific gene promoter, *Bgp1* isolated from *B. campestris* showed different patterns of expression when introduced into *A. thaliana*, a closely related genus in the family Brassicaceae, and *Nicotiana tabacum*, a member of the unrelated family Solanaceae (Xu et al., 1993). In *Arabidopsis*, the *Bgp1* promoter directs *gus* gene expression in both pollen and tapetum, correlating precisely with the endogenous *Bgp1* gene expression in *B. campestris*. In tobacco, however, the same promoter directs *gus* gene expression in pollen, but not in tapetum.

Attempts have been made to define the regulatory elements that lie within the promoter regions of anther-specific genes and presumably regulate anther-specific gene expression by interacting with specific transcriptional factors. Twell et al. (1989, 1991) analysed three pollen-specific promoters, *LAT 52, LAT 56,* and *LAT 59,* in tomato using a series of promoter deletion fragments. These promoter deletions were fused with *gus* reporter gene and subsequently analyzed for their activities in transgenic tobacco plants. The results suggested that at least three distinct regions within the *LAT 52* promoter, 52-A, 52-B, and 52-C, regulate gene expression in pollen. Regions 52-A and 52-B contain sequences that strongly enhance expression, whereas region 52-C is sufficient to direct gene expression (Twell et al., 1991). Similarly, two regions (56-A and 56-B) in *LAT 56* promoter and one region (59-D) in *LAT 59* promoter were shown to be responsible for the activities of these promoters in pollen.

Recently, a detailed analysis has been performed on an anther-specific pro-moter, *Bgp1* from *B. campestris* (Xu et al., 1993). *Bgp1* is a unique promoter active in both diploid tapetum and haploid pollen. To elucidate the way in which it controls gene expression in both diploid and haploid tissues, a series of 5' promoter deletions has been created and introduced into *Arabidopsis thaliana*. The results indicate that *Bgp1* expression in the tapetum and pollen requires the presence of different 5' sequences. It appears that the promoter sequence up to −116 nucleotides relative to the transcriptional start site is sufficient for the specific expression of the *Bgp1* in the tapetum, whereas the region up to −168 nucleotides represents the minimal promoter region for retention of pollen ex-pression.

In summary, the results obtained so far using promoter deletion analysis have defined relatively short promoter sequences, which are required for correct tissue-specific expression in various plants (Table 2-2). However, the precise regulatory elements responsible for anther-specific gene expression have not yet been identified.

Applications of Anther-Specific Genes in Biotechnology

The molecular control of plant reproduction is of increasing interest to the seed industry because the ability to manipulate fertility has important economic implications. The procedures to introduce male sterility into many plant species still remain unknown, and the development of new procedures for crop varieties is a priority. With the production of molecular tools, it has become possible to design new strategies to manipulate male fertility. Male sterility is essential for the production of F1 hybrid seed, which is of interest to seed companies because it produces plants with improved qualities such as increased yields (20 to 50% increase due to hybrid vigor) and increased uniformity when planted as field crops.

Natural and induced male sterility mutations have proved useful for producing hybrid seeds, especially for those plants in which mechanical removal of anthers from flowers is difficult and impractical. The ability to create male sterile plants by genetic engineering will greatly facilitate the production of hybrid seeds and eventually lead to increased crop productivity. Male specific genes are of importance to plant breeders as they enable breeding for desirable traits in a far more predictable and reliable way. The power of this knowledge and technology was first demonstrated in 1990. A tapetum-specific gene was isolated (*TA29* from tobacco) and its promoter used to drive the expression of the fungal ribonuclease gene (Mariani et al., 1990), or diptheria toxin gene, DTA (Koltunow et al., 1990) in the developing anthers of *B. napus* and tobacco, inducing sterile pollen. Both chimaeric *TA29*/RNase genes and *TA29*/DTA genes selectively ablated tapetal function during anther development, blocking pollen formation and leading to the production of male sterile plants.

While these killer gene systems have some limitations including difficulty in the maintenance and propagation of the line and in environmental stability, nevertheless they represent a remarkable advance. The use of herbicide is not an environmentally friendly alternative and the use of restorer genes is extremely time consuming. To overcome these problems, additional systems that use genes of plant origin are needed. One such system, described by van der Meer et al. (1992), involves switching off a key pigmentation gene, chalcone synthase (*chs*) gene in the anthocyanin biosynthesis pathway using an antisense approach. This work indicates indirectly that flavonoids may have an essential function in pollen development. However, the mode of action of the flavonoid in male gametophyte development is not known. The *chs* gene does not show anther or pollen specificity. The inactivation of this gene changed the flower color, as well as the pollen color (yellow to white). Worrall et al. (1992) utilized a tapetum-specific promoter to express a modified pathogenesis-related vacuolar 1,3 β-glucanase (callase) gene in tapetum before the callase activity appear in the anther locule. The precocious expression of this gene led to premature dissolution of the microsporo-cyte callose wall and subsequently resulted in male sterility. Recently, we have used antisense RNA technique to down regulate the expression of a key anther-specific gene *Bcpl* in *A. thaliana*. Transgenic plants harboring antisense genes are male sterile (Fig. 2-4 Xu et al. 1995). The induction of male sterility in the model plant *Arabidopsis* using *Bcp 1* antisense RNA has provided a new technol-ogy for the production of hybrid crop plants that is immediately applicable to breeding of *Brassica* vegetables (for example, cabbage, cauliflower, brussels sprouts, broccoli, kale, and oilseeds such as canola) and ultimately to other crops with homologous genes.

In conclusion, recent progress in cloning and characterizing genes that are expressed specifically in the anther tissues has significantly advanced our knowl-edge of the male haplophase in reproduction. However, to fully understand the molecular mechanisms that control pollen development, a number of questions

Figure 2-4   Male sterile phenotype of *Arabidopsis thaliana* plants carrying antisense *Bcp1* transgene compared with wild type plants. *A. thaliana* is self-fertilising and the male sterile phenotype is characterised by short empty siliques. (a) Wild type plants showing elongated siliques produced following self-pollination. (b) Male sterile transgenic plants carrying short empty siliques. Arrow indicates the junction between the siliques and pedicel. (Xu et al. 1995).

remain unanswered. For example, what is the biological function of the anther-specific genes identified and how do these genes interact during pollen development? In addition, the applications of anther-specific genes in improving plant varieties of potential agricultural or horticultural interest remain largely unexplored. Thus, the future work should be shifted from simply cloning the genes to exploit these aspects.

Acknowledgments

We thank Terryn Hough for helpful discussion and Melissa Fitzgerald for preparing Figure 2.2.

References

Albani, D., L. S. Robert, P. E. Donaldson, I. Altosaar, P. G. Arnison, and S. F. Fabijanski, 1990. Characterization of a pollen-specific gene family from *Brassica napus* which is activated during early microspore development. *Plant Mol. Biol.* 15:605–622.

Albani, D., I. Altosaar, P. G. Arnison, and S. F. Fabijanski. 1991. A gene showing sequence similarity to pectin esterase is specifically expressed in developing pollen of *Brassica napus*. Sequences in its 5′ flanking region are conserved in other pollen-specific promoters. *Plant Mol. Biol.* 16:501–513.

Albani, D., R. Sardana, L. S. Robert, I. Altosaar, P. G. Arnison, and S. F. Fabijanski. 1992. A *Brassica napus* gene family which shows sequence similarity to ascorbate oxidase is expressed in developing pollen. Molecular characterization and analysis of promoter activity in transgenic tobacco plants. Plant J. 2(3):331–342.

Blackmore, S., C. McConchie, and R. B. Knox. 1987. Phylogenetic analysis of the male ontogenetic program in aquatic and terrestrial monocotyledons. *Cladistics* 3:333–347.

Brown, S. M. and Crouch, M. L. 1990. Characterization of a gene family abundantly expressed in *Oenothera organensis* pollen that shows sequence similarity to polygalacturonase. Plant Cell 2:263–274.

Davies, S. P., M. B. Singh, and R. B. Knox. 1992. Identification and in situ localization of pollen-specific genes. *Int. Rev. Cytol.* 140:19–34.

Foster, G. D., R. P. Robinson, M. R. Blundell, R. Hodge, J. Draper, and R. J. Scott. 1992. A *Brassica napus* mRNA encoding a protein homologous to phospholipid transfer proteins is expressed specifically in the tapetum and developing microspores. *Plant Sci.* 84:187–192.

Hamilton, D. A., D. M. Bashe, J. R. Stinson, and J. P. Mascarenhas. 1989. Characterization of a pollen-specific genomic clone from maize. *Sex Plant Reprod.* 2:208–212.

Hanson, D. D., D. A. Hamilton, J. L. Travis, D. M. Bashe, and J. P. Mascarenhas. 1989. Characterization of a pollen-specific cDNA clone from Zea mays and its expression. *Plant Cell* 1:173–179.

Hodge, R., W. Paul, J. Draper, and R. Scott. 1992. Cold-plaque screening: a simple technique for the isolation of low abundance, differentially expressed transcripts from conventional cDNA libraries. *Plant J.* 2(2):257–260.

Jefferson, R. A., S. M. Burgess, and D. Hirsh. 1986. β-Glucuronidase from *Escherichia coli* as a gene-fusion marker. *Proc. Natl. Acad. Sci. USA* 83:8447–8451.

Jefferson, R. A., T. A. Kavanagh, and M. W. Bevan. 1987. GUS fusions: β-glucuronidase as a sensitive and versatile gene fusion marker in higher plants. *EMBO J.* 6:3901–3907.

Koltunow, A. M., J. Truettner, K. H. Cox, M. Wallroth, and R. B. Goldberg. 1990. Different temporal and spatial gene expression patterns occur during anther development. *Plant Cell* 2:1201–1224.

Knox, R. B. 1984. The pollen grain. In *Embryology of Angiosperms,* ed. B. M. Johri, pp. 197–272. Springer-Verlag, Berlin.

Knox, R. B. 1987. Pollen differentiation patterns and male function. In *Differentiation Patterns in Higher Plants,* ed. K. M. Urbanska, pp. 33–52, Academic Press, London.

Knox, R. B. and J. Heslop-Harrison. 1969. Cytochemical localization of enzymes in the wall of the pollen grains. *Nature* 223:92–94.

Mariani, C., M. De Blockeleer, J. Truettner, J. Leemans, and R. B. Goldberg, 1990. Induction of male sterility in plants by a chimaeric ribonuclease gene. *Nature* 347:737–741.

Mascarenhas, J. P. 1992. Pollen gene expression: Molecular evidence. *Int. Rev. Cytol.* 140:3–18.

Pacini, E. 1990. Tapetum and microspore function. In *Microspores: Evolution and Ontogeny,* eds. S. Blackmore and R. B. Knox, pp. 213–237, Academic Press, London.

Paul, W., Hodge, R. Smartt, S., Draper, J. and Scott, R. 1992. The isolation and characterization of the tapetum-specific *Arabidopsis thaliana* A9 gene. Plant Mol. Biol. 19:611–622.

Roberts, M. R., Foster, G. D. Blundell, R. P., Robinson, S. W. Kumar, A. Draper, J. and Scott, R. 1993. Gametophytic and sporophytic expression of an anther specific *Arabidopsis thaliana* gene. The Plant J. 3:111–120.

Smith, C.J.S., C. F. Watson, J. Ray, C. R. Bird, and P. C. Morris. 1988. Antisense RNA inhibition of polygalacturonase gene expression in transgenic tomatoes. *Nature* 334:724–726.

Smith, C.J.S., C. F. Watson, J. Ray, C. R. Bird, and P. C. Morris. 1990. Inheritance and effect on ripening of antisense polygalacturonase gene in transgenic tomatoes. *Plant Mol. Biol.* 14:369–379.

Stieglitz, G. and H. Stern. 1973. Regulation of β-1,3-glucanase activity of developing anthers of lilium. *Develop. Biol.* 34:169–173.

Stinson, J. R., A. J. Eisenberg, R. F. Willing, P. M. Enrico, D. D. Hanson, and J. P. Mascarenhas. 1987. Genes expressed in the male gametophyte of flowering plants and their isolation. *Plant Physiol.* 83:442–447.

Theerakulpisut, P., H. L. Xu, M. B. Singh, J. M. Pettitt, and R. B. Knox. 1991. Isolation and developmental expression of Bcp1, an anther-speciifc cDNA clone in *Brassica campestris*. *Plant Cell*. 3:1073–1084.

Twell, D., R. A. Wing, J. Yamaguchi, and S. McCormick. 1989. Isolation and expression of an anther-specific gene from tomato. *Mol. Gen. Genet*. 217:240–245.

Twell, D., J. Yamaguchi, R. A. Wing, J. Ushiba, and S. McCormick. 1991. Promoter analysis of genes that are coordinately expressed during pollen development reveals pollen-specific enhancer sequences and shared regulatory elements. *Genes Dev*. 5:496–507.

van der Krol, A. R., P. E. Lenting, J. Veenstra, I. M. van der Meer, R. E. Koes, A.G.M. Gerates, and A. R. Stuitje. 1988. An antisense chalcone synthase gene in transgenic plants inhibits flower pigmentation. *Nature* 333:866–869.

van der Meer, I. M., M. E. Stam, A. J. van Tunen, J.N.M. Mol, and A. R. Stuitje. 1992. Antisense inhibition of flavonoid biosynthesis in petunia anthers results in male sterility. *Plant Cell* 4:253–262.

Visser, R.G.F., I. Somhorst, G. J. Kuipers, N. J. Ruys, W. J. Feenstra, and E. Jacobsen. 1991. Inhibition of the expression of the gene for granule-bound starch synthase in potato by antisense constructs. *Mol. Gen. Genet*. 225:289–296.

Willing, R. P. and J. P. Mascarenhas. 1984. Analysis of the complexity and diversity of mRNAs from pollen and shoots of *Tradescantia*. *Plant Physiol*. 75:865–868.

Willing, R. P., D. Bashe, and J. P. Mascarenhas. 1988. An analysis of the quantity and diversity of messenger RNAs from pollen and shoots of *Zea mays*. *Theor. Appl. Genet*. 75:751–753.

Wing, R. A., J. Yamaguchi, S. K. Larabell, V. M. Ursin, and S. McCormick. 1989. Molecular and genetic characterization of two pollen expressed genes that have sequence similarity to pectate lyases of the plant pathogen *Erwinia*. *Plant Mol. Biol*. 14:17–28.

Worrall, D., D. L. Hird, R. Hodge, W. Paul, J. Draper, and R. Scott. 1992. Premature dissolution of the microsporocyte callose wall causes male sterility in transgenic tobacco. *Plant Cell* 4:759–771.

Xu, H. 1992. Molecular analysis of anther-specific genes from *Brassica campestris* and *Arabidopsis thaliana*. Ph.D. Thesis, University of Melbourne.

Xu, H., S. P. Davies, B.Y.H. Kwan, A. P. O'Brien, M. B. Singh, and R. B. Knox. 1993. Haploid and diploid expression of a Brassica campestris anther-specific gene promoter in *Arabidopsis* and tobacco. *Mol. Gen. Genet*. 239:58–65.

Xu, H., Knox, R. B., Taylor, P. E. and Singh, M. B. 1995. *Bcp1*, a gene required for male fertility in *Arabidopsis*. Proc. Natl. Acad. Sci. USA 92(6):2106–2110.

3

Anther- and Pollen-Specific Gene Expression in Sunflower

André Steinmetz, Rachel Baltz, Claire Domon, and Jean-Luc Evrard

Introduction

Among the living organisms plants have a unique life cycle characterized by an alternation of diploid and haploid generations. The role of the diploid sporophytic generation is to produce, following meiosis, haploid spores that develop into haploid male or female gametophytes. Specific cells of the gametophytes then differentiate into male or female gametes which fuse to yield a diploid embryo that finally grows into mature sporophyte. In flowering plants, the diploid, sporophytic generation is the dominant phase, whereas the haploid gametophyte is microscopic and consists of a very reduced number of cells: two or three in the case of the male gametophyte and seven (one of which being binucleate) in the case of the female gametophyte. The haploid male gametophytes, or pollen grains, are produced in the anthers, while the female gametophytes, or embryo sacs, develop in the ovaries (anthers and ovaries are diploid, sporophytic structures). In higher plants, the development of the male and female gametophytes is therefore closely associated with the development of sporophytic tissues. This contrasts with the situation in lower plants where the gametophytes can develop and live as independent free organisms.

The function of the pollen grain is not restricted to the production of the male gametes. Since fertilization occurs in the female gametophyte embedded in sporophytic tissues of the ovary, pollen grains must provide the structures and mechanisms that allow the gametes to reach the embryo sac. These include recognition of a compatible plant via its interactions with specific cells of the stigma, development of the pollen tube, which grows through the stylar transmitting tissue until it reaches the micropyle (an opening in the integuments of the ovules through which the pollen tube enters), and movement of the gametes through the pollen tube down to the embryo sac. These functions are provided essentially by the large vegetative cell of the pollen grain in which the gametes are enclosed.

The structural and functional diversity of cell types in a multicellular organism is the result of differential gene expression. Each cell type expresses, in addition to the many housekeeping genes, specific sets of genes not expressed in other tissues. It is estimated that 20,000–25,000 different genes are expressed in each tissue type; in anthers (including the pollen grains), 40% of these genes appear to be anther-specific (Kamalay and Goldberg, 1980), whereas in isolated pollen grains the proportion of pollen-specific genes has been estimated at 20% (Mascarenhas, 1990).

Gene expression is mainly regulated at the level of transcription by proteins (transcription factors) that bind to specific DNA sequences in the vicinity of the gene to be expressed. The characterization of these proteins and of the DNA sequences to which they bind has become an active field of research in the last few years.

Our research focuses on the regulation of gene expression in the male reproductive tissues of sunflower (*Helianthus annuus* L.). Our aims are to identify the DNA sequences responsible for promoting specific expression of genes in pollen and anther tissues, to identify the proteins implicated in pollen and anther-specific gene expression, and to understand how the interaction of DNA and proteins leads to gene activation in these tissues.

Isolation of Anther- and Pollen-Specific Probes

The study of a gene requires a preliminary isolation of a molecular probe for this gene. A common approach to isolate tissue-specific probes is differential hybridization in which mRNAs are cloned as complementary DNAs (cDNAs) to produce a cDNA library; this library is then differentially screened with labeled RNAs (or cDNAs) from different tissues. Clones hybridizing exclusively with RNAs from one tissue are considered specific for that tissue.

By differential screening of a cDNA library constructed using polyA RNAs from a sunflower inflorescence at anthesis we have isolated about 30 clones that hybridized exclusively with RNA from the inflorescence and not with RNA from roots or leaves (Herdenberger et al., 1990). These clones were further characterized by Northern experiments using RNA from the various parts of the flower, i.e., pistils, corollae, anthers, and pollen. Seven clones (initially named SF1, SF2, SF6, SF7, SF9, SF18, and SF19) hybridized exclusively with RNA from anthers (see hybridization of SF2 cDNA in Fig. 3-1a), whereas two clones (SF3 and SF16) showed exclusive hybridization with RNA from pollen. To check whether some of these clones contained the same or a closely related cDNA, the inserts were isolated from the various clones, radioactively labeled, and used as probes in Southern hybridizations with the various cDNAs. Clones SF1, SF2,

SF6, and SF7 cross-hybridized and were subsequently renamed SF2a, SF2b, SF2c, and SF2d respectively (see Table 3-1). Clones SF9 and SF19 also cross-hybridized and were renamed SF9a and SF9b. The anther-specific clone SF18 as well as the two pollen-specific cDNAs SF3 and SF16 did not cross-hybridize with any of the isolated cDNA clones and are therefore unique among these clones. The data in Table 3-1 also show that most cDNAs are much smaller than the corresponding mRNAs and are therefore considered incomplete.

Expression and Structure of the Anther-Specific Genes SF2 and SF18

Expression of the SF2 and SF18 Genes

A sunflower inflorescence at anthesis contains several hundred individual florets at various development stages, ranging from small green and closed immature florets to mature florets shedding their pollen, as well as florets that have passed maturity and where stamens and pistils have shriveled and retracted into the corolla. The cDNA library therefore contains expressed DNA sequences from the various developmental stages of each floral organ in the inflorescence. To define more precisely at which stage the anther-specific genes were expressed, Northern experiments were performed using RNAs from florets at different developmental stages: The SF2 and SF18 mRNAs were not detected in closed buds

a   **b**

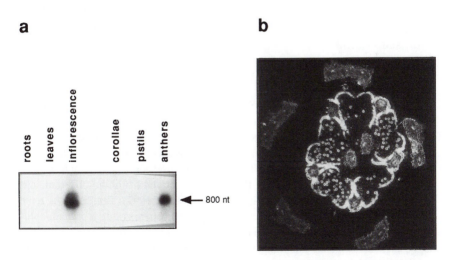

Figure 3-1   a. Northern hybridization of the SF2 cDNA with polyA RNAs from roots, leaves, inflorescence, corollae, pistils, and anthers.   *b.* In situ hybridization of the SF2 cDNA with a cross-section of a disk floret. Visible from the periphery to the center: petals (corolla), anthers with pollen grains, and the two branches of the stigma. Hybridization is visible in the peripheral cells of the anthers.

Table 3-1 Characterization of the cDNA inserts of anther- and pollen-specific cDNA clones from sunflower

	Clone		Length of cDNA	Length of mRNA
	New Name	Initial Name	(bp)	(nt)
Anther-specific	SF2a	SF1	600	800
	SF2b	SF2	600	800
	SF2c	SF6	350	ND
	SF2d	SF7	280	ND
	SF9a	SF9	1000	1700
	SF9b	SF19	600	1600
	SF18	SF18	700	900
Pollen-specific	SF3	SF3	350	1100
	SF16	SF16	400	1350

ND = not determined.

nor in florets from a closed inflorescence, but they were found simultaneously in all the florets of an open inflorescence (Evrard et al., 1991).

Northern hybridizations provide an easy approach to study the temporal expression of a gene, whereas spatial expression is best studied by in situ hybridization. Hybridization of sections through individual florets with an ^{35}S-labeled cDNA probe of SF2b and SF18 revealed that the two genes are expressed in the peripheral anther cells (Fig. 3-1b). No hybridization was detected in pollen grains, corolla, pistil, and filament.

Structure of the Anther-Specific cDNAs

The anther-specific cDNA clones, SF2a and SF2b, were both about 600 bp long, have been reverse-transcribed from an 800 nucleotide long RNA, and cross-hybridized (see Table 3-1). This strongly suggested that they derive from the same mRNA. However, when the nucleotide sequences of the two cDNAs were compared, they were found to differ by a 6 bp insertion (in SF1) and several substitutions (Fig. 3-2), indicating that they are derived from two different, but closely related copies of the same gene (Evrard et al., 1991). Both cDNAs have a typical polyadenylation signal (AATAAA) 17 bp upstream of their polyadenylation site. The unusually high sequence conservation in the untranslated 3' region suggests that the genes encoding the two mRNAs have originated by a recent duplication. Using primer extension we could show that the two SF2 cDNAs are almost full length (18 nucleotides are missing at the 5' terminus of SF2a and one nucleotide is missing in SF2b; see Fig. 3-2).

The lack of cross-hybridization between the anther-specific SF2 cDNAs and the SF18 cDNA clone indicated that the two cDNA sequences are unrelated.

```
SF2a (cDNA)    .................GAACAATTCAGTTTCTTACTTAGTATTGCTTTTGCTCGTCTTCGTTCTTGCA
SF2b (cDNA)    .gtcaactcgaaaATGGCGAACAATTCAGTTTCTTACTTAGTATTGCTTTTGCTCGTCTTCGTTCTTGCA
SF2 gene       actcaactcgaaaATGGCGAACAATTCAGTTTCTTACTTAGTATTGCTTTTGCTCGTCTTCGTTCTTGCA
                            *

                                  ↓
SF2a (cDNA)    ATATCAGAAAGCGCACCGGTACAATATTGCGACAGGGTAACCAACCTTTATCATGAAAAGTGTGACGAGA
SF2b (cDNA)    ATATCAGAAAGCGCACCGGTACAATATTGCGACAGGGTAACCAACCTTTATCATGAAAAGTGTGACGAGA
SF2 gene       ATATCAGGAAGCGCACCGGTACAATATTGCGACAGGGTAACCAACCTTTATCATGAAAAGTGTGACGAGA
                      *

SF2a (cDNA)    AGCAGTGTACTGAACACTGCAAGACAAATGAGAAGGCAGAATCAGGATACTGCCTCGTAGTTGAAAAGCA
SF2b (cDNA)    AGCAGTGTACTGAACACTGCAAGACAAATGAGAAGGCAGAATCAGGATACTGCCTCGTAGTTGAAAAGCA
SF2 gene       AGCAGTGTACTGAACACTGCAAGACAAATGAGAAGGCAGAATCAGGATACTGCCTCGTAGTTGAAAAGCA

SF2a (cDNA)    ACAACTTAGCATATGCAGCTTCGATTGTTCCAAGTATAAACCGAGTACTCCGGCTCCACCTCCACCTCCA
SF2b (cDNA)    ACAACTTAGCATATGCAGCTTCGATTGTTCCAAGTATAAACCGAGTACTCCGGCTCCACCTCCACCTCCA
SF2 gene       ACAACTTAGCATATGCAGCTTCGATTGTGCCAAGTATAAACCGAGTACTCCGGCTCCACCTCCACCTCCA
                                         *

SF2a (cDNA)    CCTCCACCAAAGTTCTTCTATTCTGGTTCATGGTTGCAGGCCAAAGTCGAGAATGTGATGCTTCCTGGGC
SF2b (cDNA)    ......CCAAAGCTCTTCTATTCTGGTTCATGGTTGCAGGCCAAAGTCGAGAATGTGATGCTTCCTGGGC
SF2 gene       ......CCAAAGCTCTTCTATTCTGGTTCATGGTTGCAGGCCAAAGTCGAGAATGTGATGCTTCCTGGGC
               ******     *

SF2a (cDNA)    AAAAGAATATGAATTGCACACAATGTCCCAAAtaatcaacagtctgcatgtttgtatgtttttgagtcttt
SF2b (cDNA)    AAAAGAATATGAATTGCACACAATGTCCCAAAtaatcaacagtctgcatgtttgtatgtttttgagtcttt
SF2 gene       AAAAGAATATGAATTGCACACAATGTCCCAAAtaatcaacagtctgcatgtttgtatgtttttgagtcttt

SF2a (cDNA)    gcatgaagtgtctttaattaaggttatatttgtcattttgagtcatattatagaaaagagtcgttgggta
SF2b (cDNA)    gcatgaagtgtcttaaattaaggttatatttgtcattttgagtcatattctagaaaagagtcgttgggta
SF2 gene       gcatgaagtgtcttaaattaaggttatatttgtcattttgactcatattctagaaaagagtcgttgggaa
                            *                              *       *                 *

SF2a (cDNA)    taactggataaagtgggtagtccatgttgtcaagtagtgggtagattttctcttgattgcatgtttgttt
SF2b (cDNA)    taactggataaagtgggtagtccatgttgtcaagtagtgggtagattttctct.gtttgcatgtttgttt
SF2 gene       taactggataaagtgggtagtccatgttgtcaagtagtgggtagattttctcttgtttgcatgtttgttt
                                                                        *  *

SF2a (cDNA)    acgtggtcatcgtgtatttacagtgtttgtgtttcccttgtgtatttaagctcttgtgttgatgaataaa
SF2b (cDNA)    atgtggccatcgtgtgtttacagtgtttgtgtttcccttgtgtatttaagctcttgtgttgatgaataaa
SF2 gene       acgtggtcatcgtgtatttacagtgtttgtgtttcccttgtgtatttaagctcttgtgttgatgaataaa
                 *     *     *

SF2a (cDNA)    attatccagttccaaca
SF2b (cDNA)    attatccagttccaaca
SF2 gene       attatccagttccaaca(attgtgagttcacttacatatctgatattaccaacaaagacttcgatgt..)
```

Figure 3-2 Comparison of the nucleotide sequences of the SF2a and SF2b cDNAs and of the exonic sequences of a genomic copy of the SF2 gene. Untranslated sequences are shown in lower case letters. Asterisks indicate positions where sequence heterogeneity is observed. Initiation codon, termination codon, and polyadenylation signal are in bold type. The arrow indicates the position of the intron in the gene. The transcription start site (nucleotide number 1) on the genomic sequence was determined by primer extension.

When the nucleotide sequences of the SF2 and SF18 cDNAs were compared, little sequence homology was observed over most of the translated sequence except for a 28 bp sequence at the 5' terminus (Fig. 3-3). In the SF2 cDNA clones this end codes for a hydrophobic amino acid sequence typical of the signal peptides of eukaryotic secreted proteins (see following section). (The sequence of a genomic clone showed that this same signal peptide is also present in the SF18 protein and indicated that the cDNA that was isolated was incomplete.)

Structure of the anther-specific proteins

A complete open reading frame (with initiation codon) of 121 amino acids (Fig. 3-4) was identified upon translation of the SF2b cDNA. The SF2a cDNA being incomplete at the 5' end, two amino acids are missing at the hydrophobic N-terminus (see Fig. 3-5c) of the polypeptide derived from this cDNA. The SF2a and SF2b polypeptides differ at three positions: SF2a has two additional proline residues (positions 91 and 92) in a proline cluster and has a phenylalanine instead of a leucine at position 94. The two polypeptides have a potential N-glycosylation site (*N*CT) at position 115. [The peptide sequence derived from the genomic clone (see following section) differs at two positions from the SF2b peptide: It has G instead of E at position 22, and A instead of S at position 76. The proline cluster contains 6 residues, as in SF2b].

As in the case of the SF2 polypeptides, the SF18 polypeptides also show a

SF2b cDNA	.gtcaactcgaaa**ATG**GCGAACAATTCAGTTTCTTACTTAGTATTGCTTTTGCTCGTCTTCGTTCTTGCA
SF18 cDNA	....TGCTCGTCTTTGTTGTTGCA

SF2b cDNA	ATATCAGAAAGCGCACCGGTACAATATTGCGACAGGGTAACCAACCTTTATCATGAAAAGTGTG......
SF18 cDNA	ATCTCAGATATTGCGACTGT.CAATGGAAAAATATGTGAA.AAACCGAGTAAGACT.TGGTTTG......

Figure 3-3 Comparison of the 5' regions of the SF2 and SF18 cDNAs. Vertical bars indicate sequence identity. Missing nucleotides at the 5' end are indicated by dots. For better alignment, gaps have been introduced into the nucleotide sequence of the SF18 cDNA.

```
                        22
SF2a       ..NNSVSYLVLLLLVFVLAIS E SAPVQYCDRVTNLYHEKCDEKQCTEHCKTNEKAESGYCLVVEKQQLSICSFDC
SF2b       MANNSVSYLVLLLLVFVLAIS E SAPVQYCDRVTNLYHEKCDEKQCTEHCKTNEKAESGYCLVVEKQQLSICSFDC
SF2 (gene) MANNSVSYLVLLLLVFVLAIS G SAPVQYCDRVTNLYHEKCDEKQCTEHCKTNEKAESGYCLVVEKQQLSICSFDC

            76            91
SF2a       SKYKPSTPAPPPPPP PPPK FFYSGSWLQAKVENVMLPGQKNMNCTQCPK
SF2b       SKYKPSTPAPPPPPP.. KLFYSGSWLQAKVENVMLPGQKNMNCTQCPK
SF2 (gene) AKYKPSTPAPPPPPP.. KLFYSGSWLQAKVENVMLPGQKNMNCTQCPK
```

Figure 3-4 Comparison of the peptide sequences derived from the nucleotide sequences of the SF2a and SF2b cDNAs and a genomic clone. Positions of sequence divergence are shown in bold type. The arrow indicates a potential N-glycosylation site.

a

```
cDNA   ..............LVFVVAISDIATVNGKICEKPSKTWFGNCKDTKCDKRCIDWEGAKHGACHQREAKHMCFCYFDC
                     ||||||||:|:||||||||||||||||||:||||||||| |||||||||      |||||||||
gene   MANNSVSYLVLVLLVFVVAISEIGTVNGKICEKPSKTWFGDCKDTDKCDNRCIDWEG         KHMCFCYFGF

cDNA   DPQKNPGPPPGAPGTPGTPPAPPGKGEGDAPHPPPTPSPPGGDGGSGPAPPAGGGSPPPAGGDGGGGAPPPAGGDGGG
       || |||||||||| |   | ||||||||||||||||||||||||| ||||||||||||||||||||||||||:|||
gene   DPKKNPGPPPGAREHQG PPAPPGKGEGDAPHPPPTPSPPGGDGGSTPAPPAGGGSPPPAGGDGGGGAPPPAGGEGGG

cDNA   GAPPPAGGDGGGAPPPGA
       |        |||||
gene   G        GGAPPAS
```

b

```
SF2    MANNSVSYLVLLLLVFVLAISESAPVQYCDRVTNLYHEKCDEKQCTEHCKTNEKAESGYCLVVEKQQLSICSFDCSKY
SF18   MANNSVSYLVLVLLVFVVAISEIGTVNGKICEKPSKTWFGDCKDTDKCDNRCIDWEGKHMCFCYFGFDPKKNPGPPPG

SF2    KPSTPAPPPPPPKLFYSGSWLQAKVENVMLPGQKNMNCTQCPK
SF18   AREHQGPPAPPGKGEGDAPHPPPTPSPPGGDGGSTPAPPAGGGSPPPAGGDGGGGAPPPAGGEGGGGGGGAPPAS
```

c

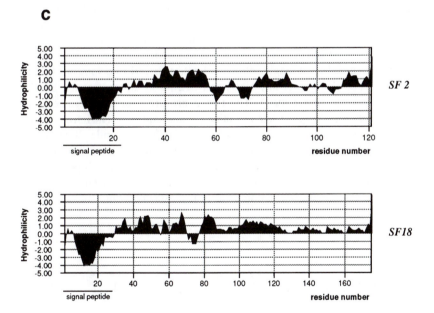

Figure 3-5 a. Comparison of the polypeptides derived from the SF18 cDNA and a genomic copy. Vertical bars indicate sequence identity. Conservative substitutions are indicated by colons. *b.* Comparison of the SF2 and SF18 polypeptides: Signal peptide at the N-terminus, cysteine residues in the central region, and proline residues at the C-terminus are written in bold type. As the SF18 cDNA was incomplete, the peptide sequence derived from the genomic copy was used in this comparison. *c.* Hydropathy plot of the SF2 and SF18 polypeptides derived using Kyte and Dolittle alogrithm. (Kyte and Doolittle, 1982).

heterogeneity in size when the peptide sequences derived from the cDNA sequence (174 amino acids) and the genomic sequence (152 amino acids) are compared (Fig. 3-5a). This heterogeneity is due to three internal deletions in the sequence derived from the genomic clone (11, 9, and 1 residue-long). While their N-terminus is hydrophobic and very similar to that of the SF2 polypeptides (see Fig. 3-5b and c), their C-terminus is characterized by a glycine- and proline-rich dodecapeptide repeat (PPPAGGDGGGGA). The presence of a hydrophobic N-terminus (signal peptide) and of a proline-rich domain suggests that the SF2 and SF18 polypeptides are extracytoplasmic, possibly located in the cell wall (of the peripheral anther cells) where they could carry out a similar, possibly structural function (Domon et al., 1990).

Study of the Anther-Specific Genes SF2 and SF18

Genomic clones containing the SF2 and SF18 genes were isolated by plaque hybridization from a nuclear DNA library constructed in Charon 40 phage. The genes were mapped on the inserts (see Fig. 3-6), and restriction fragments containing the coding region were subcloned and sequenced.

By comparing the cDNA and genomic sequences, the two genes were found to be interrupted by a single intron; this intron is 2615 bp long in the case of the SF2 gene (Domon et al., 1991) and 212 bp in the case of the SF18 gene (Baltz et al., in preparation). The two genes are split at the same position, immediately following the signal peptide coding sequence. The intron-exon junctions are typical of eukaryotic mRNA introns (GT .. AG rule; Breathnach and Chambon, 1981). As these dinucleotides occur quite frequently in introns, and more particularly in the long intron of the SF2 gene (over 100 potential donor and acceptor sites), as well as in exons, we have no idea yet as to how the correct splice sites are selected in plant genes, especially since plant introns lack

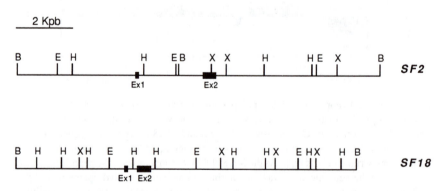

Figure 3-6   Structure of the SF2 and SF18 genes. Ex1 and Ex2 indicate exons. Restriction sites for BamHI (B), EcoRI (E), HindIII (H), and Xbal (X) are shown.

conserved sequences (like the branch point involved in animal and yeast mRNA splicing).

From the comparisons of the cDNA and genomic sequences we can infer that both the SF2 and SF18 genes are multicopy genes. This has been confirmed by Southern hybridization of restricted nuclear DNA with the corresponding radiolabeled cDNA probes. In the case of SF2, an intron-specific probe also produced several bands, indicating that at least some of these genomic copies are split by a closely related intron (Domon et al., 1990).

Expression and Structure of the Pollen-Specific Gene SF3

Expression of the SF3 Gene

The 350 bp SF3 cDNA, isolated initially from the cDNA library, hybridized to a 1100 nucleotide-long mRNA from mature pollen (Table 3-1 and Fig. 3-7). Only a very weak hybridization signal was detected with poly(A) RNA from styles and stigmas and was most likely due to a contamination by pollen grains sticking to the stigmas. Surprisingly, in the experiment shown in Figure 3-7, no hybridization signal was observed with RNAs from closed florets, suggesting that the SF3 gene is a very late pollen gene.

Structure of the SF3 cDNA

The cDNA initially isolated from the cDNA library was only 350 bp long, i.e., only about one-third the size of the corresponding mRNA (see Table 3-1). Two SF3 cDNA clones with a 950 bp long insert were subsequently isolated following

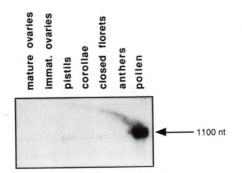

Figure 3-7 Northern hybridization of the SF3 cDNA clone with RNAs from various floral parts.

a second screening of the cDNA library. Although these were the longest inserts detected among the positive clones tested, they were still smaller than the corresponding mRNA whose size was estimated at 1100 nucleotides. One of the two clones was sequenced; its sequence analysis showed that it contained a complete open reading frame of 219 codons preceded by a 104 bp-long and followed by a 140 bp-long untranslated region (Baltz et al., 1992). That translation is unlikely to start at an earlier methionine codon is confirmed by the fact that a stop codon is present 14 codons upstream of the initiation codon. Primer extension showed that this cDNA is complete at the 5' terminus.

Structure of the SF3 Protein

In contrast to the anther-specific proteins SF2 and SF18, the pollen-specific protein SF3 lacks the hydrophobic amino terminus (Figs. 3-8a and 3-8b) and is therefore unlikely to be secreted. It has a repeated structure in which each repeated element (98 amino acids long) contains a cysteine-rich region as well as a basic domain. At the C-terminus the protein contains six copies of a pentapeptide repeat [(A,T,S) (D,E) T Q N].

A computer search for other proteins with similar structures revealed that the SF3 protein belongs to the LIM family of transcriptional regulators, which is characterized by a special arrangement of cysteine and histidine residues into a double finger motif (Baltz et al., submitted). This motif had first been recognized in the proteins LIN-11 from *Caenorhabditis elegans* (Freydt et al., 1990), *I*SL-1 from rat (Karlsson et al., 1990), and *M*EC-3 from *Caenorhabditis elegans* (Way and Chalfie, 1988). Within this family, the pollen-specific SF3 protein shows highest homology with the human cysteine-rich protein hCRP (Liebhaber et al., 1990) and the rat cysteine-rich intestinal protein CRIP (Birkenmeyer and Gordon, 1986) (Fig. 3-8c). We propose that the SF3 protein is a developmentally regulated transcription factor, which might be required for the expression of a set of late pollen genes encoding proteins involved in the formation of the pollen tube, or maybe even in fertilization.

Structure of the SF3 Gene

A genomic clone isolated from the library was restriction mapped and the gene was localized on the 14 kbp insert by Southern hybridization. The nucleotide sequence of a 2200 bp segment containing the gene was determined.

When the SF3 cDNA and genomic sequences were aligned, four introns were found to interrupt the coding sequence (see Fig. 3-9). In contrast to the long intron in the SF2 gene, these introns are short (80 to 132 bp) and AT-rich. Exon-intron junctions are typical of those of eukaryotic introns (GT..AG rule; see Fig. 3-10). No common nucleotide motif could be found near the intron-exon junctions, suggesting that splice site selection is very complex, involving possibly a considerable number of different protein factors that recognize different sequence motifs.

a

MKSFTGTTQK**CTVCEKTVYLVDKLVANQRVYHKACFRCHHCNSTLKLSNFNSFDGVVYCRHH**FD

QLFKRTGSLEKSFDGTPKFKPERTFSQETQSANRLSSFFEGTRDK**CNACAKIVYPIERVKVDGTAYHRA**

CFKCCHGGCTISPSNYIAHEGRLYCKHHHIQLFKKEGNYSQLEVEETVAAPAESETQNTETQNAETQN

ADTQNADTQNTETQNSSV

b

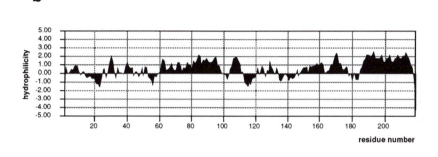

residue number

c

LIM-1

```
1           10           20           30           40           50        56
CTvCEKTVYLVDKLvanqrvYHKaCFRCHHCNsTLkLSNFNsFD..G.vvYCRH.H    SF3
CGvCQKTVYFAEEVQCEGNsFHKsCFLCMVCKKNLD.STTVAV.H.GEEIYCKS.C    hCRP
CPKCDKEVYFAERVTSLGKDWHRPCLKCEKCGKTLT.SGGH.AEHEG.KPYCNHPC    CRIP
```

LIM-2

```
1           10           20           30           40           50  52
CNaCAKIVYPIERVKVDGTAYHRACFKCCHGGCTIsPSNYIAHEGRLYCKHH      SF3
CPRCSQAVYAAEKVIGAGKSWHKACFRCAKCGKGLESTTLADKDGEIYCKGC      hCRP
```

Figure 3-8   a. Amino acid sequence of the SF3 polypeptide. The putative zinc fingers are underlined. LIM domains are written in bold type (each LIM domain consists of two zinc fingers separated by two amino acid residues).   *b.* Hydropathy plot of the SF3 polypeptide.   *c.* Comparison of the LIM motifs of the SF3 protein with the two LIM motifs of the human cysteine-rich protein (hCRP) and the unique LIM motif of the rat cysteine-rich intestinal protein (CRIP). Conserved residues are in bold type. Ligands for the zinc ions in the putative zinc fingers are underlined.

Southern hybridization on genomic blots revealed that the SF3 gene is a multicopy gene, as is the anther-specific SF2 gene. Whether all these copies are expressed or whether some are pseudogenes still has to be determined. But the sequence identity between the sequenced genomic clone and the full length cDNA is at lest a good indication that the genomic copy described here is transcribed.

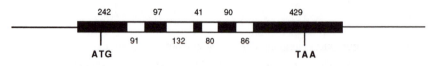

Figure 3-9 Structure of the SF3 gene. Exons are shown in black, introns in white boxes. Size of the exons and introns are indicated in bp above, respectively below, the boxes. Initiation and termination codon are shown.

Discussion and Conclusion

The anther-specific genes SF2 and SF18, as well as the pollen-specific gene SF3, are developmentally regulated and are abundantly expressed at late stages in flower development. The anther-specific proteins have no similarities with already published proteins and there is no experimental evidence yet concerning their function in anthers; there is good reason however to believe that the SF2 and SF18 proteins are components of the anther wall. The pollen-specific protein shares striking homologies with members of the LIM family of developmental regulators identified so far exclusively in animals: it is therefore reasonable to propose that this protein is a developmentally regulated transcription factor involved in the activation of a set of late pollen genes. If this proves to be the case, four questions can be asked: (1) How many and which genes are regulated by the SF3 protein? (2) Does the SF3 protein interact with DNA, and if so, what is the structure of its target site? (3) What is the molecular mechanism of gene regulation by SF3, i.e., with which other component of the transcription initiation complex does it interact? (4) How is the expression of the SF3 gene, which is itself pollen-specific, regulated?

Expression studies have shown that the SF2 and SF18 mRNAs appear at the same time in the same type of anther cells. The two genes therefore could be activated by a very similar mechanism, in which case it wouldn't be surprising to find common nucleotide motifs associated with these genes. This was observed recently (C. D., manuscript in preparation) when the complete nucleotide sequence of the SF18 gene was determined, This gene is not only split by an intron at the same position as is the SF2 gene, but the SF2 and SF18 genes also share

```
CCCTCAGGgtcatatatact ....... 66 bp ............... ccatatgttgcagCTTA (1)
TTTTGATGgtaaagttatat .... 107 bp ............... tgttgtgttttagGAAC (2)
GTCAAGAGgtagttttccac ....... 55 bp ............... ttctttgtaccagACTC (3)
TCGAGCGGgtaaacaaacat ....... 61 bp ............... taatgaaataaagGTTA (4)
AATATCAGgtgcataatcaa .. 2590 bp ............... atgacgtgaacagGAAG (SF2)
```

Figure 3-10 Comparison of the intron-exon junctions in the SF3 and SF2 genes. Exonic sequences are shown in upper-case letters. Bold lower-case letters indicate the conserved dinucleotides at the intron-exon junctions.

common sequence stretches in their immediate upstream region, as well as in the first exon (hence the conserved signal peptide in the two proteins) and in the intron. The possible implication of these conserved sequences in the expression of the two genes is presently being investigated in transient expression assays using GUS (bacterial beta-glucuronidase) as reporter gene. The identification of the sequence motif(s) specifying expression of a gene in peripheral anther cells would allow a subsequent isolation of the protein factors (by screening of an expression library with a labeled double-stranded oligonucleotide probe) and the study of the interaction of these factors with their target sites.

References

Baltz, R., C. Domon, D.T.N. Pillay, and A. Steinmetz. 1992. Isolation and characterization of a pollen-specific gene from sunflower encoding a zinc finger protein. *Plant J.* 2:713–721.

Baltz, R., J. L. Evrard, C. Domon, and A. Steinmetz. (1992) A LIM motif is present in a pollen-specific protein. Plant cell 4:1465–1466.

Birkenmeyer, E. H. and J. I. Gordon. 1986. *Proc. Natl. Acad. Sci. USA*, 83:2516–2520.

Breathnach, R. and P. Chambon. 1981. Organization and expression of eukaryotic split genes coding for proteins. *Annu. Rev. Biochem.* 50:349b–383.

Domon, C., J. L. Evrard, F. Herdenberger, D.T.N. Pillay, and A. Steinmetz. 1990. Nucleotide sequence of two anther-specific cDNAs from sunflower (*Helianthus annuus* L). *Plant Mol. Biol.* 15:643–646.

Domon, C., J. L. Evrard, D.T.N. Pillay, and A. Steinmetz. 1991. A 2.6 kb intron separates the signal peptide coding sequence of an anther-specific protein from the rest of the gene in sunflower. *Mol. Gen. Genet.* 229:238–244.

Evrard, J. L., C. Jako, A. Saint-Guily, J. H. Weil, and M. Kuntz. 1991. Anther-specific, developmentally regulated expression of genes encoding a new class of proline-rich proteins in sunflower. *Plant Mol. Biol.* 16:271–281.

Freydt, G., S. K. Kim, and H. R. Horwitz. 1990. Novel cysteine-rich motif and homeodomain in the product of the *Caenorhabditis elegans* cell lineage gene *lin-11*. *Nature* 344:876–879.

Herdenberger, F., J. L. Evrard, M. Kuntz, L. H. Tessier, A. Klein, A. Steinmetz, and D.T.N. Pillay. Isolation of flower-specific cDNA clones from sunflower (*Helianthus annuus* L). *Plant Sci.* 69:111–122.

Liebhaber, S. A., J. G. Emery, M. Urbanek, X. Wang, and N. E. Cooke. 1990. Characterization of a human cDNA encoding a widely expressed and highly conserved cysteine-rich protein with an unusual zinc-finger motif. *Nucleic Acids Res.* 18:3871–3879.

Kamalay, J. C. and R. B. Goldberg. 1980. Regulation of structural gene expression in tobacco. *Cell* 19:935–946.

Karlsson, O., S. Thor, T. Norberg, H. Ohlsson, and T. Edlund. 1990. Insulin gene enhancer binding protein Isl-1 is a member of a novel class of proteins containing both a homeo- and Cys-His domain. *Nature* 344:879–882.

Kyte, J. and R. F. Doolittle. 1982. A simple method for displaying the hydropathic character of a protein. *J. Mol. Biol.* 157:105–132.

Mascarenhas, J. 1990. Gene activity during pollen development. *Annu. Rev. Plant Physiol. Plant Mol. Biol.* 41:317–338.

Way, J. C. and M. Chalfie. 1988. *mec-3*, a homeobox-containing gene that specifies differentiation of the touch receptor neutrons in *C. elegans. Cell* 54:5–16.

4

Gene Expression in Developing Maize Pollen
Douglas A. Hamilton

Introduction

A pollen grain represents the male portion of the gametophytic stage in the angiosperm plant life cycle. At its maturity, the pollen grain is only three celled and is certainly a very reduced system in a morphological sense. Yet pollen clearly contains all the genetic information necessary for the formation of a complete plant as is shown by the fact that morphologically normal haploid plants can be generated by changing the normal pathway of development of immature pollen (Raghavan, 1976). The normal pathway of development in pollen is obviously a result of gametophytic, and not sporophytic, gene expression, but little is known about the differences between the two. Are they fundamentally the same, or different? Much work has been presented regarding gene expression in sporophytic tissues, but what is known about gene expression in pollen? Work from a number of labs is beginning to provide information regarding the genetic program in the pollen of several species. This chapter attempts to summarize what has been ascertained concerning gene expression during gametophytic development as exemplified in maize pollen.

Maize Pollen Development

In the normal course of *Zea mays* pollen development (reviewed in Kiesselbach, 1949; Goss, 1968; Mascarenhas, 1989), sporogenous cells in the sporophyte anther begin to differentiate within the nutritive tapetal tissue. The importance of this tapetal tissue in the correct development of pollen is supported by the observation that cases of male sterility have been associated with incorrect development of the tapetum itself (Laughnan and Gabay-Laughnan, 1983; Lonsdale, 1987). This observation is further supported by the finding that experimental disruption of anther function by the expression or inhibition of specific genes can result in male sterility (Mariani et al., 1990; van der Meer, 1992).

Following the differentiation of sporogenous cells into pollen mother cells (microsporocytes), these cells then divide meiotically into a tetrad of four haploid cells (microspores). After meiosis there is a long interphase stage during which the microspores separate and enlarge greatly in size and become highly vacuolate. Following this stage there is a very unequal mitotic division (microspore mitosis) which forms two cells, the vegetative and generative cells; the generative cell lies within the cytoplasm of the vegetative cell. The young pollen grain now begins to accumulate starch. In maize a second mitotic division of the generative cell occurs before anthesis, producing the two sperm cells. Thus the mature male gametophyte of maize consists of only three cells. At the point of anthesis the pollen grains are relatively dehydrated and metabolically inactive.

The next phase in the life of the maize pollen grain occurs when the pollen is deposited, normally by wind, upon one of the receptive stigmas (silks) of a female flower. At this point the pollen germinates and extrudes a pollen tube which grows quickly down the style; growth rates of as much as 240 μm/min have been recorded in maize (Heslop-Harrison and Heslop-Harrison, 1982). The two sperm move down the growing pollen tube as a unit (McConchie et al., 1987), and are deposited in the vicinity of the female gametophyte where one sperm joins with the egg and the other joins with the diploid central cell to form the primary endosperm nucleus. The major biosynthetic activities within the growing pollen tube are those associated with cell membrane and cell wall synthesis (Mascarenhas, 1975).

Clearly the activity associated with pollen germination is very different from the type of development seen prior to anthesis, and it might be reasonable to expect that two different sets of genes are expressed during these two periods. Research aimed at examining the genes expressed during maize pollen development has shed some light on the numbers, types, and timing of these genes, as well as the structure and expression of individual pollen-specific genes.

Patterns of Transcription and Translation in Maize Pollen

Research in several species of plants regarding gene expression in pollen has produced clear evidence that transcription and translation do occur in developing and germinating pollen (for a general review, see Mascarenhas, 1989, 1990). For example, in maize, the *Adh1* (alcohol dehydrogenase) gene is expressed in both the sporophyte and gametophyte and is responsible for ADH activity in the gametophyte (Freeling and Schwartz, 1973). The ADH enzyme is dimeric and may consist of two electrophoretic variants, fast (F) or slow (S). Thus plants heterozygous for the variants show dimers of FF, SS, or FS. However, extracts of pollen from heterozygous plants show only the presence of FF or SS homodimers, but no FS heterodimers (Schwartz, 1971). This would be the expected result if these enzymes were the result of gene expression only in the pollen, since individual pollen grains would inherit only one or the other *Adh1* allele, but not both. Thus the dimeric ADH product is dependent only upon the pollen genome and must be a product of pollen transcription and translation. Similar results have been achieved with several other isozymes (Sari-Gorla et al., 1986). In addition, individual members of gene families may be expressed primarily in pollen. In maize pollen, candidates include ADP-glucose pyrophosphorylases (Bryce and Nelson, 1979) and β-glucosidases (Frova et al., 1987). Recent work at the molecular level has shown that, in *Arabidopsis*, at least six α-tubulin genes are expressed in various tissues of the plant, however, one α-tubulin, the product of the *TUA1* gene, was expressed primarily in pollen (Carpenter et al., 1992).

It is assumed that the genes expressed in mature pollen are almost entirely those of the vegetative cell genome. Experimental results to date provide conflicting evidence as to whether transcription or translation is occurring in either the generative cell or sperm cells. In rye, sperm cells have been shown to incorporate ^{3}H-uridine into RNA (Haskell and Rogers, 1985). In contrast, in *Hyoscyamus niger*, neither the generative cell nor its daughter cells, the sperm cells, incorporated detectable amounts of ^{3}H-uridine into RNA (Reynolds and Raghavan, 1982). Perhaps one of the number of published methods for isolating maize sperm cells will provide suitable material for answering this question in maize.

What is known about the transcriptional and translational machinery in pollen? Ribosomal RNA (rRNA) is actively transcribed in the microspore, increasing up to the point of microspore mitosis. After this the rRNA genes decrease in their transcriptional activity and become inactive in the latest stages of pollen maturation. These genes are not transcribed during pollen germination; rather, stored pools of ribosomes and tRNAs are utilized in the protein synthesis that occurs during pollen germination and subsequent pollen tube growth (reviewed in Mascarenhas, 1975, 1988).

Mature pollen grains also contain large stores of presynthesized messenger RNAs (mRNAs), which are translated during germination, as shown by inhibitor studies. In *Tradescantia* pollen, in vitro translation of the mRNAs stored at the point of anthesis results in the production of a set of proteins that appear electrophoretically similar to proteins produced during normal pollen germination (Frankis and Mascarenhas, 1980; Mascarenhas et al., 1984). The mature ungerminated maize pollen grain has been shown to contain between 352 and 705 pg of total RNA, of which 8.9–17.8 pg is poly(A)RNA (Mascarenhas et al., 1984). These figures represent amounts somewhat larger than figures obtained for *Tradescantia* and tobacco pollen grains using similar analyses (Mascarenhas and Mermelstein, 1981; Tupy, 1982).

Thus pollen grains appear to contain all the necessary ribosomes, mRNAs, and proteins necessary for the immediate resumption of transcriptional and translational activity following deposition and hydration at a receptive stigma. This adaptation is consistent with the fact that pollen exists in a generally hostile environment between anthesis and pollination, and yet must be able to germinate rapidly and produce a pollen tube that can penetrate the style, returning it to a more favorable environment in which further development can occur.

Numbers and Types of Genes Expressed

To get an idea of the activity of genes that are expressed in pollen, the numbers of genes active in the pollen of several species has been directly estimated. In maize, the mRNAs of both mature pollen and vegetative shoots have been analyzed by comparing the kinetics of hybridization of homologous reactions of ^{3}H-cDNA to poly(A)RNA in excess (Willing et al., 1988). An analysis of this kind can generate information regarding the general numbers and types of abundance classes of messenger RNAs. For maize pollen the total complexity was calculated to be 2.4×10^7 nucleotides, which is equivalent to about 24,000 different mRNAs. These break down into three abundance classes of messages. Thirty-five percent of the mRNAs are abundant, with about 240 sequences present in roughly 32,000 copies each per pollen grain. The middle abundance class, which is the largest (49%), comprises about 6,000 different sequences present in about 1,700 copies each per grain. The final fraction is the least abundant (15%), with about 17,000 sequences present in only about 200 copies each per grain.

This analysis can be compared with a similar analysis of maize vegetative shoot mRNAs (Willing et al., 1988). Again there appear to be three abundance classes. The highest abundance class (19%) consists of about 60 sequences present in about 250 copies each per vegetative cell. The middle abundance class (54%) consists of about 2,300 sequences each present in about 20 copies per cell. The least abundant fraction (26%) consists of about 29,000 diverse sequences, each present on average only once per cell. The total complexity of the messages in maize shoot tissue comprised 4.0×10^7 nucleotides, or roughly 31,000 diverse mRNAs.

These values for the complexity of mRNAs from pollen and vegetative tissues in maize are roughly similar to those obtained by similar studies with *Tradescantia* (Willing and Mascarenhas, 1984). The number of genes active in whole tobacco anthers has been estimated to be about 26,000 different genes (Kamalay and Goldberg, 1980).

Thus, it is clear that many genes are active in the morphologically simple male gametophyte, especially when compared with the vegetative tissue, which is comprised of multiple cell and tissue types. It is also striking that the individual sequences in mature pollen are present in much greater abundances than the corresponding components in maize shoot tissue. Even the least abundant fraction in pollen contains sequences that are much more abundant (200 copies) than in the corresponding fraction in shoots (approx. 1 copy). This is presumably related to their being stored and needed in large quantities for rapid use at pollen germination.

How many of the genes expressed in pollen are pollen-specific? Many of the genes expressed in pollen are certainly cellular "housekeeping genes" that are also expressed in the sporophyte, but a proportion of them are specific to pollen. A direct estimate of the extent of gametophytic-sporophytic gene expression in maize has been made using heterologous hybridizations between pollen cDNA and shoot poly (A) RNA (Willing et al., 1988). This study suggests that roughly 65% of genes expressed in mature maize pollen are also expressed in vegetative tissues. This estimate is in line with others obtained in this manner for *Tradescantia* (at least 64%; Willing and Mascarenhas, 1984). Other estimates have been made in maize pollen by comparisons of sporophytic versus gametophytic expression of isozymes (Sari-Gorla et al., 1986). Of 34 isozymes analyzed, 29 were found to be expressed both in the sporophyte and the gametophyte, resulting in an estimate of 72% haplo-diploid expression. While all of these estimates are roughly similar, they also may have a rather large degree of uncertainty (Willing et al., 1988).

Another estimate of the percentage of pollen-specific genes expressed in pollen can be made by screening cDNA clones. In a maize cDNA library it was found that approximately 10% of the clones represented messages expressed in pollen and not in vegetative tissues (Stinson et al., 1987). This represents a smaller fraction than was predicted by the hybridization experiments, but the cloning procedure probably selects for the more abundant mRNAs, and it is possible that the pollen sequences might constitute a larger fraction of the rarer mRNAs. Indeed, a recent report suggests that limitations in obtaining "positive" plaques or colonies during differential screening may preclude the isolation of low abundance mRNAs (Hodge et al., 1992). This report indicated that when 100 randomly selected "cold" (i.e. apparently nonhybridizing) plaques from a *Brassica napus* anther cDNA library were isolated and the inserts used as probes for Northern analysis, 34% of the isolated inserts hybridized specifically to pollen or anther RNA. This result confirms that estimates of the numbers of tissue-specific genes derived from standard screening might yield low values, perhaps especially in pollen or anther tissue.

It must be noted that these estimates arise from messages isolated from mature pollen, and so primarily represent genes activated late in pollen development and which may be involved in pollen germination and tube growth. As yet we have little information about the numbers of genes active just after meiosis and during the early stages of microspore development. Whether these are largely similar or different from those expressed during later pollen development is also unclear. It seems reasonable, however, to assume that many of the early-expressed pollen specific genes would be different from those expressed later, reflecting the differences in the types of growth and development occurring at these stages.

The large degree of overlap between the genes expressed both in the gameto-phyte and the sporophyte has prompted the proposal that enhancement of traits in the diploid plant could be achieved by genetic selection for favorable traits at the gametophyte level (Ottaviano et al., 1980; Mulcahy and Mulcahy, 1987; reviewed in Mascarenhas, 1990). Several correlations of this type have been reported in maize (Ottaviano et al., 1980), and screening of this type has appar-ently resulted in the generation of chilling resistant (Barnabas and Kovacs, 1988) and herbicide tolerant maize plants (Sari-Gorla et al., 1989). Plant breeders will no doubt continue to show interest in the possibilities of haploid selection in maize.

Isolation and Sequence Characterization of Genes

A number of pollen-specific cDNA clones from libraries prepared to maize pollen have been isolated and characterized (Stinson et al., 1987; Hanson et al., 1989; Niogret et al., 1991). In addition, at least one of the corresponding genomic clones has also been characterized (Hamilton et al., 1989). Several pollen-specific cDNA clones from a cDNA library made to mature maize pollen have been characterized with respect to the timing of their appearance (Stinson et al., 1987). Developmental Northern analysis showed that the messages corresponding to the identified clones were found to appear just after microspore mitosis and increase thereafter, reaching their maximum accumulation at pollen maturity. Similar results were obtained for several pollen-specific *Tradescantia* cDNA clones (Stin-son et al., 1987). These were compared to the pattern of hybridization to the mRNA of the non-pollen-specific gene actin. The message for this housekeeping gene was found to be first detectable prior to microspore mitosis, reaching a maximum at late pollen interphase and decreasing thereafter. Alcohol dehydroge-nase expression in maize pollen (Stinson and Mascarenhas, 1985) and β-galactosi-dase expression in *Brassica campestris* (Singh et al., 1985) also probably repre-sent genes primarily involved with microspore development and have been termed "early" (Stinson et al., 1987). Thus in contrast, the timing of expression of the maize cDNA clones mentioned above may be characterized as "late."

Another late group of genes, apparently coding for a pollen polygalacturonase, has also been isolated from a mature maize pollen cDNA library (Niogret et al., 1991). These clones were similar to those of Stinson et al. (1987) in that they hybridized to a message which was undetectable in late vacuolated immature pollen but which reached maximum abundance at pollen maturity. As yet, no group has reported the isolation of pollen-specific genes of the early class from maize, although several genes of this type have been reported in other species (Albani et al., 1990; Smith et al., 1990; Roberts et al., 1991).

It has been hypothesized that late genes probably encode for proteins with functions associated with pollen maturation and/or germination and growth of the pollen tube (Stinson et al., 1987; Mascarenhas, 1988). Additional experimental support for a biphasic (early/late) type of gene expression during microsporogenesis has been provided by examination of the proteins produced by developmentally staged microspores (Bedinger and Edgerton, 1990). One-dimensional SDS-PAGE electrophoresis of proteins extracted from pollen at various developmental stages shows a transition in the positions and intensities of numerous bands at around the time of the first microspore mitosis. In vitro translation of RNA isolated from staged microspores also reflects this change. Similar findings were reported by Mandaron et al. (1990). These results correlate well with those of Stinson et al. (1987) and with earlier results in tobacco (Tupy et al., 1983), indicating that there does appear to be a genuine developmental switch occurring in microspore gene expression around the time of microspore mitosis. However, in spite of this evidence it must be noted that the division of gene activity into early and late categories is somewhat artificial and cannot describe all microspore-expressed genes, especially since several pollen-specific genes have shown intermediate or continuous expression during pollen development (Albani et al., 1991, 1992). The roles of genes of this type can only be speculated upon.

Further characterization of one maize cDNA clone, termed Zm13, has been reported. This clone was shown by Northern analysis and transformation studies to be pollen-specific (Hanson et al., 1989; Guerrero et al., 1990). Zm13 is a full length cDNA consisting of 929 nucleotides. This includes a 5'-untranslated region of 127 nucleotides and a 3'-untranslated region of 292 nucleotides. The mRNA codes for a predicted polypeptide of 170 amino acid residues with a molecular mass of 18.3 kD. A strongly hydrophobic region in the hydropathy profile of the predicted polypeptide meets the major criteria for a signal sequence, indicating that the protein is probably secreted across a membrane. There is a possible glycosylation site (NXT). A canonical AATAAA polyadenylation signal motif resides 180 base pairs upstream from the poly(A) addition site (Hanson et al., 1989).

Using the Zm13 clone as a probe, a genomic library of maize has been screened and the corresponding genomic clone has been isolated (Hamilton et al., 1989). The gene has been fully sequenced, including extensive regions of the 5' and 3' flanking regions. No introns are present. There is a single nucleotide difference between the cDNA and genomic clones. This is a G-to-A transition, which is in the 3' untranslated portion of the message. This difference probably arises because the cDNA and genomic libraries were made from the tissues of different strains of maize: the cDNA from Goldcup, a hybrid, and the genomic from W-22, a pure breeding line. This sequence change is not surprising given the high degree of polymorphism in maize (Evola et al., 1986). It is interesting to note that the nucleotide transition occurred in a nontranslated portion of the gene, while the translated portion was entirely conserved.

Zm13 shows significant homology with a pollen- and anther-expressed gene (*lat52*) isolated from tomato pollen (Twell et al., 1989a). The predicted amino acid sequences of these two clones show a 32% identity, including the presence of six conserved cysteine residues. In both cases, Southern analysis appears to indicate that the genes are present in only one or a few copies in the genome. Neither of the two clones shows significant homology with any known protein on record in any protein or DNA sequence databank searched to date. Although it has been reported that *lat52* and Zm13 show partial sequence homology to several proteinase inhibitors (McCormick, 1991), the pollen proteins do not show conservation of the amino acids at the active site of the inhibitors. Recently, a clone has been isolated from a sorghum cDNA library using a fragment of Zm13 as probe (E. Pe, personal communication). This is a partial cDNA, but it shows a 90% sequence homology with Zm13, indicating significant conservation of this gene between these two monocots.

In situ hybridization studies using RNA probes have localized the message for Zm13 to the cytoplasm of the vegetative cell (Hanson et al., 1989). There dos not appear to be any association with the sperm cells. Hybridization with the Zm13 message is also present within the cytoplasm of the germinating pollen tube. Antibodies raised to an 18-amino acid fragment of the predicted peptide have shown reactivity to several bands on a Western blot of proteins isolated from mature pollen (A. Scheewe, T. Reynolds and J. P. Mascarenhas, unpublished data). The antibodies identify two proteins of 27 and 29 kD, both of which are considerably larger than the predicted protein size of 18.3 kD. This size discrepancy is most likely explained by glycosylation of the protein, as pretreatment of the proteins with glycanase results in the enhancement of a band at approximately the predicted size. The amino acid sequence does show a potential glycosylation site on the protein.

The antibodies have been used to determine that the Zm13 protein is translated during pollen germination and tube growth and is not just a remnant from an earlier developmental program. This theory argues that the Zm13 protein may play an important role in pollen germination and tube growth and is supported by the finding that proteins that cross react to the Zm13 antibody are found in the pollens of other plant species (A. Scheewe, T. Reynolds and J. P. Mascarenhas, unpublished data).

Identification of the Sequence Elements Required for Pollen-Specific Expression of Zm13

In order to identify the regions of the 5' flanking region of Zm13 that produce its pollen-specific expression, various deletion fragments of the promoter have been transcriptionally fused to a β-glucuronidase (GUS) reporter gene combined with the polyadenylation region of the nopaline synthase (NOS) gene (Guerrero et al., 1990). These constructs have been introduced into tobacco via *Agrobacterium*-mediated transformation. Promoter fragments were used which extended from about -3000, -1002, -585, and -314, to $+61$, relative to the start of transcription. All of these fragments were successful in directing pollen-specific expression of the GUS gene in tobacco as analyzed both by histochemical staining for GUS expression and by Northern analysis. Transgenic plants containing either the -1002 or -314 constructs showed the appearance of GUS activity beginning after the first mitotic division of the developing pollen grains, corresponding with the known appearance of the Zm13 message in maize pollen. This transcriptional activity indicates that the necessary *cis*-acting elements for the correct spatial and temporal expression of Zm13 reside within the -314 to $+61$ region. It also shows that a monocot pollen promoter can be recognized by the transcriptional apparatus of a dicot (Guerrero et al., 1990).

Recently, the -1002 and -260 constructs have been stably integrated into *Arabidopsis* by *Agrobacterium*-mediated transformation (J. Rueda and J. P. Mascarenhas, unpublished results). Histochemical staining for the GUS enzyme has shown results similar to those with tobacco, i.e., correct temporal and spatial expression. This staining again shows that the Zm13 pollen-specific promoter elements are recognized in both monocots and dicots.

To analyze the Zm13 promoter in greater detail, a transient expression system has been employed (Hamilton et al., 1992), similar to that used with tomato pollen (Twell et al., 1989b). Constructs containing smaller fragments of the promoter fused to the GUS coding region have been introduced into pollen via high-velocity particle bombardment. *Tradescantia* pollen has generally been used for these transformations because of the ease with which it can be reliably stored and germinated; maize pollen is very difficult to use.

The relative expression produced by the various promoter fragments was measured visually by histochemical staining for GUS activity and more quantitatively by measuring production of fluorescent methylumbelliferone (Jefferson, 1987). Controls included bombardment of *Tradescantia* leaves and tobacco suspension cells, as well as bombardment with GUS under the control of the CaMV 35S promoter. Constructs for these experiments consisted of fragments extending from -1001, -260, -100, -54 and -38, all ending at $+61$, fused to GUS/NOS. The highest levels of expression were obtained with the -260 construct. Expression of the other constructs were approximately 52%, 6%, 1%, and undetectable (for the -1001, -100, -54, and -38 fragments, respectively) relative to the expression shown by the -260 construct. The decrease in expression obtained when using the larger (-1001) fragment suggests the presence of a negative regulatory element in this region. The -100 fragment showed a marked decrease in GUS activity, yet retained its pollen-specificity of expression. The -54 fragment showed almost total loss of GUS expression, and expression with the -38 fragment (only 4 bp 5' to the TATA box) was entirely undetectable. These results seem to indicate that a quantitative element(s) responsible for efficient expression in pollen resides in the -260 to -100 region, while a pollen-specific element(s) is present in the -100 to -54 region.

Comparisons of the sequence between −100 and −54 show no obvious consensus sequences with other pollen- and anther-specific genes, although some of the regions presumed to contain specificity elements are fairly small (Twell et al., 1991; van Tunen et al., 1990). Several elements within the −260 to −100 region are similar to the TGTGG motif that Twell et al. (1991) showed could act as an enhancer to a truncated pollen promoter or to a heterologous CaMV 35S promoter. Motifs similar to this have been found within numerous plant promoters such as maize and *Arabidopsis* Adh genes (Ferl and Nick, 1986; McKendree et al., 1990), light regulated genes (Gilmartin et al., 1990; Green et al., 1988; Kuhlemeier et al., 1988), ABA regulated genes (Guiltinan et al., 1990), and is similar to the Hex sequence located in the promoters of the CaMV 35S and *Agrobacterium* nopaline synthase (NOS) genes (Katagiri et al., 1989). Thus a possible model for the positions of active sequences in the Zm13 promoter would have a pollen-specific element core promoter between −100 and −54, with the −260 to −100 region containing one or more enhancerlike elements that amplify the expression of the core promoter. Experiments are underway using these promoter fragments to determine if they conform to this model. Further deletion experiments will locate more precisely the important regulatory elements.

One interesting detail that became apparent during the analysis of Zm13 expression and translation studies was the presence of a temporal delay between the detection of message and detection of the protein. The mRNA for Zm13 is first detectable just after microspore mitosis and accumulates in maize pollen until it reaches a maximum in the pollen grain at anthesis (Guerrero et al., 1990; unpublished data). The protein, however, is not detected in maize until just before anthesis, suggesting the presence of some form of translational control. This temporal delay in translation appears to occur in at least one other published pollen-specific gene (Brown and Crouch, 1990). In contrast, when constructs containing various Zm13 promoter fragments fused to GUS were analyzed in transgenic tobacco plants, no delay in translation was observed (Guerrero et al., 1990). One explanation for this might be that the constructs used are somehow missing additional regions necessary for translational regulation. All of the constructs used in these experiments included the first 61 bases of 5′ untranslated sequence, while the actual Zm13 message has a 128 base 5′ untranslated region. Thus it is possible that regulatory elements act on translation within the +61 to +128 region. Translational repression is known in several systems but is not well understood—several other possibilities exist besides *cis*-acting regulatory elements (Hershey, 1991).

Expression of Specific Genes

Another pollen-specific cDNA clone isolated from maize, designated Zm58 (D. Hamilton, D. Bashe, and F. P. Mascarenhas, unpublished results), shows sequence homology with two tomato clones expressed in tomato anthers and pollen, *lat56* and *lat59* (McCormick et al., 1989). All three of these show a further similarity in their amino acid sequences to pectate lyases known from the plant bacterial pathogen *Erwinia* (McCormick et al., 1989; Wing et al., 1989). This commonality with a plant pathogenic bacteria is suggestive because the invasive growth of pollen within the female tissue in order to fertilize the egg and central cells has been compared to a pathogen-host interaction (Clarke and Gleeson, 1981; Wing et al., 1989). When pollen grains germinate, a pollen tube emerges through the pollen wall and then grows through the tissues of the style. Both of these functions would require both wall degradation and synthesis. During tube growth the principal component of the vesicles that fuse with the growing tube walls is pectic in nature (Heslop-Harrison and Heslop-Harrison, 1982; Pressey and Reger, 1989). Because pectin is a primary component of the cell wall, pollen tube growth would necessitate reallocation of pectin molecules to form new walls. Thus the presence of large quantities of enzymes associated with wall synthesis and degradation would be expected in pollen. In addition to pectate lyase, several cDNA clones corresponding to polygalacturonase have been isolated in pollen from maize (Niogret et al., 1991) and *Oenothera organensis* (Brown and Crouch, 1990), and a cDNA has been isolated which shows sequence similarity to pectin esterase in *Brassica napus* (Albani et al., 1991). The fact that enzymes of this type have been isolated from at least five pollen cDNA libraries suggests that the messages coding for pectin-degrading enzymes are abundantly expressed in pollen.

There are several other reports of genes expressed in maize pollen whose activity has been monitored. In most cases these genes are not pollen specific. Some of the earliest reports of this type were those concerning the *waxy (wx)* locus in maize. This gene encodes the major starch granule-bound ADP-glucose-glucosyl transferase responsible for the synthesis of amylose and is expressed during starch accumulation in pollen as well as in the embryo sac and endosperm (Klosgen et al., 1986; Mascarenhas, 1990). This gene has been characterized at the molecular level (Klosgen et al., 1986). Another body of work has grown around the expression of the *Adh1* gene. Again, this gene, which codes for the enzyme alcohol dehydrogenase, is expressed in pollen as well as several other tissues in the plant (Schwartz, 1971; Felder et al., 1973; Freeling and Schwartz, 1973). The timing of expression of the *Adh1* gene has been described earlier in this chapter, as has that for actin. Other isozyme families where one member is expressed only in maize pollen include ADP-glucose pyrophosphorylases (Bryce and Nelson, 1979) and β-glucosidases (Frova et al., 1987).

The expression of the catalase and superoxide dismutase enzymes have been examined in mature maize pollen. In the maize plant, of the three genes encoding catalase isozymes, the *Cat1* gene is the only one expressed in mature pollen and is transcribed and translated after tetrad formation (Acevedo and Scandalios, 1990). There are six superoxide dismutase (Sod) isozymes expressed in several maize tissues, and at least four of these are detectable in mature pollen at levels similar to those of the scutellum. Thus Sod gene expression is not tissue-specific in maize (Acevedo and Scandalios, 1990).

One area of particular interest involves the expression of heat shock genes and proteins (HSPs) in pollen. When most organisms are subjected to elevated temperatures, they produce a unique set of proteins which, among other functions, apparently help to protect their cells against damage. Although this response is relatively universal (Lindquist and Craig, 1988), some developmental stages of various organisms (often involving reproductive cell types) do not respond to heat shock in a typical fashion (e.g., developing zygotes of *Xenopus*; Bienz, 1984). Mature pollen does not appear to mount a normal heat shock response. While heat shocked immature pollen does synthesize several of the HSPs found in heat shocked vegetative tissues, this ability to respond diminishes during pollen maturation (Frova et al., 1989). Mature and germinating pollen appear incapable of synthesizing normal HSPs (Altschuler and Mascarenhas, 1982; Cooper et al., 1984; Dupuis and Dumas, 1990; Frova et al., 1989). Coupled to this fact is the intriguing fact that maize pollen is relatively sensitive to high temperatures (Dupis and Dumas, 1990). The questions must be asked: Why does mature maize pollen fail to produce normal HSPs? Are HSPs somehow detrimental to late pollen functions? Does the absence of these HSPs relate in any way to the heat sensitivity observed in maize pollen? A recent work has examined levels of several HSP mRNAs in mature maize pollen and found no increase in those transcripts as a result of heat shock, suggesting that the block is at the level of transcription (Hopf et al., 1992). Since high field temperatures are known to depress kernel yields, apparently because pollen viability and germination are inhibited (Schoper et al., 1987), answers to these questions may provide information leading to strategies for improving heat tolerance in maize pollen.

Summary and Conclusions

The angiosperm microgametophyte, as exemplified by maize pollen in this chapter, displays a complex genetic program during its development. There appear to be two main classes of genes expressed during pollen development, with the transition occurring roughly at microspore mitosis: The earlier class of genes is presumed to be associated with immature microspore formation; the later genes probably represent those associated with pollen maturation, germination, and tube growth. This hypothesis is supported by the observation that several pollen-specific clones of the late class have been identified whose sequences are similar to enzymes associated with cell wall synthesis and degradation—a necessary and major requirement for pollen tube growth (Wing et al., 1989).

The genetic promoter elements required for a pollen response continue to elude our understanding, but the seemingly unrestricted interchangeability of pollen promoters among a wide range of host plants argues for some universality within them. Pollen-specific genes have been isolated at an increasing rate in the last several years, and commonalities among the promoters will prove harder to overlook. Eventually, characterization of genes involved in key regulatory processes in pollen should provide insight into the nature of haploid gene expression and its relation to that of the diploid plant. Findings of this nature should point the way toward practical applications of pollen molecular biology in the areas of plant breeding, biotechnology, and basic science.

Acknowledgments

The author gratefully acknowledges the helpful advice and critical reviews of Mike Turcich, Lynne Nicolson, and Joseph Mascarenhas.

References

Acevedo, A. and J. G. Scandalios. 1990. Expression of the catalase and superoxide dismutase genes in mature pollen in maize. *Theor. Appl. Genet.* 80:705–711.

Albani, D., L. S. Robert, P. A. Donaldson, I. Altosaar, P. G. Arnison, and S. F. Fabijanski. 1990. Characterization of a pollen-specific gene family from *Brassica napus* which is activated during early microspore development. *Plant Mol. Biol.* 15:605–622.

Albani, D., I. Altosaar, P. G. Arnison, and S. F. Fabijanski. 1991. A gene showing sequence similarity to pectin esterase is specifically expressed in developing pollen of *Brassica napus*. Sequences in its 5' flanking region are conserved in other pollen-specific promoters. *Plant Mol. Biol.* 16:501–513.

Albani, D., R. Sardana, L. S. Robert, I. Altosaar, P. G. Arnison, and S. F. Fabijanski. 1992. A *Brassica napus* gene family which shows sequence similarity to ascorbate oxidase is expressed in developing pollen. Molecular characterization and analysis of promoter activity in transgenic tobacco plants. *Plant J.* 2:331–342.

Altschuler, M. and J. P. Mascarenhas. 1982. The synthesis of heat shock and normal proteins at high temperatures and their possible roles in survival under heat stress. In *Heat Shock: From Bacteria to Man*, eds. M. J. Schlessinger, M. Ashburner, and M. A. Tissieres, pp. 291–297. Cold Spring Harbor Lab, Cold Spring Harbor, New York.

Barnabas, B. and G. Kovacs. 1988. Perspectives of pollen and male gamete selection in cereals. In *Fertilization and Embryogenesis in Ovulated Plants*, ed. O. Erdelska), pp. 137–147. VEDA, Bratislava.

Bedinger, P. A. and M. D. Edgerton. 1990. Developmental staging of maize microspores reveals a transition in developing microspore proteins. *Plant Physiol.* 92:474–479.

Bienz, M. 1984. Developmental control of the heat shock response in *Xenopus*. *Proc. Natl. Acad. Sci. USA* 81:3138–3142.

Brown, S. M. and M. L. Crouch. 1990. Characterization of a gene family abundantly expressed in *Oenothera organensis* pollen that shows sequence similarity to polygalact-uronase. *Plant Cell* 2:263–274.

Bryce, W. H. and O. E. Nelson. 1979. Starch synthesizing enzymes in the endosperm and pollen of *Zea mays*. *Plant Physiol.* 63:312–317.

Carpenter, J. L., S. E. Ploense, D. P. Snustad, and C. Silflow. 1992. Preferential expression of an α-tubulin gene of *Arabidopsis* in pollen. *Plant Cell* 4:557–571.

Clarke, A. E. and P. A. Gleeson. 1981. Molecular aspects of recognition and response in the pollen-stigma interaction. *Rec. Adv. Phytochem.* 15:161–211.

Cooper, P., T.-H. D. Ho, and R. M. Hauptmann. 1984. Tissue specificity of the heat-shock response in maize. *Plant Physiol.* 75:431–441.

Dupuis, I. and C. Dumas. 1990. Influence of temperature stress on in vitro fertilization and heat shock protein synthesis in maize. (*Zea mays* L.) reproductive tissues. *Plant Physiol.* 94:665–670.

Evola, S. V., F. A. Burr, and B. Burr. 1986. Evidence for the inclusion of controlling elements within the structural gene at the *waxy* locus in maize. *Genetics* 99:275–284.

Felder, M. R., J. G. Scandalios, E. H. Liu (1973) Purification and partial characterization of two genetically defined alcohol dehydrogenase isoenzymes in maize. Biochem Biophys Acta 317:149–159.

Ferl, R. J. and H. S. Nick. 1986. In vivo detection of regulatory factor binding sites in the 5′ flanking region of maize *Adh1*. *J Biol. Chem.* 262:7947–7950.

Frankis, R. C. and J. P. Mascarenhas. 1980. Messenger RNA in the ungerminated pollen grain: a direct demonstration of its presence. *Ann. Bot.* 45:595–599.

Freeling, M. and D. Schwartz. 1973. Genetic relationships between the multiple alcohol dehydrogenases of maize. *Biochem. Genet.* 8:27–36.

Frova, C., G. Binelli, and E. Ottaviano. 1987. Isozyme and HSP gene expression during male gametophyte development in maize. In *Isozymes: Current Topics in Biological and Medical Research, Vol. 15, Genetics, Development and Evolution,* eds. M. C. Rattazzi and J. G. Scandalios, pp. 97–120. Liss, New York.

Frova, C., G. Taramino, and G. Binelli. 1989. Heat shock proteins during pollen development in maize. *Dev. Genet.* 10:324–332.

Gilmartin, P., L. Sarokin, J. Memelink, and N-H. Chua. 1990. Molecular light switches for plant genes. *Plant Cell* 2:369–378.

Goss, J. A. 1968. Development, physiology, and biochemistry of corn and wheat pollen. *Bot. Rev.* 34:333–358.

Green, P. J., M-H Yong, M. Cuozzo, Y. Kano-Murakami, P. Silverstein, and N-H. Chua. 1988 Binding site requirements for pea nuclear protein factor GT-1 correlate with sequences required for light-dependent transcriptional activation of the rbcS-3A gene. *EMBO J.* 7:4035–4044.

Guerrero, F. D., L. Crossland, G. S. Smutzer, D. A. Hamilton, and J. P. Mascarenhas. 1990. Promoter sequences from a maize pollen-specific gene direct tissue-specific transcription in tobacco. *Mol. Gen. Genet.* 224:161–168.

Guiltinan, M. J., W. R. Marcotte, and R. S. Quatrano. 1990. A plant leucine zipper protein that recognizes an abscisic acid response element. *Science* 250:267–271.

Hamilton, D. A., D. M. Bashe, J. R. Stinson, and J. P. Mascarenhas. 1989. Characterization of a pollen-specific genomic clone from maize. *Sex. Plant Reprod.* 2:208–212.

Hamilton, D. A., M. Roy, J. Rueda, R. K. Sindu, J. Sanford, and J. P. Mascarenhas. 1992. Dissection of a pollen-specific promoter from maize by transient transformation assays. *Plant Mol. Biol.* 18:211–218.

Hanson, D. D., D. A. Hamilton, J. L. Travis, D. M. Bashe, and J. P. Mascarenhas, 1989. Characterization of a pollen-specific cDNA clone from *Zea mays* and its expression. *Plant Cell* 1:173–179.

Haskell, D. W. and O. M. Rogers. 1985. RNA synthesis by vegetative and sperm nuclei of trinucleate pollen. *Cytologia* 50:805–809.

Hershey, J.W.B. 1991. Translation control in mammalian cells. *Ann. Rev. Biochem.* 60:717–755.

Heslop-Harrison, J. and Y. Heslop-Harrison. 1982. The growth of the grass pollen tube: 1. Characteristics of the polysaccharide particles ("P-particles") associated with apical growth. *Protoplasma* 112:71–80.

Hodge, R., W. Paul, J. Draper, and R. Scott. 1992. Cold-plaque screening: a simple technique for the isolation of low abundance, differentially expressed transcripts from conventional cDNA libraries. *Plant J.* 2:257–260.

Hopf, N., N. Plesofsky-Vig, and R. Brambl. 1992. The heat shock response of pollen and other tissues of maize. *Plant Mol. Biol.* 19:623–630.

Jefferson, R. A. 1987. Assaying chimeric genes in plants: the GUS gene fusion system. *Plant Mol. Biol. Rep.* 5:387–405.

Kamalay, J. C. and R. B. Goldberg. 1980. Regulation of structural gene expression in tobacco. *Cell* 19:934–946.

Katagiri, F., E. Lam, and N-H. Chua. 1989. Two tobacco DNA-binding proteins have homology to CREB. *Nature* 340:727–730.

Kiesselbach, T. A. 1949. In *The Structure and Reproduction of Corn*. pp. 37–50. Univ. of Nebraska Press, Lincoln, NE.

Klosgen, R. B., A. Gierl, Z. Schwartz-Sommer, and H. Saedler. 1986. Molecular analysis of the *waxy* locus of *Zea mays*. *Mol. Gen. Genet.* 203:237–244.

Kuhlemeier, C., M. Cuozzo, P. J. Green, E. Goyvaerts, K. Ward, and N-H. Chua. 1988. Localization and conditional redundancy of regulatory elements in rbcS-3A, a pea gene encoding the small subunit of ribulose-bisphosphate carboxylase. *Proc. Natl. Acad. Sci. USA* 85:4662–4666.

Laughnan, J. R. and S. Gabay-Laughnan. 1983. Cytoplasmic male sterility in maize. *Annu. Rev. Genet.* 117:27–48.

Lindquist, S. and E. A. Craig. 1988. The heat shock proteins. *Ann. Rev. Genet.* 22:631–677.

Lonsdale, D. M. 1987. Cytoplasmic male sterility: a molecular perspective. *Plant Physiol. Biochem.* 25:265–271.

Mandaron, P., M. F. Niogret, R. Mache, and F. Moneger. 1990. In vitro protein synthesis in isolated microspores of *Zea mays* at several stages of development. *Theor. Appl. Genet.* 80:134–138.

Mariani, C., M. DeBeuckeleer, J. Truettner, J. Leemans, and R. B. Goldberg. 1990. Induction of male sterility in plants by a ribonuclease gene. *Nature* 347: 737–741.

Mascarenhas, J. P. 1975. The biochemistry of angiosperm pollen development. *Bot. Rev.* 41:259–314.

Mascarenhas, J. P. 1988. Anther- and pollen-expressed genes. In *Temporal and Spatial Regulation of Plant Genes,* eds. D.P.S. Verma and R. B. Goldberg, pp. 97–115. Springer-Verlag, New York.

Mascarenhas, J. P. 1989. The male gametophyte of flowering plants. *Plant Cell* 1:657–664.

Mascarenhas, J. P. 1990. Gene activity during pollen development. *Ann. Rev. Plant Physiol. Plant Mol. Biol.* 41:317–338.

Mascarenhas, J. P. and J. Mermelstein. 1981. Messenger RNAs: their utilization and degradation during pollen germination and tube growth. *Acta Soc. Bot. Pol.* 50:13–20.

Mascarenhas, N. T., D. Bashe, A. Eisenberg, R. P. Willing, C. M. Xiao, and J. P. Mascarenhas. 1984. Messenger RNAs in corn pollen and protein synthesis during germination and pollen tube growth. *Theor. Appl. Genet.* 68:323–326.

McConchie, C. A., T. Hough, and R. B. Knox. Ultrastructural analysis of the mature sperm cells of mature pollen of maize. *Zea mays. Protoplasma* 139:9–19.

McCormick, S. 1991. Molecular analysis of male gametogenesis in plants. *Trends Genet.* 6:298–303.

McCormick, S., D. Twell, R. Wing, V. Ursin, J. Yamaguchi, and S. Larabell. 1989. Anther-specific genes: molecular characterization and promoter analysis in transgenic plants. In *Plant Reproduction: From Floral Induction to Pollination,* eds. E. Lord and G. Bernier, pp. 128–135. American Society of Plant Physiology, Rockville, MD.

McKendree, W. L., A-L. Paul, A. J. DeLisle, and R. J. Ferl. 1990. In vivo and in vitro characterization of protein interactions with the dyad G-box of the *Arabidopsis* Adh gene. *Plant Cell* 2:207–214.

Mulcahy, D. L. and G. B. Mulcahy. 1987. The effects of pollen competition. *Am. Sci.* 75:44–50.

Niogret, M-F., M. Dubald, P. Mandaron, and R. Mache. 1991. Characterization of pollen polygalacturonase encoded by several cDNA clones in maize. *Plant Mol. Biol.* 17:1155–1164.

Ottaviano, E., M. Sari-Gorla, and D. L. Mulcahy. 1980. Pollen tube growth rates in *Zea mays:* implications for genetic improvement of crops. *Science* 210:437–438.

Pressey, R. and B. J. Reger. 1989. Polygalacturonase in pollen from corn and other grasses. *Plant Science* 59:57–62.

Raghavan, V. 1976. *Experimental Embryogenesis in Vascular Plants.* Academic Press, New York.

Reynolds, T. L. and V. Raghavan. 1982. An autoradiographic study of RNA synthesis during maturation and germination of pollen grains of *Hyoscyamus niger*. *Protoplasma* 111:177–182.

Roberts, M. R., F. Robson, G. D. Foster, J. Draper, and R. C. Scott. 1991. A *Brassica napus* mRNA expressed specifically in developing microspores. *Plant Mol. Biol.* 17:295–299.

Sari-Gorla, M., C. Frova, G. Binelli, and E. Ottaviano. 1986. The extent of gametophytic-sporophytic gene expression in maize. *Theor. Appl. Genet.* 72:42–47.

Sari-Gorla, M., E. Ottaviano, E. Frascaroli, and P. Landi. 1989. Herbicide-tolerant corn by pollen selection. *Sex. Plant Reprod.* 2:65–69.

Schoper, J. B., R. J. Lambert, B. L. Vasilas, and M. E. Westgate. 1987. Plant factors controlling seed set in maize. *Plant Physiol.* 83:121–125.

Schwartz, D. 1971. Genetic control of alcohol dehydrogenase—A competition model for regulation of gene action. *Genetics* 67:411–25.

Singh, M. B. and Knox, R. B. (1985) Gene controlling beta-galactosidase deficiency in pollen of oilseed rape. J. Hered. 76:199–201.

Smith, A. G., C. G. Gasser, K. A. Budelier, and R. T. Fraley. 1990. Identification and characterization of stamen- and tapetum-specific genes from tomato. *Mol. Gen. Genet.* 222:9–16.

Stinson, J. R. and J. P. Mascarenhas. 1985. Onset of alcohol dehydrogenase synthesis during microsporogenesis in maize. *Plant Physiol.* 77:222–224.

Stinson, J. R., A. J. Eisenberg, R. P. Willing, M. E. Pe, D. D. Hanson, and J. P. Mascarenhas. 1987. Genes expressed in the male gametophyte of flowering plants and their isolation. *Plant Physiol.* 83:442–447.

Tupy, J. 1982. Alterations in polyadenylated RNA during pollen maturation and germination. *Biol. Plant.* 24:331–340.

Tupy, J., J. Suss, E. Hrabetova, and L. Rihova. 1983. Developmental changes in gene expression during pollen differentiation and maturation in *Nicotiana tabacum* L. *Biol. Plant* 25:231–237.

Twell, D., R. Wing, J. Yamaguchi, and S. McCormick. 1989a. Isolation and expression of an anther-specific gene from tomato. *Mol. Gen. Genet.* 247:240–245.

Twell, D., T. M. Klein, M. E. Fromm, and S. McCormick. 1989b. Transient expression of chimeric genes delivered into pollen by microprojectile bombardment. *Plant Physiol.* 91:1270–1274.

Twell, D., J. Yamaguchi, R. A. Wing, J. Ushiba, and S. McCormick. 1991. Promoter analysis of genes that are coordinately expressed during pollen development reveals pollen-specific enhancer sequences and shared regulatory elements. *Genes Devel.* 5:496–507.

van der Meer, I. M., M. E. Stam, A. J. van Tunen, J.N.M. Mol, and A. R. Stuitje, 1992. Antisense inhibition of flavonoid biosynthesis in petunia anthers results in male sterility. *Plant Cell* 4,253–62.

van Tunen, A. J., L. A. Mur, G. S. Brouns, J-D. Rienstra, R. E. Koes, and J.N.M.

Mol. 1990. Pollen- and anther-specific promoters from petunia: Tandem promoter regulation of the *chiA* gene. *Plant Cell* 2:393–401.

Willing, R. P. and J. P. Mascarenhas. 1984. Analysis of the complexity and diversity of mRNAs from pollen and shoots of *Tradescantia*. *Plant Physiol.* 75:865–868.

Willing, R. P., D. Bashe, and J. P. Mascarenhas. 1988. An analysis of the quantity and diversity of messenger RNAs from pollen and shoots of *Zea mays*. *Theor. Appl. Genet.* 75:751–753.

Wing, R. A., J. Yamaguchi, S. K. Larabell, V. M. Ursin, and S. McCormick, 1989. Molecular and genetic characterization of two pollen-expressed genes that have sequence similarity to pectate lyases of the plant pathogen *Erwinia*. *Plant Mol. Biol.* 14:17–28.

5

In Vitro Pollen Cultures:
Progress and Perspectives

Erwin Heberle-Bors, Eva Stöger, Alisher Touraev,
Viktor Zarsky, and Oscar Vicente

Introduction

Today a number of cell culture and molecular techniques are used or are under development to manipulate sexual reproduction. Protoplast fusion is used not only to create somatic hybrids but also to produce cybrids that regenerate into cytoplasmic male sterile plants. Unreduced gametophytes are selected for polyploidization or distant hybridization. In vitro pollination and in vitro fertilization after isolation of egg and sperm cells are used to overcome pre- and postfertilization crossing barriers (Shivanna and Johri, 1985).

On the molecular level, DNA fingerprinting and Restriction Fragment Length Polymorphism analysis are used to screen for rare hybrids. A recently developed modification of the polymerase chain reaction (PCR) allows the generation of specific molecular markers termed randomly amplified polymorphic DNAs (RAPDs). Since this procedure is technically simple, quick to perform, and requires only small amounts of DNA, the routine use of RAPD markers in large-scale crop improvement programs is now feasible (Waugh and Powell, 1992). One of the most spectacular achievements of molecular methods has been, however, the creation of male sterile and restorer lines by transforming plants with the barnase/barstar genes cloned behind anther-specific promoters (Mariani et al., 1990, 1992). The molecular analysis of self-incompatibility has produced fascinating insight into the communication of the two sexes (Thompson and Kirch, 1992) but until now no molecular technique to manipulate self-incompatibility has developed from this analysis.

In most cases, pollen is the target for manipulation of sexual plant reproduction because male gametogenesis is less protected than female gametogenesis in vivo. Pollen development can be manipulated in a further way. In in vitro cultures, microspores can develop to mature pollen grains that are fertile. Since they do this without interaction with the surrounding anther tissue, they are free from controls exerted in vivo by the sporophyte. This allows separation of the processes under the command of the pollen genome from the processes in the sporophytic anther tissues and also for some potential applications in plant breeding. When short-term stress treatments are applied on microspores or young pollen grains, they change their developmental direction in a fundamental way. Instead of pollen grains, i.e., male gametophytes, they develop into haploid embryos, i.e., sporophytes. Again, such haploids are of interest for applications in plant breeding, and the investigation of the processes governing induction of pollen embryogenesis provide an alternative to somatic embryogenesis. This chapter presents the progress that has been achieved with these two techniques and the perspective that these techniques have.

Pollen Development

Pollen represents the male gametophytic generation in the life cycle of higher plants. Its development depends totally on the sporophyte since it takes place within tissues of the sporophyte, i.e., in the anther. Following flower initiation and the formation of anther primordia the archesporial cells divide to form tapetal and pollen mother cells. The tapetal cells undergo rapid cell division and polyploidization and develop into a typical nutritional tissue, which can be an amoeboid or a cellular, secretory tapetum, as in tobacco. The pollen mother cells are separated from the surrounding tapetum by a callose wall and undergo meiosis. The microspores are released from the tetrads through the action of a tapetum-derived callase. The microspores form the characteristic pollen wall, consisting of intine and exine. Exine formation depends largely on the tapetum and is completed before first pollen mitosis. Intine formation in tobacco starts during microsporogenesis but it is only after first pollen mitosis when the intine becomes more prominent, particularly at the apertures (Kastler, 1990). Tobacco microspores pass quite synchronously through one cell cycle and produce a large vacuole that "pushes" the microspore nucleus from its initially central position to a peripheral position. Toward the end of microsporogenesis the tapetum starts to degenerate.

First pollen mitosis is an asymmetrical division resulting in a small generative cell and a large vegetative cell (early binucleate stage). The vegetative cell engulfs the generative cell so that the two cells are separated by two plasma membranes and a thin cell wall in between (mid-binucleate stage). The generative cell undergoes a rapid cell cycle immediately after first pollen mitosis and—in tobacco and other species—is arrested in G2 phase. In the vegetative cell, the vacuole is rapidly absorbed by a highly active cytoplasm, as indicated by the formation of a large nucleolus in the vegetative nucleus. The vegetative cell does not undergo a further cell cycle but remains arrested in G0 (Zarsky et al., 1992).

It terminally differentiates into what makes up most of the mature pollen grain. Nutrients are taken up from the anther and many genes are switched on at the mid-binucleate stage, encoding for functions required during pollen development and during pollination and tube growth (Mascarenhas, 1990). Most of the latter functions occur on the basis of mRNAs and proteins that are synthesized during pollen development and are stored in the mature pollen (Mascarenhas, 1988, 1989, 1990). During the latest stages, the pollen undergoes dehydration (Hoekstra et al., 1988) and prepares for shedding and pollination. The mature tobacco pollen is filled with storage material (starch grains and lipid droplets). The exine cavities are filled with material derived from the tapetum, and the pollen grain is covered with the remnants of the tapetum (pollenkitt and tryphin, Pacini, 1990).

The interest to study pollen dates back to the work of Amici in 1824, and pollen structure, development, and function has been described in many species since then. Descriptive work is ongoing, and fascinating insights are still produced, now mostly on the ultrastructural level. The analytical period probably started with the experiments of Shimakura (1934) and Gregory (1940) when for the first time anthers were excised from the plant and cultured under more or less controlled conditions in vitro. A first important result from such studies was that meiosis can proceed relatively independently from the tapetum in isolated meiocytes, at least starting with the zygotene stage of the meiotic prophase (Ito and Stern 1967).

In similar experiments, Guha and Maheshwari (1964, 1966) cultured immature *Datura innoxia* anthers in vitro. To their surprise, embryos emerged from the cultured anthers that developed into haploid plants. Since then, pollen-derived plants have been produced in anther cultures of many species.

Anther Culture Versus Pollen Culture

Anther cultures are technically simple and are presently used in many commercial labs for dihaploid production (for details see section on dihaploids for plant breeding). For studies on the mechanism of embryogenic induction, they are, however, hardly applicable since the sporophytic anther wall tissue interferes in the process. Even if it usually does not proliferate to produce diploid and heterozygote regenerants, the anther wall affects pollen embryogenesis by providing both beneficial and inhibitory conditions and substances that are not yet well characterized. The position within the anther may also affect the fate of the individual pollen grain simply by controlling access to nutrients from the medium. More importantly, the heterogeneous cell population in the anther (containing only a very small fraction of embryogenic pollen from which embryos and eventually haploid plants will develop) does not allow the use of standard biochemical and molecular methods to study the process of induction of pollen embryogenesis. In contrast, in isolated pollen cultures the pollen grains are (more or less) evenly distributed in the culture medium, they can easily be watched under the microscope and, under optimal conditions, highly homogeneous populations of embryogenic cells can be obtained. On the other hand, embryo formation from pollen cultures is usually technically more difficult. Appropriate methods to isolate the pollen from the anthers are required, specific culture vessels have to be used, and usually, medium changes have to be performed.

Pollen Cultures as Experimental Systems of Plant Generative Development

Over the years a plethora of pollen culture techniques for embryo formation have been developed (Heberle-Bors, 1985, 1989). Today, there exist two true ab initio pollen cultures that produce embryos in good yields. In tobacco, mid-binucleate pollen is isolated and cultured for a few days in a starvation medium. In this medium, embryogenic pollen grains are formed that develop into embryos after transfer to a rich medium (Kyo and Harada 1986, Benito Moreno et al., 1988a; Garrido et al., 1991). In *Brassica napus,* microspores at around the first pollen mitosis are isolated and cultured in a rich medium directly. A short-term, mild heat shock of about 32°C initiates embryogenesis that proceeds at lower temperatures (Lichter, 1985). In both systems, no further pretreatment of plants or flower buds is required. Both systems are used to study the molecular events occurring during embryogenic induction.

In cereals, whole plants have been recovered from isolated microspores of rice (Datta et al., 1990a,b; Bolik and Koop, 1991), and maize (Pescitelli et al., 1990; Gaillard et al., 1991; Abele et al., 1992) but yields are still low in relation to the number of microspores cultured. In maize, in contrast, no specific treatment but growth substances in the medium seem to sufficient for embryogenic induction (Pescitelli et al., 1990; Mitchell and Petolino, 1991).

Microspores or young pollen grains have another choice for developing in in vitro cultures. Tobacco pollen grains at the early binucleate stage that can be induced to form embryogenic pollen grains develop into mature, fertile pollen when they are cultured directly in a rich medium (Benito Moreno et al., 1988b). Younger stages can also be cultured to maturity. Microspores require a more complex medium for the early stages of in vitro maturation, and a medium change has to be performed in order to complete maturation. In wheat, binucleate pollen grains can be cultured to maturity in vitro (Stauffer et al., 1991). Genetic tests proved that the in vitro matured pollen pollinated and fertilized the mother plant and not some accidental contaminating in vivo pollen (Benito Moreno et al., 1988b; Stauffer et al. 1991).

In these initial experiments, the fraction of microspores that acquired full gametophytic function in vitro was relatively low. Pollen developed not as synchronously as in vivo, as seen in some size variation of the in vitro matured pollen, and germination frequency in vitro and seed set after in situ pollination were lower than with in vivo pollen. An improvement of in vitro maturation has been achieved by a better simulation of the conditions that normally prevail inside the anther. After addition of nucleosides to the first medium and of proline and more sucrose to the second medium, pollen developed much more synchronously, higher yields of seeds were obtained, fruit set was much more reproducible and, in addition, the in vitro matured pollen could be stored maintaining its gametophytic function (Tupy et al., 1991, A. Touraev, unpubl. res.). This indicates that indeed pollen culture under these conditions is a reasonable simulation of in vivo pollen development inside the anther.

Actually, these pollen cultures are one of the very few plant in vitro systems of cytodifferentiation, in which an explant cell remains committed to the developmental pathway it was committed to before isolation. In most other plant in vitro systems in contrast, explant cells dedifferentiate and form calli that can be induced to regenerate into organs or whole plants via somatic embryogenesis or shoot/root organogenesis. With the two types of pollen culture we now have an experimental system that allows us to study one of the most fundamental developmental processes in plants, i.e., the alternation of generations. In a rich medium the microspores and pollen grains develop into gametophytes as they would in vivo, and a short stress treatment induces the development of sporophytes (Heberle-Bors, 1989).

Molecular Analysis of Embryogenetic Induction in Pollen Cultures

As mentioned previously, there are at present two efficient systems of embryogenic pollen culture developed in *Nicotiana tabacum* and in *Brassica napus*. In both cases a stress treatment (starvation and heat shock, respectively) applied at a specific moment during gametophytic development triggers the induction of embryogenic development. In rapeseed premitotic and mitotic microspores, as well as early postmitotic pollen grains, are competent for induction by the heat shock treatment. This results in two different pathways for embryogenesis initiation, either by equal division of the microspore or by division of the vegetative cell in the young pollen grain (Pechan and Keller, 1988; Telmer et al., 1992). The tobacco system has the advantage of a single pathway for embryogenesis induction. The developmental stage from which pollen embryogenesis can be most efficiently induced is the mid-binucleate stage. Bud selection combined with Percoll density gradient centrifugation allows preparation of a highly homogeneous population of mid-binucleate pollen that is cultured under starvation conditions. After transfer to a sugar-containing medium, the vegetative pollen cell divides to form eventually an embryo (Kyo and Harada, 1986; Garrido et al., 1991).

Analysis of cell cycles during normal and embryogenic tobacco pollen development has shown that one of the crucial events of embryogenic induction is the reactivation, during the starvation treatment, of the cell cycle in the vegetative cell which is arrested in G1 throughout gametophytic development (Zarsky et al., 1990, 1992). To confirm the data produced by classical cytometry after Feulgen staining, thymidine-labeling and autoradiography were used on the pollen cultures to study pollen cell cycles.

A similar analysis has been now performed during embryogenic induction from *B. napus* early binucleate pollen, showing also in this case reactivation of the cell cycle in the vegetative nucleus that enters S-phase during the heat shock treatment (Zaki and Dickinson 1991, Binarova, Zarsky, Pechan, in prep.).

The importance of cytoskeleton dynamics in the regulation of plant development is well documented, and changes in the cytoskeleton also seem to be important during induction of pollen embryogenesis. Kyo and Harada (1986) described a massive reorganization of cellular structures in embryogenic pollen grains of tobacco. In rapeseed, while no preprophase band is present in gametophytic microspores (see Baskin and Cande 1990, for a review) one of the first cytological changes observed during embryogenic induction is the formation of a preprophase band that divides the microspore in two halves (D. Simmonds, pers. comm.). Changes in the actin network have also been recently detected during induction of embryogenesis in this system (Binarova, pers. comm.).

Tobacco and rapeseed embryogenic pollen (microspore) cultures are being used to study the molecular mechanisms underlying the developmental switch from gametophytic toward sporophytic pollen development. Changes in gene expression associated with the inductive phase of pollen embryogenesis have been investigated. In *Brassica,* several specific mRNAs and proteins are synthesized in response to the heat shock (Pechan et al., 1991). It is not yet clear whether these proteins are really involved in embryogenic induction or are simply the products of heat shock genes required to survive the heat stress. In tobacco, on the contrary, no new proteins reproducibly associated with formation of embryogenic pollen have been detected (Kyo and Harada, 1990a; Garrido et al., 1992). However, in vitro translation of RNAs isolated from mid-binucleate and from embryogenic pollen allowed detection of at least two abundant mRNA species, coding for low molecular weight proteins that are transcriptionally induced during the starvation treatment and accumulate in embryogenic pollen but presumably are not translated in vivo (Capkova et al. 1988, Garrido et al., 1992). Experiments with inhibitors suggested that transcription during starvation is required for subsequent formation of embryos in the rich medium (Garrido et al., 1992).

Transcriptional activation of genes during the starvation treatment has been confirmed in one specific case. A tobacco cDNA clone coding for a small heat shock protein (18 kDa) has been isolated from a mid-binucleate pollen cDNA library, and Northern blot analysis has revealed that much higher levels of its mRNA are present in embryogenic pollen as compared to mid-binucleate pollen (Vicente, et al. 1992). These results are particularly interesting since they suggest a common molecular mechanism for induction of pollen embryogenesis in tobacco and *Brassica:* homologous genes seem to be induced in the two systems by starvation and heat shock, respectively, although the signal transduction pathway of the two stress treatments may be different. By differential screening of specific pollen cDNA libraries, other genes activated during the inductive phase of pollen embryogenesis no doubt will be isolated and characterized in the near future.

The cell cycle studies previously mentioned suggest that genes involved in control of the cell cycle may play an essential role in the process of induction of pollen embryogenesis. Therefore, the study of the expression of specific cell cycle genes during the heat shock (in rapeseed) or starvation (in tobacco) treatments can be used as an alternative approach to investigate the molecular basis of embryogenic induction. We have recently isolated a tobacco cDNA clone coding for the p34cdc2 protein kinase, the major regulator of cell cycle in eukaryotic cells. RNAs hybridizing with the cdc2 probe have been detected in mid-binucleate and embryogenic tobacco pollen, but the transcripts are clearly different (Wilson et al. 1993). Although alternative explanations cannot be ruled out, it is possible that the starvation treatment induces the expression of an embryo-specific cdc2 gene. Experiments are in progress to confirm this hypothesis.

Post-translational modification, and specifically phosphorylation, of proteins also seems to be involved in the process of embryogenic induction. After in vivo labeling with $^{32}P_i$ added to the pollen cultures, specific protein phosphorylation patterns associated with induction of pollen embryogenesis have been observed in *N. tabacum* and *N. rustica:* about six proteins, presumably localized in the plasma membrane, are specifically phosphorylated in embryogenic pollen grains (Kyo and Harada, 1990a, 1990b; Kyo and Ohkawa, 1991). In vitro phosphorylation assays in extracts of mid-binucleate and embryogenic tobacco pollen also showed qualitative and quantitative changes on protein kinase activities during the starvation treatment (Garrido et al., 1992). Therefore, protein kinases are likely to be involved in the transduction of the starvation signal, mediating its effects on gene expression and cell cycle regulation. Characterization of protein kinase activities during normal and embryogenic pollen development could also give valuable insights into the mechanisms of induction of pollen embryogenesis.

These studies in the model plants tobacco and rapeseed will hopefully reveal the basic molecular mechanisms regulating the induction of pollen embryogenesis and eventually open the possibility of controlling and even manipulating this process in these and other species.

Dihaploids for Plant Breeding

For some field crops, the production of dihaploid plants from in vitro anther culture is now an established technique for producing pure lines in one step

(Morrison and Evans, 1988). The primary regenerants are usually haploid and can easily be diploidized by colchicin treatment. In most cereals, spontaneous diploids are formed that can be used directly in breeding programs (Henry and de Buyser, 1990). Commercial varieties have been produced using dihaploid breeding such as the wheat variety Florin in France (Henry and de Buyser, 1990).

Dihaploids from anther culture are commonly produced from preselected breeding material in the F3 to F4 generation to achieve full homozygosity. With better pollen plant yields, the population of dihaploids will much better represent the full scope of meiotic recombination, and it will be possible to select directly from the dihaploids of segregating F1 plants. Despite the progress in application, the processes that lead to the formation of an embryo from an immature pollen are still largely unknown and cannot be controlled to a sufficient degree. In all crop species from which pollen plants can be produced in anther cultures, there exists a strong genotypic dependence, and somaclonal variation represents a serious threat to the applicability of the technique. In tobacco, for example, it has been shown that DNA amplification takes place in haploids derived from pollen, but not in those derived from ovules (Reed and Wernsman 1989). Another limitation of the technique is the possibility of formation (e.g., in potato) of heterozygous diploids from unreduced microspores.

In cereals, the formation of albino plantlets is a big problem, and in durum wheat, no or only very few green plantlets can be produced (Hadwiger and Heberle-Bors, 1986). A screen of all tetraploid wheats with the AABB genome identified only one species (*T. dicoccon*) which gave rise to green plantlets (Löschenberger and Heberle-Bors, unpubl. results). An enigma is the formation of spontaneous diploids in variable frequency that regenerate in cereal anther cultures. In hexaploid wheat, normal-looking plants that are aneuploids and show reduced fertility at seed set regenerate in a disturbingly high frequency (Youssef et al., 1989). The frequency of albinos and of spontaneous diploids is genotype-dependent (Löschenberger and Heberle-Bors, 1992).

Prospects of Dihaploid Production from Pollen Cultures

In some *Brassica oleracea* subspecies, dihaploids cannot be produced in anther cultures although it is possible in pollen cultures (Duijs et al. 1989; Takahata and Keller, 1991). This illustrates that one of the prospects of pollen culture is the production of dihaploids from pollen in cases in which anther culture fails. In anther cultures only a minute fraction of the pollen grains start to divide and a fraction of these form embryos and even fewer form plants. In good tobacco pollen cultures in contrast, the majority of pollen grains can be induced to divide depending on the quality of the donor plants, the synchrony of pollen development, and appropriate conditions during starvation. Much still must be done to define the culture conditions that allow all the dividing pollen grains to complete embryonic development. Only once all or a high fraction of pollen grains can be regenerated to plants will it be possible to approach the ideal that represents a population of pollen plants with all viable meiotic recombinants. Dihaploids could then be produced in large numbers from F1-hybrids, and selection on the haploids or doubled haploids would produce superior lines that are true breeding.

High yields of pollen plants are also a prerequisite for the use of pollen cultures for in vitro selection. In most cases so far, diploid cell or tissue cultures have been used for in vitro selection (van den Bulk, 1991). Only dominant genes can be detected in such an approach. In a few cases only, haploid cell cultures have been used (Carlson, 1973; Sacristan, 1985). Embryogenic pollen cultures of *Brassica napus* have been used to exploit haploidy after a mutagenic treatment (Polsoni et al., 1988; Swanson et al., 1988).

Pollen cultures may also solve the problem of albino formation, spontaneous diploidization and other forms of somaclonal variation found in cereal anther-culture derived plants. In one-step regeneration anther cultures, less albinos and spontaneous diploids are produced (Liang and Tang, 1987) indicating that the culture regime has a strong influence on albino formation and spontaneous diploidization. Only in pollen cultures can the culture conditions be controlled so stringently that derivations from the normal and wanted form of development can be avoided.

Pollen Cultures to Study Normal Pollen Development and Pollen Function

The experiments on in vitro pollen maturation in tobacco (Benito Moreno et al., 1988b; Tupy et al., 1991) and wheat (Stauffer et al., 1991) helped to settle a dispute to which Mascarenhas has pointed (1990). To which degree does pollen development depend on the sporophyte, or what is the developmental autonomy of the pollen? A clear example of dependency from the sporophyte is the production of callase in the tapetum that dissolves the callose around the tetrads (Worrall et al., 1992). Our experiments showed that in vitro tobacco pollen development is possible with microspores that have a fully formed exine at the time of isolation. This stage coincides with the degeneration of the tapetum. This result indicates that once the microspore is fully developed, gametophyte development is autonomous in the sense that it requires only nutrients from the sporophyte. At a time when the bulk of pollen-specific gene expression takes place, there seems to exist no control by the sporophyte.

Early and mid uninucleate microspores invariably died in liquid culture (A. Touraev, unpubl. res.). Although it cannot be excluded that better knowledge of in vitro culture conditions may allow earlier stages to be cultured, we feel that this will be possible only in "artificial anthers" that simulate pollen development to a much higher degree. Only in such artificial anthers will it be possible to find out whether young microspores can produce an exine on their own and whether the presence of an exine is a defining property of a pollen grain.

Gene expression as measured by two-dimensional gel electrophoresis of ^{35}S-methionine-labeled proteins indicated that in vitro matured pollen differ to some extent from in vivo matured pollen (van Herpen et al., 1992). Since in vivo pollen passes through dehydration, differences in gene expression are to be expected, and the comparison of in vivo and in vitro matured pollen may allow the identification of genes involved in pollen dehydration. These authors were also able to show that in vitro matured pollen acquires first the ability to set seed in situ and then the ability to germinate in vitro. This finding may indicate that the in vitro pollen receives stimulatory substances from the stigma which it lacks in vitro.

This is in fact what we recently found. There exists hormonal control of pollen development by the sporophyte. In the early experiments, even with the improvements of Tupy et al. (1991), seed set after pollination with in vitro matured pollen never was as high as with in vivo pollen although synchrony of pollen development in vitro was high. Addition of a diffusate from in vivo pollen to the germination medium of in vitro matured pollen strongly increased seed set and also germination frequency of the in vitro matured pollen. We then found that flavonols are present in the diffusate, and addition of flavonols such as quercetin, kaempferol, or myricetin at low concentration had a strong enhancing effect on pollen germination, pollen tube growth, and seed set (Ylstra et al., 1992). Other flavonoids had no effect. Molecular work had shown that the genes coding for flavonoid biosynthesis enzymes are active only in the anther tissues but not in the pollen (van Tunen and Mol, 1987). These findings indicated that we had detected a new class of plant signal substances that fit the classical definition of hormones. They are produced at a site different from the site of action, they are active at low concentration, and they show chemical specificity.

The flavonols were active not only when added to the germination medium. When added to the maturation media, pollen developed faster, i.e., it germinated earlier as compared to control cultures without flavonols (Ylstra et al., 1992).

Molecular genetic experiments confirmed this finding. Transgenic petunia plants deficient in the first flavonoid biosynthetic enzyme, due to antisense inhibition of the chalcon synthase gene, resulted in male sterility (van der Meer et al., 1992). Similar experiments by Taylor and Jorgensen (1992), also on petunia, showed that the sterile pollen in their case is fertile when used to pollinate wild-type plants, and the term conditional male fertility was created to distinguish this phenomenon from true male sterility (arrest during pollen development) and self-incompatibility (allele-specific inhibition of pollen function). These results indicated that the pollen lacked substances that in backcrosses, but not in selfings, could be substituted by the stigma. Addition of flavonols directly to the stigma of the transformants also rescued the sterile pollen in selfings (Mo et al., 1992).

In the experiments of van der Meer et al. (1992), it was not possible to rescue the sterile pollen, and the pollen was apparently blocked in its development at the early binucleate stage. This coincided with our finding that the addition of flavonols to the pollen maturation medium enhanced pollen development. More recent experiments showed that indeed only binucleate pollen grains were responsive to added flavonols whereas microspores were not (A. Touraev, unpubl. results).

These data indicate a two-step mechanism of flavonol action. In the first step, flavonols synthesized by the anther wall affect pollen development directly. In the second step, flavonols are incorporated into or onto the exine and their effect occurs later, during pollination and tube growth. Flavonoids have been described to be present in relatively large quantities in pollen of many higher plants, including gymnosperms and dicot and monocot angiosperms (Stanley and Linskens, 1974; Wiermann and Vieth, 1983).

Inhibition of flavonoid biosynthesis is also a natural mechanism of male sterility. In peach, several male sterility loci that are involved in flavonoid biosynthesis have been identified (D. Werner, pers. comm.). Also in this plant, pollen development is arrested at around first pollen mitosis.

The finding that in vitro matured pollen produces fewer seeds than in vivo pollen points to another interesting aspect. In vitro matured pollen obviously has reduced reproductive success and requires signals from the sporophyte to achieve full fertility. It can therefore be termed "minimal pollen" exclusively through the action of its own genome. It develops without interference from the sporophyte and lacks substances from the sporophyte that increase reproductive success after pollen shedding. Flavonols may not be the only substances affecting reproductive success. In fact, proteins present in pollen diffusates have been postulated to enhance pollen germination and seed set (Kirby and Vasil, 1979).

The sporophyte may also provide the pollen with other properties that affect reproductive success such as defense mechanisms against pathogens or mechanisms that make the pollen attractive for pollinators. The comparison of in vivo and in vitro pollen may establish where and how these properties emerge during pollen development but also whether it is the pollen itself or the sporophyte that provides these properties.

Potential Plant Breeding Applications of In Vitro Pollen Maturation

Inbred Lines from Male Sterile Lines, Clone Crosses

Rescue of sterile pollen is one obvious application of in vitro pollen maturation. In the case of the peach, male sterility is a mechanism that enforces out-crossing. However, the breeder is interested in inbreeding to create defined lines that are of value either directly or after hybridization. Since, in peach as well as in the petunias of van der Meer et al. (1992), pollen development is blocked around first pollen mitosis. The only way to rescue the sterile pollen would be to isolate the microspores and to culture them in the presence of flavonols. Also in other cases of male sterility, pollen development is blocked around the first pollen mitosis and in all these cases, it should be possible to rescue the pollen by isolation of the microspores and in vitro maturation. "Self"-pollination with this pollen would produce homozygotes that would otherwise be difficult to obtain.

Similarly it may be possible to restore sexual reproduction in plants in which pollen development is impaired such as banana or garlic that reproduce only vegetatively. With restored male sexual function, it should be possible to perform clone crosses that should create a lot of new genetic variability.

A particularly interesting application is to use flavonol deficiency in anthers for hybrid seed production. Creation of 100% hybrid seed requires in the case of nuclear male sterility that the male sterile line is homozygous for the sterility trait. Male sterile plants are, however, usually heterozygous for the sterility trait. There are two routes possible to achieve homozygosity. Either flavonol deficiency does not inhibit the formation of mature pollen, then chemical complementation (addition of flavonols to the stigma) allows selfing, and homozygous male sterile lines can be produced. Or, flavonol deficiency blocks pollen development, then immature pollen has to be isolated, rescued in vitro, and used for self-pollination. The homozygous male sterile plants can then be pollinated with pollen from a second, male fertile, inbred line. All resulting seeds collected from the male sterile plants will be hybrid and male sterile. If seed or fruit production on the hybrid is wanted, male fertility restoration is required. Again, chemical restoration by flavonols treatment may lead to seed set. Or, genetic engineering may provide means to overcome the block of flavonol biosynthesis in the anthers of the hybrids (for example, neutralization of the antisense transcript used to suppress chalcone synthase gene expression by an appropriate construct in the male fertile parent, J. Mol, A. van Tunen, B. Ylstra, pers. comm.).

Inbred Lines from Self-Incompatible Plants

Genetic and cell biological evidence indicates that in sporophytic self-incompatibility, the self-incompatibility (S)-proteins of the pollen that are produced by the tapetum should be localized in the exine of the pollen wall (Heslop-Harrison et al., 1973). If the pollen develops in vitro, without contact with the tapetum, no S-proteins should be present in the exine cavities and no SI-reaction should take place since communication with the female allele is interrupted. An ultrastructural study showed that indeed the cavities of in vitro matured pollen are empty (Kastler, 1990). Such pollen should be self-fertile. As with male sterility, this pollen could be used for "selfings" that could lead to inbred lines, which at a later stage can be used for normal hybrid crosses since the SI systems stays intact.

Other procedures exist to break self-incompatibility but they are not applicable in all crop species. In vitro pollen maturation could be an alternative in these cases.

Pollen Selection

Pollen selection in vivo has been proposed as an important factor of angiosperm evolution and also as a breeding tool (Ottaviano and Mulcahy, 1989). As compared to other organs, a high number of genes are expressed in pollen (Willing and Mascarenhas, 1984). There exists a haplo-diploid genetic overlap. Pollen is haploid and is produced in large numbers. Genetic variability to be used under selection pressure is produced by meiotic recombination and can be easily predicted on the basis of parental selection. No regeneration is involved that may introduce somaclonal variation.

Efficient use of in vivo pollen selection requires suitable selection procedures, particularly the application of selective pressures. Selective forces such as environmental stress can be easily applied but selection experiments involving pathotoxins or herbicides are more difficult to set up. They easily damage the anther tissues and would simply arrest pollen development. In pollen cultures, no sporophytic tissue would interfere in pollen selection. Physical and chemical stress can be applied in pollen cultures and the selected pollen can be used in backcrosses.

Pollen Transformation

For the creation of transgenic individuals of higher eukaryotes, there are some obvious advantages for using male gametes as "supervectors." They can easily be manipulated since they are free-living cells, usually produced in large quantity. They are "designed" to transfer DNA, i.e., the male nucleus, to the target, i.e., the egg cell. Furthermore, the egg cell is the optimal target cell for the creation of a transgenic organism. In higher plants, the pollen grain represents the actual free-living phase that is active to transfer the sperm cells to the egg cell whereas the sperm cells are passive entities that, in a more or less passive way, fuse with the egg and the polar cell to form the zygote and the primary endosperm nucleus, respectively.

Pollen transformation has attracted the interest of scientists for a long time because no real success has been achieved. Throughout the many years of experimentation on pollen transformation, a bewildering multitude of techniques has appeared (for a review see Heberle-Bors et al., 1990, Heberle-Bors, 1991). They can be divided into techniques involving DNA uptake or delivery into pollen before pollination and techniques in which the DNA is applied to stigmas before or after pollination (pollen tube pathway).

Most commonly genomic or plasmid DNA has been mixed directly with pollen and applied to stigmas (Hess, 1969, 1987; Ohta, 1986). DNA delivery into pollen has been attempted using *Agrobacterium tumefaciens,* electroporation, microinjection, and particle guns (see Heberle-Bors et al., 1990). In some cases, successful production of transgenic plants has been claimed using these methods. However, convincing proof has not been presented in these cases, both with respect to expression and transmission of the transgenes and reproduciblity of results.

DNA Uptake Before Pollination

In the earliest attempts to use pollen as a vector, genomic DNA was incubated with pollen, and the incubated pollen was then used for pollination (Hess, 1969; de Wet et al., 1985); later, cloned DNA was used (Sanford and Skubik, 1986; Booy et al., 1989; Stöger et al. 1992). Although in theory the DNA should be rapidly degraded by nucleases released from the pollen grains (Matousek and Tupy, 1983, 1984), in some cases spectacular transformation frequencies have been reported (Ohta, 1986). A problem with such claims is the evidence presented. A number of criteria have to be fulfilled in order to verify the transgenic nature of a plant. First, the marker gene that has been used to select (e.g., by antibiotic resistance) or identify (e.g., by enzymatic assays) a transgenic plant among the nontransformed ones has to be detected physically by Southern blotting. Second, for nuclear transformation, Southern blotting must be performed in such a way as to demonstrate covalent linkage of the transferred DNA to the nuclear DNA of the recipient and that it is not present in extrachromosomal or contaminating DNA (for plastid or mitochondria transformation, integration in the respective genomes). Third, the gene has to be shown to pass through meiosis by demonstrating Mendelian segregation in the offspring of the primary transformant (maternal inheritance for plastid or mitochondria transformation). For all these criteria, the appropriate controls have to be presented, particularly to exclude false positives.

None of the published reports on pollen transformation holds up to these criteria, and attempts to reproduce claims of successful pollen transformation by other authors have failed (Sanford et al., 1985; Sanford and Skubik, 1986; Booy et al., 1989; Stöger et al., 1992). In this confusing situation, we thought it important to unequivocally settle one crucial point in pollen transformation: Can DNA be transferred into mature pollen by the published methods? Constructions were made involving a pollen-specific promoter and the *gus*-gene as a reporter gene (Stöger et al., 1992). This construct was transferred into tobacco via *Agrobacterium*-mediated leaf-disk transformation, and the stable transformants showed the expected specific expression in late pollen stages and no expression in somatic tissues. This construct was also expressed at the same stages when delivered into pollen via a DNA gun.

When the same construct in solution was mixed with tobacco pollen under a variety of conditions, including conditions that have been used for pollen transformation or for DNA uptake in other targets such as dry seeds, as well as conditions that eliminate nuclease activity (Negrutiu et al., 1985), no GUS$^+$ pollen were found. Also with germinated pollen, no GUS–expression was found. Electroporation (square-wave, low-voltage pulse, as well as discharge, high-voltage pulse) has also been tried, again without success (Stöger, unpubl. results). It appears that the pollen wall is an effective barrier of DNA uptake that can only be overcome with brute force, i.e., with a DNA gun. Recently, it has been claimed that tobacco pollen can be transformed by electroporation and that transformed tobacco plants have been recovered after pollination (Saunders et al., 1991). However, no full proof with respect to the transgenic nature of the presumed transformants has been presented, and, particularly, data on the segregation in the next generation are lacking.

Agrobacterium tumefaciens has also been claimed to be able to transfer DNA into pollen (Hess, 1987; Hess et al., 1991). To test this claim, a construct with a pollen-specific promoter that contained an intron in the coding region of the *gus*-gene was cloned into the T-DNA of *A. tumefaciens* and was used in cocultivation experiments with mature pollen. The intron was correctly spliced in pollen, as observed in stable transformants as well as in transient expression experiments after pollen bombardment and effectively prevented synthesis of the GUS protein in the agrobacteria (Stöger et al., 1992). Inronless constructs, however, were efficiently expressed in agrobacteria and led to artefactual blue pollen in histochemical GUS assays, due to diffusion of either the enzyme or, more likely, diffusion of the intermediate product of the enzymatic reaction into the pollen grains.

Still with some *A. tumefaciens* strains a high number of GUS$^+$ pollen could be found after cocultivation. This cleavage of the GUS-substrate could be attributed to the presence of a plant endogenous β-glucuronidase (Alwen et al., 1992). The endogenous β-glucuronidase was found to have an activity optimum at ph 5 while the *E. coli* β-glucuronidase used as the reporter enzyme, has an activity optimum at pH 7. The activity of the endogenous β-glucuronidase could simply be inhibited by maintaining a pH of 7 in the culture medium and under these conditions no GUS$^+$ pollen could be found after cocultivation with any *A. tumefaciens* strain containing the *gus*-intron construct (Stöger et al., 1992). This finding contrasted earlier claims that reproductive tissues contain specifically an endogenous GUS-like activity (Plegt and Bino, 1989; Hu et al., 1990).

DNA delivery into pollen by a DNA gun has also been reported by Twell et al. (1989), van der Leede-Plegt et al. (1992) and Hamilton et al. (1992) and is to date the only reproducible method effective to transfer exogenous DNA into pollen grains. In all these cases, including our own, until now no transgenic plants could be found after pollination with the transformed pollen (J. Mascarenhas, S. McCormick, A. van Tunen, pers. comm.). A very simple reason for this failure may be that the transferred and expressed DNA is actually only present in the vegetative cell but not in the generative or sperm cells, depending on the species, of the pollen. The vegetative cell occupies most of the space within the pollen wall while the generative and later the sperm cells are suspended in the cytoplasm of the pollen grains, located at unpredictable positions, and consist of a nucleus with not much cytoplasm. During fertilization, the contents of the vegetative cell is left behind in the tube cell and the synergid, and only the contents of the sperm cells or the sperm nucleus alone are transferred to the egg cell at fertilization. Any DNA present in the vegetative cell or integrated into the vegetative nucleus will therefore not be transmitted to the egg cell. Still, in some cases, the microprojectiles may also penetrate the generative or sperm cells, and in these cases a transmission to the egg cell is possible. The number of such pollen grains will, however, be extremely low, and there is the danger that the nucleus will be damaged by the tungsten or gold particles. Ongoing experiments in a number of labs will show whether this method will prove to be successful.

Pollen Tube Pathway

The first attempt using the pollen tube pathway has been to apply DNA into wheat florets at the time of fertilization (Picard et al., 1988), or the stigmas have been cut immediately after fertilization and the DNA placed on the cut surface (Zhou et al., 1988; Luo and Wu, 1988). A variation of this technique was to apply agrobacteria instead of DNA or plasmid on wheat stigmas (Hess et al., 1991).

Recently, Langridge et al. (1992) have critically assessed transformation via *Agrobacterium* and the pollen tube pathway. Although antibiotic-resistant plants could be recovered, no evidence was found for transmission of the gene detected by hybridization in the progeny of the putative transgenic plants nor could enzyme activity associated with the antibiotic-resistance genes be found in plant extracts. The authors rather assumed that *A. tumefaciens* transformed an endophyte that was prevalent in the greenhouse at the time when these experiments were performed. Such "surrogate transformation" has indeed been performed in perennial ryegrass by transformation of the *Acremonium* endophyte (Murray et al., 1992).

In conclusion, pollen transformation has a long but unfortunately totally unsuccessful and dubious history. This unsuccessfulness and dubiousness is shared with analogous attempts to create transgenic animals via sperm transformation (Lavitrano et al., 1989; Brinster et al., 1989). One may conclude that male gametes, which at first seem to be the most promising vector for DNA transfer, are protected from DNA uptake and transfer. It even makes sense that male gametes, unlike female gametes that are enveloped with various protective tissues in animals and plants, should be provisioned with mechanisms to be protected from pollution with foreign DNA that they may encounter on their way after release from the male sex organs to the egg cell. Pollution with foreign DNA may well occur after landing of the pollen on a stigma. DNA from decaying pollen of other species, from plant pathogenic insects, microorganisms, or viruses may accumulate, particularly on wet stigmas. The female side is active in defense reactions indicated by the presence of PR-proteins in the style. Such PR-proteins may not only be involved in preventing entry of living pathogens but also of their decay products, including DNA. On the male side, nucleases are an effective mechanism to prevent unwanted DNA uptake (Matousek and Tupy, 1983, 1984). If only these nucleases or through other additional mechanisms, in any case mature tobacco pollen did not take up functional DNA in our experiments (Stöger et al., 1992). Similar results have been obtained by Sanford et al. (1985) and Booy et al. (1989).

Microspore Transformation and In Vitro Maturation

As mentioned previously, the only proven way to deliver DNA into pollen is via the DNA gun, and the critical point with this approach is whether the generative or sperm cells can be transformed. Theoretically, a simple way to get DNA into the generative or sperm cells would be to transfer DNA into microspores and to let the rest be done by the pollen itself. After first pollen mitosis the transferred DNA would be partitioned into the generative and vegetative cell and after second pollen mitosis into the two sperm cells (Alwen et al. 1990). DNA replication during the cell cycle of the generative cell could even contribute to the integration of the transferred DNA into the genome of the generative cell.

This approach is possible since our discovery that isolated microspores when cultured in appropriate conditions in vitro, will develop into mature pollen that is functional (Benito Moreno et al., 1988b). DNA transferred into tobacco microspores by the biolistic approach has been shown to be expressed in the vegetative cell at maturity (Stöger et al., 1992) but no transgenic plants have been found to date after pollination with this pollen (E. Stöger, A. Touraev, unpublished results). Numbers are strongly at odds with this approach at present. Pollen transformation frequency are on average 2.5×10^{-5} (Stöger et al., 1992) using a self-made gunpowder DNA gun that allows only single shots because of the toxicity of the burned gunpowder. If one assumes a 1:100 ratio for transient to stable transformation as calculated for other targets, a chance of 2.5×10^{-7} to get a transformed plant is obtained. Maturation efficiency is on average 50%, to yield a probability of 10^{-7}. This is far beyond feasibility and may explain why until now we have not succeeded in finding a transgenic plant. However, the system has a great potential for optimization.

To our satisfaction, these optimization attempts resulted and may still result in side products that shed light on important open questions concerning sexual reproduction of higher plants (see flavonols above). First, the transformation frequency can be improved by using a better DNA gun that does not produce toxic waste, thus allowing multiple shots into a culture. Second, in vitro maturation of tobacco microspores has been optimized (Tupy et al., 1992, confirmed by own unpublished data). Third, preliminary results indicate that the transformed pollen can be selected during in vitro maturation. Transformation with resistance genes against a number of selective agents and the *gus*-gene was tried and resulted in a low number of viable mature pollen that were GUS$^+$. Fourth, we were able to optimize seed set after pollination with in vitro matured pollen by adding flavonols to the last maturation medium and to the pollination medium (Ylstra et al., 1992). Experiments are underway to determine the degree of pollen selection during pollen tube growth in the style and whether the transformed pollen may be at a selective disadvantage.

Transformation of Pollen Embryos

In both systems of embryogenic pollen cultures, i.e., in tobacco and rapeseed, relative high numbers of embryos can be produced, and attempts have been made to use these systems for transformation. Microinjection into about eight-celled *B. napus* pollen embryos resulted in the formation of genetic chimera but these chimera were rescued by secondary embryo formation (Neuhaus et al., 1987). Pollen embryos have also been used in the case of cereals to transform them by microinjection (Potrykus, 1991). Antibiotic-resistant plants have been recovered that seemed to contain the transferred gene constructs but in the next generation, no transgenic plants were found. Bombardment of embryogenic pollen is currently being pursued in several labs.

Pollen embryos can be used for transformation in another, more indirect, way. Embryogenic cell cultures have been produced from pollen embryos (Datta et al., 1990b, Mitchell and Petolino, 1991), and protoplasts from microspore embryos have been transformed using standard transformation techniques (Datta et al., 1990a). This approach is particularly rewarding in cereals in which pollen embryos from anther cultures are convenient sources of embryogenic suspension cultures.

References

Abele, J. C., L. W. Kannenberg, R. Keats, G. Sohota, and E. B. Swanson. 1992. Increased induction of microspore embryos following manipulation of donor plant environment and culture temperature in corn (*Zea mays* L.). *Plant Cell Tiss. Org. Cult.* 28:87–90.

Alwen, A., N. Eller, M. Kastler, R. M. Benito Moreno, and E. Heberle-Bors. 1990. Potential of in vitro pollen maturation for gene transfer. *Physiol. Plant* 79:194–196.

Alwen, A., R. M. Benito Moreno, O. Vicente, and E. Heberle-Bors. 1992. Plant endogenous β-glucuronidase activity: how to avoid interference with the use of the *E. coli* β-glucuronidase as a reporter gene in transgenic plants. *Transgenic Res.* 1:63–70.

Baskin, T. I., and W. Z. Cande. 1990. The structure and function of the mitotic spindle in flowering plants. *Ann. Rev. Plant Physiol. Plant Mol. Biol.* 41:277–315.

Benito Morreno, R. M., A. Alwen, M.-T. Hauser, and E. Heberle-Bors. 1988a. Sporophytes and male gametophytes from in vitro cultured tobacco pollen. In *Sexual Reproduction in Higher Plants*, eds. M. Cresti, P. Gori, and E. Pacini, pp. 137–142. Springer-Verlag, New York.

Benito Moreno, R. M., F. Macke, A. Alwen, and E. Heberle-Bors. 1988b. In situ seed production after pollination with in vitro matured, isolated pollen. *Planta* 176:145–148.

Bolik, M. and H. U. Koop. 1991. Identification of embryogenic microspores of barley (*Hordeum vulgare* L.) by individual selection and culture and their potential for transformation by microinjection. *Protoplasma* 162:61–68.

Booy, G., F. A. Krens, and H. J. Huizing. 1989. Attempted pollen-mediated transformation of maize. *J. Plant Physiol.* 135:319–324.

Brinster, R. L., E. P. Sandgren, R. R. Behringer, and R. D. Palmiter. 1989. No simple solution for making transgenic mice. *Cell* 59:239–241.

Capkova, V., E. Hrabetova, and J. Tupy. 1988. Protein synthesis in pollen tubes: preferential formation of new species independent of transcription. *Sex. Plant Reprod.* 1:150–155.

Carlson, P. S. 1973. Methionine sulfoxcimine-resistant mutants of tobacco. *Science* 180:1366–1368.

Datta, S. K., K. Datta, and I. Potrykus. 1990a. Embryogenesis and plant regeneration from microspores of both "Indica" and "Japonica" rice (*Oryza sativa*). *Plant Sci.* 67:83–88.

Datta, S. K., A. Peterhans, K. Datta, and I. Potrykus. 1990b. Genetically engineered fertile Indica-rice recovered from protoplasts. *Bio/Techn.* 8:736–740.

De Wet, J.M.J., R. R. Bergquist, J. R. Harlan, D. E. Brink, C. E. Cohen, C. A. Newell, and A. E. De Wet. 1985. Exogenous gene transfer in maize (*Zea mays*) using DNA treated pollen. In *Experimental Manipulation of Ovule Tissue*, eds. G. P. Chapman, S. H. Mantell, and W. Daniels, pp. 197–209. Longman, London.

Duijs, J. G., R. E. Voorrips, and J.B.M. Custer. 1989. Microspore culture in Brassica oleracea vegetables. *Acta Bot. Neerl.* 38:343–344.

Gaillard, A., P. Vergne, and M. Beckert. 1991. Optimization of maize microspore isolation and culture conditions for reliable plant regeneration. *Plant Cell Rep.* 10 55–58.

Garrido, D., B. Charvat, R. M. Benito Moreno, A. Alwen, O. Vicente, E. Heberle-Bors. 1991. Pollen culture for haploid plant formation in tobacco. In *A Laboratory Guide for Cellular and Molecular Plant Biology*, eds. I. Negrutiu and G. Gharti-Chhetri, pp. 59–69. Birkhäuser, Basel.

Garrido, D., N. Eller, E. Heberle-Bors, and O. Vicente. 1992. De novo transcription of specific mRNAs during induction of tobacco pollen embryogenesis. *Sex. Plant Reprod.* in press.

Gregory, W. C. 1940. Experimental studies on the cultivation of excised anthers in nutrient solution. *Am. J. Bot.* 27:687–692.

Guha, S. and S. C. Maheshwari. 1964. In vitro production of embryos from anthers of *Datura. Nature* 204:497.

Guha, S., and S. C. Maheshwari. 1966. Cell division and differentiation of embryos in the pollen grains of *Datura* in vitro. *Nature* 212:97–98.

Hadwiger, M. A., and E. Heberle-Bors. 1986. Pollen plant production in *Triticum turgidum* ssp. *Durum. Proceedings of International Symposium of Nuclear Techniques and in-vitro Culture for Plant Improvement.* pp. 213–220. FAO/IAEA, Vienna.

Hamilton, D. A., M. Roy, J. Rueda, R. K. Sindhu, J. Sanford, J. P. Mascarenhas. 1992. Dissection of a pollen-specific promoter from maize by transient transformation assays. *Plant Mol. Biol.* 18:211–218.

Heberle-Bors, E. 1985. In vitro haploid formation from pollen: a critical review. *Theor. Appl. Genet.* 71:361–374.

Heberle-Bors, E. 1989. Isolated pollen cultures in tobacco: plant reproductive development in a nutshell. *Sex. Plant Reprod.* 2:1–10.

Heberle-Bors, E. 1991. Germ line transformation in higher plants. *IAPTC Newsletter* 64:2–10.

Heberle-Bors, E., R. M. Benito Moreno, E. Alwen, E. Stöger, and O. Vicente. 1990. Transformation of pollen. In eds. H.J.J. Nijkamp, L.H.W. Van der Plas, J. Van Aartrijk. *Progress in Plant Cellular and Molecular Biology.* pp. 244–251. Kluwer, Dordrecht Boston London.

Henry, Y., and J. de Buyser. 1990. Wheat anther culture: agronomic performance of doubled haploid lines and the release of a new variety 'Florin'. In *Biotechnology in Agriculture and Forestry*, Vol. 13, Wheat. ed. Y.P.S. Bajaj, Springer, Berlin.

Heslop-Harrison, J., Y. Heslop-Harrison, R. B. Knox, and B. Howlett. 1973. Pollen-wall proteins: Gametophytic and sporophytic fractions in the pollen walls of the Malvaceae. *Ann. Bot.* 37:403–412.

Hess, D. 1969. Versuche zur Transformation an höheren Pflanzen: Induktion und konstante Weitergabe der Anthocyansynthese bei Petunia hybrida. *Z. Pflanzenphysiol.* 60:348–358.

Hess, D. 1987. Pollen based techniques in genetic manipulation. *Int. Rev. Cytol.* 107:169–190.

Hess, D., K. Dressler, R. Nimmrichter. 1991. Transformation experiments by pipetting *Agrobacterium* into the spikelets of wheat (*Triticum aestivum* L.). *Plant Sci.* 72:233–244.

Hoekstra, F. A., T. Van Roekel, and N. Ten Pas. 1988. Pollen maturation and desiccation tolerance. In *Sexual Reproduction in Higher Plants*, pp. 291–296. eds. M. Cresti, P. Gori, and E. Pacini. Springer-Verlag, Berlin.

Hu, C., P. Chee, R. Chesney, J. Zhou, and P. Miller. 1990. Intrinsic GUS-like activities in seed plants. *Plant Cell Rep.* 9:1–5.

Ito, M., and H. Stern. 1967. Studies of meiosis in vitro. I. In vitro culture of meiotic cells. *Dev. Biol.* 16:36–53.

Kastler, M. 1990. Ultrastrukturelle Unterschiede zwischen in-vivo und in-vitro gereiften Tabakpollen. Diploma Thesis, University of Vienna.

Kirby, E. G., and I. K. Vasil. 1979. Effect of pollen-protein diffusates on germination of eluted pollen samples of *Petunia hybrida* in vitro. *Ann. Bot.* 44:361–367.

Kyo, M., and H. Harada. 1986. Control of the developmental pathway of tobacco pollen in vitro. *Planta* 168:427–432.

Kyo, M., and H. Harada. 1990a. Specific phosphoproteins in the initial period of tobacco pollen embryogenesis. *Planta* 182:58–63.

Kyo, M., and H. Harada. 1990b. Phosphorylation of proteins associated with embryogenic dedifferentiation of immature pollen grains of *Nicotiana rustica. J. Plant Physiol.* 136:716–722.

Kyo, M., and T. Ohkawa. 1991. Investigation of subcellular localization of several phosphoproteins in embryogenic pollen grains of tobacco. *J. Plant. Physiol.* 137:525–529.

Langridge, P., R. Brettschneider, P. Lazzeri, and H. Lorz. 1992. Transformation of cereals via Agrobacterium and the pollen pathway: a critical assessment. *Plant J.* 2:631–638.

Lavitrano, M., A. Camaioni, V. M. Fazi, S. Dolci, M. G. Farace, and A. C. Spadafora. 1989. Sperm cells as vectors for introducing foreign DNA into eggs: genetic transformation of mice. *Cell* 57:717–723.

Liang, G. H., A. Xu, and H. Tang. 1987. Direct generation of wheat haploids via anther culture. *Crop Sci.* 27:336–339.

Lichter, R. 1985. From microspores to rape plants: a tentative way to low glucosinolate strains. In *Cruciferous Crops: Production, Utlization, Description*. Vol. 2, pp. 268–277. ed. H. Sorensen, Nijhoff/Junk, Dordrecht.

Löschenberger, F. and E. Heberle-Bors. 1992. Anther culture responsiveness of austrian winter wheat. (*Triticum aestivum* L.) cultivars. *Die Bodenkultur* 43:115–122.

Luo, Z. X., and R. Wu. 1989. A simple method for the transformation of rice via the pollen-tube pathway. *Plant Mol. Biol. Rep.* 7:69–77.

Mariani, C., M. De Beuckeleer, J. Truettner, J. Leemans, and R. G. Goldberg. 1990. Induction of male sterility in plants by a chimaeric ribonuclease gene. *Nature* 347:737–741.

Mariani, C., V. Gossele, M. De Beudkeleer, M. De Block, R. B. Goldberg, W. De Greef, and J. Leemans. 1992. A chimaeric ribonuclease-inhibitor gene restores fertlity to male sterile plants. *Nature* 357:384–397.

Mascarenhas, J. P. 1988. Anther- and pollen-expressed genes. In *Temporal and Spatial Regulation of Plant Genes,* eds. D.P.S. Verma, and R. B. Goldberg, pp. 97–115. Springer-Verlag, New York.

Mascarenhas, J. P. 1989. The male gametophyte of flowering plants. *Plant Cell* 1:657–664.

Mascarenhas, J. P. 1990. Gene activity during pollen development. *Ann. Rev. Plant Physiol. Plant Mol. Biol.* 41:317–338.

Matousek, J. and J. Tupy. 1983. The release of nucleases from tobacco pollen. *Plant Sci. Lett.* 30:83–89.

Matousek, J. and J. Tupy. 1984. Purification and properties of extracellular nuclease from tobacco pollen. *Biol. Plant* (Praha) 26:62–73.

Mitchell, J. C. and J. F. Petolino. 1991. Plant regeneration from haploid suspension and protoplast cultures from isolated microspores of maize. *J. Plant Physiol.* 137:530–536.

Mo, Y., C. Nagel, and J. P. Taylor. 1992. Biochemical complementation of chalcone synthase mutants defines a role for flavonols in functional pollen. *Proc. Natl. Acad. Sci. USA* 89:7213–7217.

Morrison, R. A., and D. A. Evans. 1988. Haploid plants from tissue culture: new plant varieties in a shortened time. *Bio/Technology* 6:684–690.

Murray, F. R., G.C.M. Latch, and D. B. Scott. 1992. Surrogate transformation of perennial ryegrass, *Lolium perenne,* using genetically modified *Acremonium* endophyte. *Mol. Gen. Genet.* 233:1–9.

Negrutiu, I., E. Heberle-Bors, and I. Potrykus. 1985. Attempts to transform tobacco pollen by direct gene transfer. In *Biotechnology and Ecology of Pollen,* eds. D. L. Mulcahy, Bergamini, G. Mulcahy, and E. Ottaviano, pp. 65–70. Springer-Verlag New York.

Neuhaus, G., G. Spangenberg, O. Mittelsten-Scheid, H. G. Schweiger. 1987. Transgenic rapeseed plants obtained by the microinjection of DNA into microspore-derived embryoids. *Theor. Appl. Genet.* 75:30–36.

Ohta, Y. 1986. High efficiency genetic transformation of maize by a mixture of pollen and exogenous DNA. *Proc. Natl. Acad. Sci. USA* 83:715–719.

Ottaviano, E., and D. L. Mulcahy. 1989. Genetics of angiosperm pollen. *Adv. Genetics* 26:1–64.

Pacini, E. 1990. Tapetum and microspore function. In *Microspores. Evolution and Ontogeny,* eds. S. Blackmore, and R. B. Knox, pp. 213–237. Academic Press, London.

Pechan, P. M., and W. A. Keller. 1988. Identification of potentially embryogenic microspores in *Brassica napus. Physiol. Plant* 74:377–384.

Pechan, P. M., D. Bartels, D.C.W. Brown, and J. Schell. 1991. Messenger-RNA and protein changes associated with induction of *Brassica* microspore embryogenesis. *Planta* 184:161–165.

Pescitelli, S. M. C. D. Johnson, and J. F. Petolino. 1990. Isolated microspore culture in maize: effects of isolation technique, reduced temperature, and sucrose level. *Plant Cell Rep.* 8:628–631.

Picard, E., J. M. Jacquemin, F. Granier, M. Bobin, and P. Forgeois. 1988. Genetic transformation of wheat (*Triticum aestivum*) by plasmid DNA uptake during pollen tube germination. In *7th International Wheat Genetics Symposium.* pp. 779–787. Cambridge. Cambridge University Press.

Plegt, L., and R. Bino. 1989. β-glucuronidase activity during development of the male gametophyte from transgenic and non-transgenic plants. *Mol. Gen. Genet.* 216:311–327.

Polsoni, L., L. S. Kott, and W. D. Beversdort. 1988. Large scale microspore culture technique for mutation-selection studies in Brassica napus. *Can. J. Bot.* 66:1681–1685.

Potrykus, I. 1991. Gene transfer to plants: Assessment of published approaches and results. *Ann. Rev. Plant Physiol. Plant Mol. Biol.* 42:205–225.

Reed, S. M., and E. A. Wernsman. 1989. DNA amplification among anther-derived doubled haploid lines of tobacco and its relationship to agronomic performance. *Crop Sci.* 29:1072–1076.

Sacristan, M. D. 1985. Selection for disease resistance in Brassica cultures. *Hereditas Suppl.* 3:57–63.

Sanford, J. C., and K. A. Skubik. 1986. Attempted pollen mediated plant transformation using Ti plasmid. In *Biotechnology and Ecology of Pollen,* eds. D. L. Mulcahy, Mulcahy G. Bergamini, and E. Ottaviano, pp. 71–76. Springer-Verlag, New York.

Sanford, J. C., K. A. Skubik, and B. I. Reisch. 1985. Attempted pollen mediated plant transformation employing genomic donor DNA. *Theor. Appl. Genet.* 69:571–574.

Saunders, J. A., B. F. Matthews, and S. L. Van Wert. 1991. Pollen electrotransformation for gene transfer in plants. In *Guide to Electroporation and Electrofusion* eds. D. C. Chang, B. M. Chassy, J. A. Saunders, and A. E. Sower, pp. 227–247. Academic Press.

Shimakura, K. 1934. The capability of continuing divisions of the *Tradescantia* pollen mother cell in saccharose solution. *Cytologia* 5:363–373.

Shivanna, K. R. and Johri, B. M. 1985. The angiosperm pollen: structure and function. New Delhi, Wiley Eastern, xv. 374 p.

Stanley, R. G., and H. F. Linskens. 1974. *Pollen—Biology, Biochemistry, Management.* Springer-Verlag, Berlin.

Stuaffer, C., R. M. Benito Moreno, and E. Heberle-Bors. 1991. In situ pollination with in vitro matured pollen of *Triticum aestivum. Theor. Appl. Genet.* 81:576–580.

Stöger, E., R. M. Benito Moreno, B. Ylstra, O. Vicente, E. Heberle-Bors. 1992. Comparison of different techniques for gene transfer into mature and immature tobacco pollen. *Trans. Res.* 1:71–78.

E. B. Swanson, M. P. Coumans, G. L. Brown, J. D. Patel, W. D. Beversdore. 1988. The characterization of herbicide tolerant plants in Brassica napus L. after in vitro selection of microspores and protoplasts. *Plant Cell Rep.* 7:83–87.

Takahata, Y., and W. A. Keller. 1991. High frequency embryogenesis and plant regeneration in isolated microspore culture of *Brassica oleracea* L. *Plant Sci.* 74:235–242.

Taylor, L. P., and R. Jorgensen. 1992. Conditional male fertility in chalcone synthase-deficient petunia. *J. Hered.* 83:11–17.

Telmer, C. A., D. H. Simmonds, and W. Newcomb. 1992. Determination of developmental stage to obtain high frequencies of embryogenic microspores in Brassica napus. *Physiol. Plant* 84:417–424.

Thompson, R. D., and H. H. Kirch. 1992. The S locus of flowering plants: when self-rejection is self-interest. *Trends in Genetics* 8:381–387.

Tupy, J., L. Rihova, V. Capcova, and V. Zarsky. 1992. Differentiation and maturation of in situ and in suspension culture. Angiosperm pollen and ovules, E. Ottaviano (eds.), New York: Springer Verlag, p. 309–314.

Twell, D., T. M. Klein, M. E. Fromm, S. McCormick. 1989. Transient expression of chimeric genes delivered into pollen by microprojectile bombardment. *Plant Physiol.* 91:1270–1274.

Van Den Bulk, R. W. 1991. Application of cell and tissue culture and in vitro selection for disease resistance breeding—a review. *Euphytica* 56:269–285.

Van Der Leede-Plegt, L. M., B.C.E. Van De Ven, R. J. Bino, T.P.M. Van Der Salm, and A. J. Van Tunen. 1992. Introduction and differential use of various promoters in pollen grains of *Nicotiana glutinosa* and *Lilium longiflorum*. *Plant Cell Rep.* 11:20–24.

Van Der Meer, I. M., J. E. Stam, A. J. Van Tunen, J.N.M. Mol, and A. R. Stuitje. 1992. Inhibition of flavonoid biosynthesis in petunia anthers by antisense approach results in male sterility. *Plant Cell* 4:253–262.

Van Herpen, M.M.A., P.F.M. De Groot, J.A.M. Schrauwen, K.J.P.T. Van Den Heuvel, K.A.P. Wetterings, G. J. Wullems, 1992. In-vitro culture of tobacco pollen: gene expression and protein synthesis. *Sex Plant Reprod.* 5:304–309.

Van Tunen, A. J., and J.M.M. Mol. 1987. A novel purification procedure for chalcone flavanone isomerase from Petunia hybrida and the use of its antibodies to characterize the Po mutation. *Arch. Biochem. Biophys.* 257:85–91.

Vincente, O., D. Garrido, V. Zarsky, N. Eller, L. Rihova, M. Berenyi, J. Tupy. 1992. Heberle-Bors, E. Induction of embryogenesis in isolated pollen cultures of tobacco. *Angiosperm pollen and ovules* (E. Ottaviano , et al. eds.) New York, Springer Verlag, p. 279–284.

Waugh, R., and W. Powell. 1992. Using RAPD markers for crop improvement. *Trends Biotech* 10:186–191.

Wiermann, R., and K. Vieth. 1983. Outer pollen wall, an important accumulation site for flavonoids. *Protoplasma* 118:230–233.

Willing, R. P., and J. P. Mascarenhas. 1984. Analysis of the complexity and diversity of mRNAs from pollen and shoots of *Tradescantia*. *Plant Physiol.* 75:865–868.

Wilson, C., N. Eller, A. Gartner, O. Vincente, E. Heberle-Bors. 1993. Isolation and characterization of a tobacco cDNA clone encoding a putative map kinase. *Plant Mol. Biol* 23:543–551.

Worrall, D., D. L. Hird, R. Hodge, W. Paul, J. Draper, and R. Scott. 1992. Premature dissolution of the microsporocyte callose wall causes male sterility in transgenic tobacco. *Plant Cell* 4:759–771.

Ylstra, B., A. Touraev, R. M. Benito Moreno, E. Stöger, A. J. Van Tunen, O. Vicente, J.N.M. Mol, and E. Heberle-Bors. 1992. Flavonols stimulate development, germination, and tube growth of tobacco pollen. *Plant Physiol.* 100:902–907.

Youssef, S. S. R. Morris, P. S. Baenziger, and C. M. Papa. 1989. Cytogenetic studies of progenies from crosses between "Centurk" wheat and its doubled haploids derived from anther culture. *Genome* 32:622–628.

Zaki, M.A.M., and H. G. Dickinson. 1991. Microspore-derived embryos in *Brassica:* the significance of division symmetry in pollen mitosis I to embryogenic development. *Sex Plant Reprod.* 4:48–55.

Zarsky, V., L. Rihova, and J. Tupy. 1990. Biochemical and cytological changes in young tobacco pollen during in vitro starvation in relation to pollen embryogenesis. In *Progress in Plant Cellular and Molecular Biology,* eds. H.J.J. Nijkamp, L.H.W. Van der Plas, J. Van Aartrijk, pp. 228–233. Kluwer, Dordrecht.

Zarsky, V., D. Garrido, L. Rihova, J. Tupy, O. Vicente, and E. Heberle-Bors. 1992. Derepression of the cell cycle by starvation is involved in induction of tobacco pollen embryogenesis. *Sex. Plant Reprod.* 5:189–194.

Zhou, G. Y., J. Weng, Z. Z. Gong, Y. S. Zhen, W. X. Yang, W. F. Shen, Z. F. Wang, Q. Z. Tao, J. G. Huang, S. Y. Qian, G. L. Lin, M. C. Ying, D. Y. Xue, A. H. Hong, Y. J. Xu, S. B. Chen, and X. L. Duan. 1988. A technique for introducing exogenous DNA into plants after self pollination. *Scientia Agricultura Sinica* 21:1–6.

PART II
Allergen Characterization*

*Note: During the compilation of this volume, the allergen nomenclature system was revised. For an updated nomenclature of allergens, the readers are referred to an article published in *J. Allergy Clin Immunology,* 46: 5–14, 1995.

6

Molecular Characterization of Allergens of Kentucky Bluegrass Pollen

Egil Olsen, Ming Yang and Shyam S. Mohapatra

Introduction

Grass pollen is one of the major causes of allergy worldwide and since the 1960s many different grass species have been extensively studied for their allergenic proteins by biochemical and immunological methods. Grass pollen extracts are complex mixtures of allergenic and nonallergenic proteins and may consist of up to 14 different IgE-binding components (Ford and Baldo, 1986). Several allergens have been defined and partly characterized, but only a few have been isolated in pure form from pollen extract (Ansari et al., 1989a, b). Comparison of pollen proteins from different grass species have revealed considerable cross-reactivity (Leiferman and Gleich, 1986; Bernstein et al., 1976). Employing monoclonal antibodies, groups of allergens have been defined with interspecies cross-reactivity (Standring et al., 1987; Singh and Knox, 1985). Because of the difficulties in isolating sufficient amounts of pure allergens from pollen extracts, the availability of structural information of grass allergens was limited until recombinant DNA techniques were introduced (Fang et al., 1988).

Kentucky bluegrass (KBG) pollen is one of the main causes of respiratory allergies in the prairies and other parts of North America and in Europe and Australia. Previous studies on the aqueous extracts of KBG pollen revealed a multiplicity of antigenic and allergenic components with a wide range of molecular size and charge. However the molecular properties of these proteins remained uncharacterized. Recently, isolation of complementary DNA clones coding for IgE-binding proteins allowed the determination of the primary structures of the allergens and subsequently enabled studies of the structure-function relationships of these molecules. This chapter reviews the current state of knowledge of these allergens from KBG pollen.

Isolation and Characterization of cDNA Clones Coding for IgE-Binding Proteins

A cDNA library was prepared in λgt11 based on mRNA isolated from lyophilized KBG pollen (Mohapatra et al., 1990a, b). *Escherichia coli* 1090 was infected with the recombinant λgt11, and IgE-binding plaques were identified by incubating replicas of the plaques on nitrocellulose with pooled sera of grass pollen allergic individuals. The individual clones were further purified by subcloning. Detection of expressed recombinant allergenic molecules by employing allergen-specific monoclonal antibodies was not achieved because of pronounced unspecific binding to the plaques. The clones identified by this immunoscreening procedure were classified according to three main sequence species as defined in Table 6-1. The most abundant group of transcripts, as revealed by the number of cDNA clones, encoded a major group of allergens designated as Poa p IX. We discuss here the molecular immunologic characteristics of these cloned proteins, *Poa p* IX proteins, and another group of recently characterized proteins.

Molecular Characterization of *Poa p* IX Allergens

Primary Structures of Group IX Allergens

The major group of allergens were designated as *Poa p* IX, in accordance with the then-used international nomenclature system. A number of cDNA clones including three different full-length cDNAs of the isolated recombinant phages were sequenced by using single strand plasmid templates (Silvanovich et al., 1991). The amino acid sequences deduced from the nucleotide sequences represented three homologous basic proteins of M_r 28–35 kD, which shared a 255-residue-long conserved region (Fig. 6-1). These results indicated the presence of isoallergens of mw 28–35 kD in KBG pollen.

Table 6-1 Important IgE-binding proteins of Kentucky bluegrass pollen

Molecular mass (kDa)	pI	No. of cDNAs characterized	Classical nomenclature
28–34	>9.5	8	Group IX[1]
	5.2–5.7	–	Group V[1]
36	6.0–6.6	1	Group I[2]
58	>9.5	1	Group IV

[1]Group IX allergens share about 75% amino acid sequence homology with that of Group V allergens from other grass pollens.

[2]Group I allergens share ˜57% amino acid sequence homology with Group II and Group III grass pollen allergens.

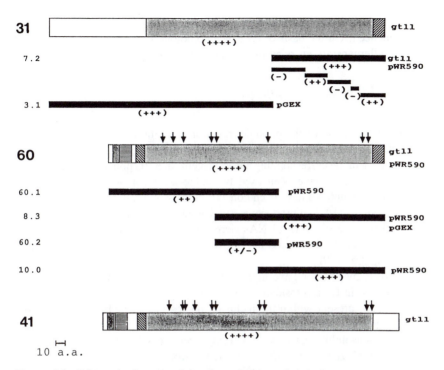

Figure 6-1 Schematic diagram of the three cDNAs and their fragments produced by different expression vectors. 31, 60 and 41 refer to full-length cDNAs encoding *Poa p* IX allergens. Arrows indicate single amino acid substitutions. Different shaded areas represent homologous stretches in these allergens. The solid bars represent fragments expressed in different vectors, i.e., λgt 11, pWR 590 or pGEX and their ability to bind human IgE antibodies as examined by dot/slot blotting. −, −/+, 1+, 2+, 3+, 4+ indicate relative IgE binding ranging from no binding to highest binding.

The deduced amino acid sequences were all different from the known sequences of major allergens in grass pollen by sharing marginal similarity to the fully sequenced *Lol p* II and III (Ansari et al., 1989a, 1989b) in ryegrass and to the internal peptides isolated from group I allergens in different grasses (Esch and Klapper, 1989) as well as the total amino acid sequence deduced from genes coding for *Lol p* I in ryegrass pollen (Perez et al., 1990). Interestingly, the previously mentioned consensus region shared a stretch of five amino acids in common with the N-terminal peptides isolated from the Group V allergens of both timothy and KBG pollen (Klysner et al., 1990). Because of this difference and the isoelectric point of the deduced proteins, which implied that these were basic proteins as opposed to the acidic Group V proteins, these cloned proteins were designated as Group IX (Olsen et al., 1991; Yang et al., 1991; Olsen and

Mohapatra, 1992). Recently however the proteins of Group V have been cloned and sequenced. On the basis of sequence homologies, it can be inferred that all of these allergens are encoded by a family of related genes.

Expression of Recombinant Allergens

Further examination of the possible importance of the molecules coded for by these cDNA clones warranted the expression of the recombinant proteins and identification of the corresponding native proteins in KBG pollen extracts. Several expression systems have been used to synthesize recombinant allergens, and limitations encountered have been non-IgE binding and water-insoluble products (Tovey et al., 1989), and products partly degraded by bacterial proteases (Rafnar et al., 1991). The *Poa p* IX cDNAs were expressed by employing the pWR 590 vector (Yang et al., 1991) and later by a modified pGEX vector system (Olsen and Mohapatra, 1992).

The pWR 590 vector gave high yield of recombinant proteins fused to the bacterial protein β-galactosidase (Guo et al., 1984). The fusion products were insolubilized, present within the inclusion bodies inside the bacteria, and were not subjected to degradation by bacterial proteases. The fusion proteins, which were water-insoluble and need denaturing agents (SDS, urea, or guanidin) to become solubilized, were difficult to purify from the other bacterial proteins present in the inclusion bodies (Olsen et al., 1991). The fusion recombinant proteins produced with the aid of the pWR590 vectors enabled us to identify native proteins in the KBG extract, which corresponded to the three homologous cDNA clones from KBG pollen.

To further investigate the use of recombinant allergens for in vitro as well as in vivo diagnostic assays (Mohapatra 1992a, Mohapatra and Sehon, 1992b), we explored the synthesis of these allergens utilizing the pGEX expression vector system, which yielded water soluble fusion products (Smith and Johnson, 1988). The fusion partner, the glutathione-S-transferase (GST), enable the isolation of the fusion proteins by absorption onto sepharose-glutathione gel matrix. The fusion protein can either be eluted from the gel by soluble glutathione or the nonbacterial part of the protein can be cleaved off by the enzyme thrombin at a specific site recognized by thrombin, positioned in between the GST and the recombinant pollen polypeptide, and coded for within the vector itself (Smith and Johnson, 1988). Fusion proteins eluted from the column may be degraded by bacterial proteases (Olsen and Mohapatra, 1992). This problem can be resolved by synthesis of recombinant protein free of GST by cleavage of the fusion protein when it is bound to the column (Olsen and Mohapatra, 1992) (Fig. 6-2). These studies led to the production of water-soluble, nonfused recombinant proteins, which was crucial in the demonstration of the clinical significance of the recombinant proteins.

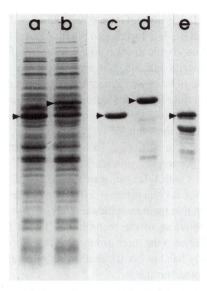

Figure 6-2 SDS-Polyacrylamide gel electrophoretogram of proteins extracted from the *E. coli* containing pGEX-2T1 vector plasmid (a) or pGEX 2T1 vector ligated to the allergen cDNA, KBG 8.3. The lower and upper arrowheads indicate the position of GST and GST-allergen fusion protein respectively. These proteins were purified by Glutathione agarose (c, and d) and thrombin cleavage (e).

Ig Binding Properties of Poa p *IX Allergens*

The Ig binding by itself only demonstrates the existence of epitope(s) similar to those present on naturally occurring allergen proteins. It also demonstrates the extent to which the cloned allergens actually are expressed in the pollen. The lack of detection of these allergens in previous investigations of KBG pollen (Standring et al., 1987) is most likely caused by the batch to batch variations in the amount of these allergens in extracts of KBG pollen (Olsen et al., 1991; Yang et al., 1991). The clinical importance of the *Poa p* IX allergens were investigated further by screening for *Poa p* IX-specific IgE antibodies in serum of Norwegian grass-pollen-allergic individuals (Olsen et al., 1992). Because more than 95% of this group of individuals were positive in this analysis, the *Poa p* IX were considered to comprise a group of major allergens of grass pollens.

We extended this analysis of IgE and IgG binding to KBG and rKBG8.3 in 978 allergic individuals from several different geographic regions. Of these 51% of total allergic patients' sera showed specific IgE binding to KBG pollen extract, whereas 39% bound to the recombinant allergen. In IgE and IgG4 correlation analysis of total allergic patient's sera, a positive correlation was found in reactivity of IgE and IgG4 antibodies between KBG extract and recombinant allergen

(IgE: $r = 0.656$, $P < 0.001$ and IgG4: $r = 0.5875$, $P < 0.0001$). Correlations between KBG IgE and KBG IgG4 were also positive ($r = 0.5384$, $P < 0.0001$). However, the correlation coefficient was lower ($r = 0.3$, $P < 0.0001$) between recombinant allergen specific IgE and IgG4 antibodies. The analysis of correlation between IgE and IgG4 antibodies using recombinant allergen specific sera showed similar correlations and the coefficients were higher compared to those seen with total allergic sera (see Table 6-2).

Further investigation of the immune responses to the grass pollen allergens and the *Poa p* IX allergens in particular, focused on the serum Ig pattern in three groups of individuals: nonatopics, atopics with IgE specific for tree pollen and grass pollen, and atopics with IgE against tree pollen but not grass pollen (Olsen et al., 1992). There were no quantitative differences in the IgG response to the individual pollen allergen in the nonatopics and any of the two groups of atopics, as judged by immunoblotting of the recombinant Poa p IX protein. The only noticeable difference between the three groups was that atopics with IgE response to grass pollen generally had antibodies against a higher number of grass pollen proteins than the two other groups.

Another important facet of immunologic characterization of these cloned allergens has been to identify the epitopes of these allergens recognized by the B and T lymphocytes. These aspects are described in detail in chapters 8. The allergenic molecules encoded by overlapping cDNA clones provide the opportunity to identify these epitopes. Direct IgE binding to the overlapping fragments of *Poa p* IX produced by rDNA methods revealed several IgE-binding epitopes located in the conserved region of the *Poa p* IX allergens (Fig. 6-1). Experiments to localize the individual epitope by overlapping synthetic peptides at the C-terminus led to identification of fewer number of IgE-binding epitopes than

Table 6-2 Correlation Analysis of IgE and IgG4 Antibody Binding to KBG and Recombinant Allergen rKBG8.3 (RA)

Comparison	Correlation[a] Coefficient (r)
Unselected allergic sera ($n = 978$)	
KBG IgE: RA IgE	0.656
KBG IgG4 : RA IgG4	0.587
KBG IgE : KBG IgG4	0.538
RA IgE : RA IgG4	0.357
RA positive allergic sera ($n = 384$)	
KBG IgE : RA IgE	0.715
KBG IgG4 : RA IgG4	0.696
KBG IgE : KBG IgG4	0.677
RA IgE : RA IgG4	0.483

[a]The probability (P) values for all correlation analysis was < 0.0001.

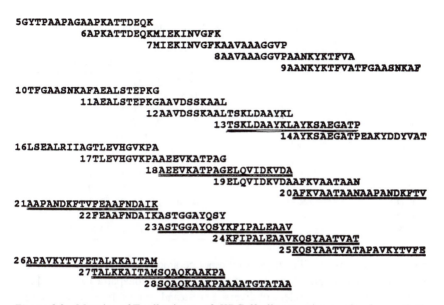

```
5GYTPAAPAGAAPKATTDEQK
      6APKATTDEQKMIEKINVGFK
            7MIEKINVGFKAAVAAAGGVP
                  8AAVAAAGGVPAANKYKTFVA
                        9AANKYKTFVATFGAASNKAF

10TFGAASNKAFAEALSTEPKG
      11AEALSTEPKGAAVDSSKAAL
            12AAVDSSKAALTSKLDAAYKL
                  13TSKLDAAYKLAYKSAEGATP
                        14AYKSAEGATPEAKYDDYVAT
16LSEALRIIAGTLEVHGVKPA
      17TLEVHGVKPAAEEVKATPAG
            18AEEVKATPAGELQVIDKVDA
                  19ELQVIDKVDAAFKVAATAAN
                        20AFKVAATAANAAPANDKFTV
21AAPANDKFTVFEAAFNDAIK
      22FEAAFNDAIKASTGGAYQSY
            23ASTGGAYQSYKFIPALEAAV
                  24KFIPALEAAVKQSYAATVAT
                        25KQSYAATVATAPAVKYTVFE
26APAVKYTVFETALKKAITAM
      27TALKKAITAMSQAQKAAKPA
            28SQAQKAAKPAAAATGTATAA
```

Figure 6-3   Mapping of T cell epitopes of rKBG 60 allergen using overlapping synthetic peptides. The number preceding indicates the designation of the particular peptide in Table 6-3. The underlined peptides were identified as T-cell and B-cell epitopes.

defined by the larger fragments (Fig. 6-3). A possible explanation to this difference is that some of the epitopes are conformational and are not exposed properly without being an integral part of the flanking regions of the molecule. This limitation in overlapping peptide strategy has also been recognized by others (Atassi, 1984). The indication of some linear epitopes fits well with the observation that even water-insoluble as well as guanidin-treated recombinant allergens expose IgE-binding epitopes.

In addition to B cell epitopes, the overlapping peptides have been used to define the T cell epitopes of these allergens utilizing peripheral blood mononuclear cell proliferation assays. A number of peptides capable of proliferation are shown in Fig. 6-3 and the order of reactivity is described in Table 6-3. Identification of these T-cell reactive peptides are important in the context of peptide immunotherapy as a modality in individuals allergic to these allergens.

Cross-Reactivities in Grass Pollen Allergens

With the availability of recombinant allergens and the information on their primary structures and their epitopes recognized by B and T cells, there has been increasing emphasis on the investigations concerning the cross-reactivities among different allergens because of obvious implications in therapy and diagnosis.

Table 6-3 Summary of PBMC proliferation studies with synthetic overlapping peptides of rKBG60

Individuals	Order of peptide reactivity
S.M.	25>23>26>28>13
E.O.	18>13>26>25>23
L.D.	18>23>25=27=14
E.D.	25>13>28>23>18
L.Z.	27>13>25>23=14
J.G.	26>18>14=13>25>28
S.J.	13>25>14
G.P.	26>27>23>28>18=25>13=14
E.H.	13>5>14>18=19=22=23>6=10=25>17
C.D.	20>13=14>17=25>28=19=18>9>5
K.M.	13>21=23>14=9>12>26
T.K.	27>13>20=19>6=9>10=23

Extensive cross-reactivities have been reported among individual groups of aeroallergens, such as among grasses, trees, and weeds (Mohapatra 1992a, b, 1993, 1994a,b,c) and among pollens of diverse genera and species, for example, among fruit pollens (Takahashi et al., 1987); olive, ash, privet, and Russian olive (Baldo, 1992); tomato, ragweed, and Japanese cedar (McCormick et al., 1991); and between parthenium and ragweed (Sriramarao and Rao, 1993) and *parietaria* and grass (D'Amato et al., 1991).

Cross-reactivities have been studied both at the B and T cell level not only among allergens belonging to the same aeroallergen source, but also among different aeroallergens (Basker, 1992; Ebner et al., 1993; Mohapatra et al., 1994a,b,c). Sequence homologies among various allergens and their epitopes appear to be a principal reason for extensive cross-reactivities observed among allergens. In addition, recently a method for the analysis of structural motifs within and among diverse groups of aeroallergens was suggested as an approach to identify the major IgE-binding motifs on allergens (Mohapatra, 1992). However, whether the motifs predicted by this method would stand the functional scrutiny remains to be elucidated. Many reports describe the existence of similar allergens within different species of grass (Lieferman and Gleich, 1976; Bernstein et al., 1976). Our analysis by employing monoclonal antibodies recognizing Group I and Group V allergens, respectively, also revealed intraspecies cross-reactive allergens as *Poa p* IX allergens expose epitopes cross-reacting to both Group I and V of allergens (Olsen et al., 1991, Astwood et al., 1995). Similar results have been observed previously but not commented (Mourad et al., 1986). Therefore, the use of monoclonal antibodies as the sole criteria for identifying an allergen may not be adequate but should be supported by physicochemical characterization as well.

The monoclonal antibodies specific for Group IV grass pollen allergens demonstrated no reactivity to the *Poa p* IX in the 1D- or the 2D-PAGE immunoblotting system, nor in direct binding to the recombinant proteins. Another noteworthy observation was that none of the other isolated clones was shown to code for any of the defined major groups of allergens, groups I, II, III, and IV, which are the abundant protein constituents in the pollen extracts (Olsen et al., 1991, Stwood et al., 1995). More recently allergens have been cloned from other grasses and their primary structures have been deduced. Comparison of these sequences support the notion of conservation of the grass allergen proteins and the molecular basis of cross-reactivity.

Characterization of Other Allergens of KBG Pollen

Two other clones selected from the KBG pollen-λgt11 library with sera of patients allergic to grass pollen led to the identification of a partial cDNA clone, 51. Nucleotide sequence analysis of clone 51 indicated that the polypeptide encoded by this cDNA is different from that of the known recombinant grass pollen ALs. In SDS-PAGE immunoblot analysis of KBG pollen proteins using the murine antiserum to this rAL produced with the aid of pGEX-2T-1 expression system, two polypeptides of about 30 and 58 kDa in size were detectable. Furthermore, this antiserum reacted with one or more polypeptides ranging in size from 30 to 80 kDa from pollen extracts of several grasses, birch, ragweed, and parietaria. Because of the broad cross reactivity, the putative protein has been designated as, CRAL51. These studies suggest that the cloned AL represents a member of a family of highly cross-reactive ALs in plant pollens. On the basis of these observations, it is believed that this group of ALs may constitute the third major group of ALs in grass pollens.

Summary

The availability of recombinant allergens and immunologically relevant synthetic peptides designed from our knowledge of the structures of allergens are expected to aid in the development of new immunotherapeutic approaches as has been discussed elsewhere (Mohapatra 1994b,c).

We have employed cDNA cloning techniques to identify and characterize one of the major groups of allergens in KBG pollen, denoted *Poa p* IX. We have demonstrated the need for the high resolution power of 2D-PAGE combined with immunoblotting to identify and characterize individual allergens in complex extracts such as grass pollen extracts. This method was also used to identify the native allergens corresponding to the recombinant molecules. Our results have emphasized the need for different criteria as amino acid sequences, size, and charge as well as antigenic and allergenic properties to identify any allergens as opposed to the characterization of allergens using monoclonal antibodies. We have demonstrated the use of the pGEX expression systems, which generate nondenatured, water-soluble recombinant molecules, to synthesize milligram amounts of individual allergens. In addition to overlapping peptide synthesis these methods have enabled the identification of IgE-binding and T-cell epitopes on the *Poa p* IX molecules and delineation of cross-reactivity among allergens.

References

Ansari, A. A., P. Shenbagamurthi, and D. G. Marsh. 1989a. Complete amino acid sequence of a *Lolium perenne* (perennial rye grass) pollen allergen, *Lol p* II. *J. Biol. Chem.* 264:11181.

Ansari, A. A., P. Shenbagamurthi, and D. G. Marsh. 1989b. Complete primary structure of a *Lolium perenne* (perennial rye grass) pollen allergen, *Lol p* III: Comparison with known *Lol p*I and II sequences. *Biochemistry* 28:8665.

Astwood, J., S. S. Mohapatra, H. Ni, and R. D. Hill. 1995. Barley allergen homologous protein in crop plants. *Clin. Exp. Allergy* 25:66.

Atassi, M. Z. 1984. Antigenic structures of proteins. *Eur. J. Biochem.* 145:1.

Baldo, B. A., R. C. Panzani, D. Bass, and R. Zerboni. 1992. Olive and privet pollen allergens. Identification and cross reactivity with grass pollen proteins. *Mol. Immunol.* 29:1209.

Baskar, S., P. Parronchi, S. S. Mohapatra, S. Romagnani, and A. A. Ansari. 1992. Human T cell responses to the purified pollen allergens of the grass, *Lolium perenne:* relationship between structural homology and T cell recognition. *J. Immunol.* 148:2378.

Bernstein, I. L., M. Perera, J. Gallagher, J. G. Michael, and S.G.O. Johansson. 1976. In vitro cross-reactivity of major aeroallergenic pollens by the radioallergosorbent technique. *J. Allergy Clin. Immunol.* 57:141.

D' Amato, G., R. De-Palma, A. Verga, P. Martucci, and G. Labefalo. 1991. Antigenic activity of non-pollen parts (leaves and stems) of allergenic plants (*Parietaria judaica* and *Dactylis glomerata*). *Ann. Allergy* 67:421.

Ebner, C., F. Ferreira, K. Hoffman, R. Hirschwehr, S. Schenk, Z. Szebfalusi, H. Breiteneder, P. Parronchi, S. Romagnani, O. Scheiner and D. Kraft. 1993. T cell clones specific for Bet V1, the major birch pollen allergen, crossreact with major allergens of hazel, coral and alder. *Mol. Immunol.* 30:1323.

Esch, R. E. and D. G. Klapper. 1989. Isolation and characterization of a major cross-reactive grass group I allergenic determinant. *Mol. Immunol.* 26:557.

Fang, K.S.Y, M. Vitale, P. Fehlner, and T. P. King. 1988. cDNA cloning and primary structure of a white-face hornet venom allergen, antigen 5. *Proc. Natl. Acad. Sci. USA* 85:895.

Ford, S. A. and B. A. Baldo. 1986. A re-examination of ryegrass (*Lolium perenne*) pollen allergens. *Internat. Arch. Allergy Appl. Immunol.* 81:193.

Guo, L., P. P. Stepien, J. Yuon Tso, R. Brousseau, S. Narang, D. Y. Thomas, and R. Wu. 1984. Synthesis of insulin gene. *Gene* 29:251.

Klysner, S., F. Matthiesen, and H. Lowenstein. 1990. Affinity purification of grass pollen group V allergens using monoclonal antibodies. *J. Allergy Clin. Immunol.* 85:279.

Leiferman, K. M. and G. J. Gleich. 1976. The cross-reactivity of IgE antibodies with pollen allergens. I. Analysis of various species of grass pollen. *J. Allergy Clin. Immunol.* 58:129.

McCormick, S., D. Twell, G. Vaneanneyt, and J. Yamaguchi. 1991. Molecular analysis of gene regulation and function during male gametrophyte development. *Symp. Soc. Exp. Biol.* 45:229.

Mohapatra, S. S. 1992. Diagnosis and therapy of IgE-mediated allergies with recombinant allergens: *Poa pratensis* IX allergen as a model. *Pharmacia Allergy Research Foundation Awards Book,* ed. M. Debelic, pp. 4–15.

Mohapatra, S. S., R. Hill, J. Astwood, A.K.M. Ekramoddoullah, E. Olsen, A. Silvanovitch, T. Hatton, T. T. Kisil, and A. H. Sehon. 1990a. Isolation and characterization of a cDNA clone encoding an IgE binding protein from Kentucky Bluegrass (*Poa pratensis*) pollen. *Internat. Arch. Allergy Appl. Immunol.* 91:362.

Mohapatra, S. S., R. D. Hill, and A. H. Sehon. 1990b. Molecular cloning of allergens: progress and perspectives. *Aerobiologia* 6: 205.

Mohapatra, S. S. 1993. Structural motifs as a basis of cross-reactivity among pollen allergens. In Proc. Int. Symp. *Molecular Biology and Immunology of Allergens,* eds. D. Kraft and A. H. Sehon, pp. 69–81. CRC Press, Boca Raton, FL.

Mohapatra, S. S. 1994. Molecular characterization of pollen allergens: Implications for immunotherapy. In: *Recent trends in Aerobiology, Allergy and Immunology,* Agashe, S. N. (ed). Oxford IBH, New Delhi, pp. 278–292.

Mohapatra, S. S. and A. H. Sehon. 1992. Therapeutic potential of recombinant allergens. *Internat. Arch Allergy Immunol.* 98:265.

Mohapatra, S. S., S. Mohapatra, M. Yang, A. A. Ansari, P. Parronchi, E. Maggi, and S. Romagnani. 1994. Molecular basis of cross-reactivity among allergen-specific human T cells: T cell receptor gene usage and epitope structure. *Immunology* 81:15.

Mohapatra, S. S., C. F. Nicodemus, C. Schou, and R. Valenta. 1994. Recombinant allergens and epitopes. *Allergy and Clin. Immunol. News* 6:45.

Mourad, W., G. Pelletier, A. Boulet, N. Islam, J.-P. Valet, and J. Hebert. 1986. Allergenicity and cross-reactivity of rye grass pollen extracts revealed by monoclonal antibodies. *J. Immunol. Methods* 89:53.

Olsen, E. and S. S. Mohapatra. 1992. Expression and Thrombin cleavage of *Poa p* IX recombinant allergens fused to glutathione S-transferase. *Internat. Arch. Allergy Appl. Immunol.* 98:343.

Olsen, E., L. Zhang, R. D. Hill, F. T. Kisil, A. H. Sehon, and S. S. Mohapatra. 1991. Identification and characterization of the *Poa p* IX group of basic allergens of Kentucky bluegrass pollen. *J. Immunol.* 147: 205.

Olsen, E., A. Fallang, and S. S. Mohapatra. 1995. Characterization of antibody responses to grass pollen allergens in non-atopic and atopic individuals with sensitization to grass and tree allergens. *Allergy* (in press).

Perez, M., G. V. Ishioka, L. E. Walker, and R. W. Chesnut. 1990. cDNA cloning and immunological characterization of the ryegrass allergen *Lol p* I. *J. Biol. Chem.* 265:16210.

Rafnar, T., I. G. Griffith, M. C. Kuo, J. F. Bond, B. L. Rogers, and D. G. Klapper. 1991. Cloning of *Amb a* I (antigen E), the major allergen family of short ragweed pollen. *J. Biol. Chem.* 266:1229.

Silvanovich, A., J. Astwood, E. Olsen, L. Zhang, F. Kisil, A. Sehon, S. Mohapatra, and R. Hill. 1991. Nucleotide sequence analysis of three cDNA's coding for major isoallergens of Kentucky bluegrass pollen. *J. Biol. Chem.* 266:1204.

Singh, M. B. and R. B. Knox. 1985. Grass pollen allergens: Antigenic relationships detected using monoclonal antibodies and dot blotting immunoassay. *Internat. Arch. Allergy Appl. Immunol.* 78:300.

Smith, D. B. and K. S. Johnson. 1988. Single step purification of polypeptides in *Escherichia coli* as fusions with glutathione S-transferase. *Gene* 67:31.

Sriramarao, P., and P. V. Subba Rao. 1993. Allergenic cross-reactivity between Parthenium and ragweed pollen allergens. *Int. Arch. Allergy and Immunology.* 100:79.

Standring, R., V. Spackman, and S. J. Porter. 1987. Distribution of a major allergen of Rye grass (*Lolium perenne*) pollen between other grass species. *Internat. Arch. Allergy Appl. Immunol.* 83:96.

Takahasi, Y., S. Katagiri, and F. Ewapo. 1987. Analysis of cross reactivities among fruit pollen allergens. *Areug* 36:377.

Thomas, W. R., G. A. Stewart, R. J. Simpson. 1988. Cloning and expression of DNA coding for the major house dust mite allergen, *Der p* I, in *Escerichia coli. Internat. Arch. Allergy Appl. Immunol.* 85:127.

Tovey, E. R., M. C. Johnson, A. L. Roche, G. S. Cobon, and B. A. Baldo. 1989. Cloning and sequencing of a cDNA expressing a recombinant house dust mite protein that binds human IgE and corresponds to an important low molecular weight allergen. *J. Exp. Med.* 170:1457.

Yang, M., E. Olsen, J. Dolovich, A. H. Sehon, and S. S. Mohapatra. 1991. Immunological characterization of a recombinant Kentucky bluegrass (*Poa pratensis*) allergenic peptide. *J. Allergy Clin. Immunol.* 87:1096.

Zhang, L., F. T. Kisil, A. H. Sehon, and S. S. Mohapatra. 1991. Allergenic and antigenic cross-reactivities of Group IX grass pollen allergens. *Internat. Arch. Allergy Appl. Immunol.* 96:28.

Zhang, L., E. Olsen, R. Hill, F. Kisil, A. Sehon, and S. S. Mohapatra. 1992. Mapping of B cell epitopes of a recombinant *Poa p* IX allergen of Kentucky Bluegrass pollen. *Mol. Immunol.* 29:1383.

7

Molecular Characterization of Group I Allergens of Grass Pollen

Penelope M. Smith, R. Bruce Knox, and Mohan B. Singh

Introduction

The Group I allergens of grass pollen are one of the most widely spread and ubiquitous groups of glycoproteins that have been characterized in pollen grains (Johnson and Marsh, 1965a). Members of this group have been found in the pollen from a number of subfamilies of the Poaceae and are found in all the clinically important grasses (Table 7-1). Group I allergens from pollen of most grasses are major allergens (Ansari et al., 1987; Matthiesen et al., 1991). The most well characterized Group I allergens come from the clinically important grasses: ryegrass, Kentucky bluegrass, timothy, and orchard grass have been designated *Lol p* I, *Poa p* I, *Phl p* I, and *Dac g* I respectively (Marsh et al., 1987). A more recently studied allergen is *Cyn d* I from Bermuda grass, which has historically been considered to have different allergens from those of other clinically important grasses (Watson and Kibler, 1922; Marsh et al., 1970).

At the amino acid level there is homology between the Group I allergens from different grasses, at least in their NH$_2$-terminal region (Smith et al., 1993a). Antigenic cross-reactivity between the different allergens also suggests similarities within the group (Kahn and Marsh, 1986; Esch and Klapper, 1987; Standring et al., 1987). Allergenic cross-reactivity between different Group I allergens has also been established (van Ree et al., 1992; Bond et al., 1993) and appears to be explained by the presence of a shared IgE epitope on many of the Group I allergens from grasses in the subfamily Pooideae (Esch and Klapper, 1989b).

To date four Group I allergens, *Lol p* I, *Cyn d* I, *Sor h* I, and *Zea m* I, from grasses in three subfamilies of the Poaceae have been cloned and their amino acid sequence deduced. Knowledge of the amino acid sequence of allergens can lead to a more complete understanding of their immunology through the identification of both B- and T-cell epitopes. This review deals with the antigenic

Table 7-1 Taxonomic Relationship of Clinically Important Grasses

Family	Subfamily	Supertribe	Tribe	Genus, Species	Common Name
Poaceae	Pooideae	Triticodae	Triticeae	*Agropyron cristatum*	quack grass
				Hordeum vulgare	barley
				Secale cereale	rye
				Triticum aestivum	wheat
			Bromeae	*Bromus inermis*	smooth brome
		Poodae	Aveneae	**Agrostis alba**	red top
				Anthoxanthum odoratum	sweet vernal
				Avena sativa	cultivated oat
				Holcus lanatus	velvet grass
				Phalaris aquatica	canary grass
				Phleum pratense	timothy
			Poeae	**Dactylis glomerata**	orchard grass
				Festuca elatior	meadow fescue
				Lolium multiflorum	Italian ryegrass
				Lolium perenne*	perennial ryegrass
				Poa pratensis	Kentucky bluegrass
	Bambusoideae	Oryzodae	Oryzeae	**Oryza sativa***	rice
	Arundinoideae		Arundineae	**Phragmites communis**	common reed
	Chlorideae		Chlorideae	*Bouteloua gracilis*	grama grass
				Cynodon dactylon*	Bermuda
				Distichlis stricta	salt grass
	Panicoideae	Panicodae	Paniceae	*Paspalum notatum*	Bahia grass
		Andropogonodae	Andropogoneae	**Sorghum halepense**	Johnson grass
			Maydeae	**Zea mays***	maize

Source: Adapted from Watson, 1990.

Notes: Allergens cross-reactive with Group I allergens have been identified in all grasses and have been isolated from those listed in bold type.

* = Deduced amino acid sequence is available.

and allergenic cross-reactivity between Group I allergens of different grasses, the use of monoclonal antibodies in epitope mapping, and the information that has been made available through cloning of these allergens. In particular it focuses on *Lol p* I and *Cyn d* I as representatives of two different subfamilies.

Characterization of Group I Allergens

Lol p I

Lol p I is a major allergen of ryegrass, producing an allergic response in at least 90% of grass pollen sensitive individuals (Johnson and Marsh, 1965a; b; Cottam et al., 1986). It is an acidic glycoprotein with approximately 5% carbohydrate (Johnson and Marsh, 1966b; Howlett and Clarke, 1981; Cottam et al., 1986). The M_r of *Lol p* I has been estimated by different methods to be between 27 kDa (by amino acid analysis) and 35 kDa [by sodium dodecyl sulfate-polyacrylamide gel electrophoresis (SDS-PAGE)]. Four isoallergens of *Lol p*I have been characterized (Johnson and Marsh, 1965b) and these share at least one common IgE epitope (Smith et al., 1993c).

   Grass pollen allergens have long been known to diffuse from their cellular sites when the pollen is moistened (Howlett and Knox, 1984), and this has created considerable problems in cellular localization of allergens. The development of an anhydrous fixation technique, (Staff et al., 1990), and the use of anti-*Lol p* I Mabs (Smart et al., 1983), enabled the localization of *Lol p* I in ryegrass pollen. Anhydrous fixation, using 2,2-dimethoxypropane which converts water into acetone and methanol, prevented the movement of the allergen as the allergens are insoluble in solvents. *Lol p* I is located in the cytosol (Fig. 7-1) and to a lesser extent at the pollen surface, in cavities in the exine wall (Staff et al., 1990; Singh et al., 1991; Taylor et al., 1993). A detailed description of the techniques for allergen localization has recently appeared (Singh et al., 1993). Thus the immunocytochemical evidence suggests that *Lol p*I is secreted to the surface of grass pollen, where it is strategically sited to interact with the eye and nasal passages after the pollen impacts there.

Cyn d I

Cyn d I is an N-linked glycoprotein (Matthiesen et al., 1991) with four isoforms ranging in pI from 6.2 to 7.2. Most studies show that at least 95% of individuals allergic to Bermuda grass pollen are allergic to *Cyn d* I (Orren and Dowdle, 1977; Ford and Baldo, 1987; Matthiesen and Løwenstein, 1991; Chang et al., 1991; Matthiesen et al., 1991). *Cyn d* I, apart from its isoforms with different pI appears to have a number of different isoforms with varying M_r (Matthiesen et al., 1991; Smith et al., 1993b; Matthiesen et al., 1992). The most abundant isoforms are 31 kDa and 32 kDa (Matthiesen et al., 1991; Smith et al., 1993a)

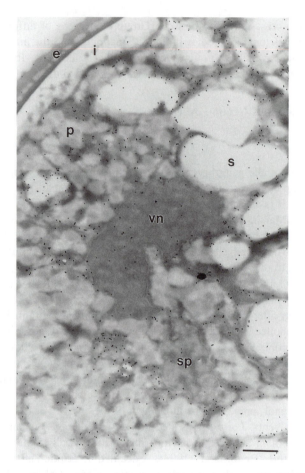

Figure 7-1 Localization of *Lol p* I on ryegrass section using mAB FMCA1 followed by gold probe and silver enhancement. Labelling is predominantly on the electron-opaque regions of the cytosol. Bar = 0.5 μm; e, exine; i, intine; vn, vegetative nucleus; sp, sperm cell; s, starch granule; p, p particle.

but 29 kDa (Smith et al., 1993a) and 23 kDa (Matthiesen et al., 1992) isoforms have been identified and the NH$_2$-terminal amino acid sequences are homologous to those of the other *Cyn d* I isoforms and Group I allergens from different grasses (Table 6-2).

Antigenic Cross-Reactivity between Different Group I Allergens

Both Mabs and polyclonal antibodies have been used in conjunction with crossed radioimmunoelectrophoresis (Matthiesen et al., 1991; Matthiesen and Løwen-

Table 7-2 Physicochemical Properties of Group I Grass Pollen Allergens

Name	M_r(kDa)	pI	No. of isoallergens	Allergenicity	CHO
Lol p I	27[a], 31[b], 32[c], 33[d,e,f], 35[g]	5.15[a], 5.25[a], 5.8[c], 5.6–6.0[c]	4[a]	100%[a], 84%[b], 99%[i]	yes (5%)[a,b]
Poa p I	27[j], 35.8/33[k]	6.4, 9.1(?)[k]	4[l]		yes[j] (3–5%)
Dac g I	34[f], 33[m]	5.9[m]		95%[m]	
Phl p I	31[n], 34[f]				
Pha a I	o				
Sec c I	29/33[f], 31[n], 28[n]				
Alo p I	32[f]				
Ave e I	35[f]				
Fes p I/	34[f], 27[j]				yes[j] (3–5%)
Fes e I					
Agr a I	27[j]				yes[j] (3–5%)
Ant o I	27[j]				yes[j] (3–5%)
Cyn d I	23[n], 29/32[p,q], 31[q,r], 32[p,q]	6.5–7.0[q], 6.2–7.2[s]	4[q]	95%[t], 76%[u]	yes[p,q]

Note: Allergens have been named according to the IUIS system of nomenclature (Marsh et al., 1987)

[a] Johnson and Marsh, 1965a, 1965b; 1966a, 1966b; Marsh et al., 1970.
[b] Cottam et al., 1986.
[c] Ansari et al., 1987.
[d] Howlett and Clarke, 1981a.
[e] Lynch and Turner, 1974.
[f] Matthiesen and Løwenstein, 1991.
[g] Griffith et al., 1991.
[h] Freidhoff et al., 1986.
[i] Singh et al., 1991.
[j] Esch and Klapper, 1987.
[k] Lin et al., 1988.
[l] Ekrammodoullah, 1990.
[m] Mecheri et al., 1985.
[n] Matthiesen et al., 1992.
[o] Suphioglu et al., 1992.
[p] Matthiesen et al., 1991.
[q] Smith et al., 1993a.
[r] Su et al., 1991.
[s] Chang et al., 1991.
[t] Ford and Baldo, 1987.
[u] Shen et al., 1988.

stein, 1991), enzyme-linked immunosorbent assay (ELISA), inhibition ELISA (Esch and Klapper, 1987), radioimmunoassay (RIA) (Kahn and Marsh, 1986), and immunoblotting (Singh and Knox, 1985; Standring et al., 1987; Matthiesen et al., 1991; Smith et al., 1993b) to study antigenic relationships between allergens of different grasses.

Use of Mabs raised against Group I allergens from (Esch and Klapper, 1987) in both direct binding ELISAs and inhibition ELISAs has shown cross-reactivity between the allergens *Lol p* I, *Poa p* I, *Fes e* I, *Ant o* I, and *Agr a* I [from ryegrass, Kentucky bluegrass (June grass), meadow fescue, redtop and sweet vernal grass] but that these Group I allergens also possess unique IgE epitopes.

Most early studies using Mabs raised against *Lol p* I suggested that *Cyn d* I is antigenically different from other grass Group I allergens. Using immunoblotting with a Mab reactive with *Lol p* I, Singh and Knox (1985) identified cross-reactive epitopes on proteins from all grasses tested (from three grass subfamilies) except Bermuda grass (which is from a fourth subfamily) (Singh and Knox, 1985). Strandring et al. (1987) produced two Mabs to allergen R7 (*Lol p* I), neither of which identified a Group I allergen in Bermuda grass. However, in the same study rabbit anti-*Lol p* I polyclonal antibodies identified a 29 to 31 kDa protein in Bermuda grass pollen extract. Only one Mab raised against *Lol p* I recognized an antigen in Bermuda grass pollen extract. This was designated Mab 3.2 and also recognized proteins in eight other grasses (Kahn and Marsh, 1986).

In more recent studies, Mabs against *Cyn d* I recognized *Lol p* I (Matthiesen et al., 1991; Smith et al., 1993b). All four Mabs produced by Smith et al. (1993b) cross-react with *Lol p* I and Group I proteins from seven other grasses. Although only one of four Mabs raised by Matthiesen et al. (1991) recognized *Lol p* I, all four recognized cross-reactive proteins in a number of other grasses including orchard grass and Kentucky bluegrass. In contrast, four Mabs raised against *Cyn d* I (*Cyn d* Bd 35K) by Chang et al. (1991) failed to recognize antigens in five other grasses (Canary grass, *Phalaris minor;* maize, *Zea mays;* annual bluegrass, *Poa annua;* cane, *Sorghum vulgare;* and ryegrass, *Lolium perenne*).

Polyclonal antibodies against *Lol p* I have failed to precipitate *Cyn d* I in CRIE (Matthiesen and Løwenstein, 1991) but those raised against *Cyn d* I precipitate *Lol p* I (Matthiesen et al., 1991). Obviously *Lol p* I and *Cyn d* I are antigenically cross-reactive to some extent. The difference in reactivity of antisera raised against the different proteins probably reflects differences in the primary structure of the two proteins but may also be related to the absence of certain immunodominant epitopes of *Lol p* I on *Cyn d* I.

Allergenic Cross-Reactivity between Grass Pollen Allergens

A number of studies have investigated allergenic cross-reactivity between grass pollen allergens from different grasses (Bernstein et al., 1976; Leiferman and Gleich, 1976; Schumacher et al., 1985; Martin et al., 1985; van Ree et al., 1992). Most of these studies have used radioallergosorbent test (RAST) inhibition with whole pollen extract, which has the disadvantage that the particular allergens that cross-react are not identified. The taxonomically related grasses such as ryegrass, orchard grass, Kentucky bluegrass, and meadow fescue (from the tribe Poeae) all tend to show similar inhibitory activity when protein extracts prepared from their pollen are used as inhibitors or solid phase antigen in RAST inhibition experiments, suggesting that they share common allergens. In contrast, Bermuda grass, Johnson grass and to a lesser extent timothy grass appear to have allergens that are distinct from those of other grasses (Bernstein et al., 1976; Schumacher et al., 1985; Martin et al., 1985).

To study cross-reactivity between Group I allergens of different grasses, van Ree et al. (1992) used purified *Lol p* I as an inhibitor of IgE binding to crude extracts of other grasses. Inhibition of IgE binding to the other pollen extracts may still be due to cross-inhibition of allergens other than Group I but the results are more specific than radioallergosorbent test (RAST) inhibition using whole pollen extracts as inhibitors. They also investigated whether inhibition was the same for serum from different individuals. They found that different sera recognized different amounts of cross-reactivity between the grasses. In general, the similarity in structure of antigens cross-reactive with *Lol p* I was, in order of decreasing similarity, (from grass pollen extracts of) orchard grass (*Dactylis glomerata*), fescue (*Festuca rubra*), timothy (*Phleum pratense*), maize (*Zea mays*), sweet vernal grass (*Anthoxanthum odoratum*) and cultivated oat (*Secale cereale*). This order does not follow taxanomic relationships, with the Group I allergen of maize from the subfamily Panicoideae being more similar than those from sweet vernal and cultivated oat which are from the same subfamily as ryegrass but a different tribe.

Inhibition between Mabs and IgE has been used to identify two IgE binding regions on *Fes e* I, the major allergen of meadow fescue (Esch and Klapper, 1989a). These were designated site I and site II. Site I was recognized by IgE from all (5) sera tested, whereas site II was recognized by IgE from only one of the five sera tested. Mabs whose epitopes overlapped site I recognized Group I allergens from four other grasses (*Lol p* I, *Poa p* I, *Ant o* I, and *Agr a* I) and so the site I determinant on the five allergens was further characterized by digestion of the Group I antigens with trypsin after protection of lysine residues by citraconylation (Esch and Klapper, 1989b). Peptides that bound to antisite A Mabs (those that overlapped site I) were isolated and the amino acid sequences identified and found to be highly conserved in all five grasses (Table 7-3). Two 28 amino acid peptides were isolated after digestion of *Lol p* I. These peptides differed by one amino acid at position 11 in the sequence (Esch and Klapper, 1989b). An epitope shared by *Lol p* I, *Poa p* I, and *Dac g* I was identified through inhibition between Mabs and IgE (Mourad et al., 1988; 1989; Lin et al., 1988). This epitope was recognized by IgE from all 12 sera tested and may represent the same epitope localized by Esch and Klapper (1989a, 1989b). Thus contrary to the findings of van Ree et al. (1992), allergenic epitopes shared by Group I allergens from different gasses do exist.

Han et al. (1993) identified two anti-*Cyn d* I Mabs (MoAb 1-61 and MoAb 4-37) that significantly inhibit the binding of IgE to *Cyn d* I, suggesting that their epitopes overlap major IgE epitopes. As these Mabs do not bind *Lol p* I and Group I allergens from a number of other grasses (Chang et al., 1991; Han

Table 7-3 NH$_2$-Terminal Sequences of *Cyn d*I Isoallergens and Other Group I Allergens

Allergen	NH$_2$-terminal amino-acid sequence
Cyn d I a[a]	A**M**G<u>D</u>KPGPXITATYGD<u>K</u>X<u>L</u>DAK<u>X</u>**AFD**
	<u>T</u>
Cyn d I b[a]	A**I**G<u>D</u>KPGPXITAXYG<u>X</u><u>K</u>X<u>L</u>XA
Cyn d I b[b]	A**I**G<u>D</u>KPGP<u>K</u>ITATYX<u>X</u><u>K</u>WLEAKA**TFY**GSNP<u>R</u>GAAXDD
Cyn d I[c]	AMG<u>D</u>KPGPXITATYGDKWLDAKATFYG
Lol p I[d]	IAKVpPG<u>p</u><u>K</u>ITAEYGDKWLDAK<u>X</u>T
Lol p I[e]	IAKVPPGPNITAEYGDKWLDAKSTWYGKPT
Po a p I[f]	IAKVpPG<u>p</u>XITATYGDKXLDAKST
Poa p I[g]	IAKVpPG<u>p</u>XITATYGD<u>KWL</u>DAKST
Phl p I[f]	IAKVpPG<u>p</u>XITATYG
Sec c Ia[f]	IPKVpPG<u>p</u>TIXAS
Sec c Ib[f]	AYGNKXLDTIT
Pha a I[h]	IAKVPPG<u>G</u>XITAEYGDKWLD

Note: **bold** identifies amino acids which differ between different *Cyn d*I isoallergens; *underlined* amino acids indicate that the designation is tentative; X indicates unknown amino acid; p indicates hydroxyproline residues. Where two acids have been identified at the same position, the second amino acid is shown on the line below.

[a] Determined after transfer to PVDF membrane.
[b] Determined after HPLC purification.
[c] Matthiesen and Løwenstein, 1991.
[d] P. M. Smith, unpublished data.
[e] Cottam et al., 1986; Griffith et al., 1991.
[f] Matthiesen et al., 1992.
[g] Ekramoddoullah, 1990.
[h] Suphioglu et al., 1992.

et al., 1993), it seems likely that the IgE epitopes identified are not shared with other Group I allergens. These results suggest that *Cyn d* I may have some distinct IgE epitopes.

Molecular Cloning of Group I Allergens

The past five years have seen an explosion in the information available about allergens. Molecular cloning has been largely responsible for this increase. Previously, amino acid sequencing of only small proteins was feasible, but with the advent of molecular cloning, isolation of cDNA clones encoding allergens and identification of their deduced amino acid sequences has become routine. Breakthroughs in expression and purification of recombinant proteins mean that pure recombinant allergens can now be produced in large amounts for use in research and with potential for immunodiagonsis and therapy. Modification of cDNAs allows the expression of truncated allergens that can be used to localize allergenic epitopes.

Lol p I

The nucleotide and deduced amino acid sequence of *Lol p* I have been determined (Perez et al., 1990; Griffith et al., 1991). Clones were obtained by cDNA cloning and PCR amplification of ryegrass pollen RNA. A partial cDNA clone encoding the C-terminal region of *Lol p* I (amino acids 96-240; 13R), which binds IgE from the serum of individuals susceptible to ryegrass pollen has been isolated using anti-*Lol p* I Mabs (Griffith et al., 1991). Full-length *Lol p* I clones have an ORF (Open Reading Frame) of 789 nucleotides and potentially encode a protein of 263 amino acids with a 23 amino acid hydrophobic leader sequence. The deduced amino acid sequence of *Lol p* I has a single N-glycosylation motif (Asn-Ile-Thr) at Asparagine 9, and the predicted molecular mass of the mature form of *Lol p* I (240 amino acids) is 26.6 kDa.

The *Lol p* I signal peptide has some motifs common to other plant signal peptides, especially those of secreted plant enzymes such as amylase. The motifs close to the cleavage site show some similarity with those that direct proteins into the lumen of the endoplasmic reticulum for glycosylation, e.g., VLL . . . FL GSAHG in *Lol p* I and VLL . . . LS ASLASG in alpha-amylase (Jones and Robinson, 1989). This molecular evidence supports the immunocytochemical findings that *Lol p* I occurs in both the cytosol and at the pollen surface, suggesting the allergen is secreted to the surface of the pollen grain (Singh et al., 1991).

Lol p I is expressed by a pollen specific gene (Griffith et al., 1991). Use of a *Lol p* I cDNA to screen an RNA gel blot of total RNA from ryegrass pollen, leaves, roots, and inflorescence identified a transcript only in the tissues containing pollen (pollen and inflorescence). These results have been confirmed using antibody and IgE probes on Western blots of proteins of these tissues separated by SDS-PAGE.

There is homology between *Lol p* I and the low molecular weight allergens *Lol p* II and III; 48% homology (33% identity) between *Lol p* I and II and 44% homology (26% identity) between *Lol p* I and III. *Lol p* II and III can be aligned with the C-terminal region of *Lol p* I (Griffith et al., 1991; Fig. 7-2). Antigenic cross-reactivity between *Lol p* II and III has been identified, but despite the homology between these proteins and *Lol p* I, most studies have not detected cross-reactivity between *Lol p* I and II and *Lol p* I and III (Ansari et al., 1987). However, a recent study has suggested that cross-reactivity between *Dac g* II and *Lol p* I/II is increased after immunotherapy (Roberts et al., 1992).

By using the deduced amino acid sequence of *Lol p* I to construct synthetic peptides which were used to stimulate peripheral blood mononucleocytes, a T-cell determinant has been shown to lie between amino acids 191 and 210 of *Lol p* I (Perez et al., 1990).

Cyn d I

A full-length cDNA clone encoding *Cyn d* I have been isolated using molecular cloning techniques and PCR amplification. Clones that encode proteins of 244 and 246 amino acids have been isolated. The mature protein has a calculated M_r of 26.8 kDa with one potential N-glycosylation site at Asparagine 9, a site shared

```
Lol p Iᵃ     GELELQFRRVKCKYPDGTKPTFHVEKASNPNYLAILVKYVDGDGDVVAVD  177
Lol p IIᵇ                   AAPVE-T---G-DEKN--LSI--NKEGDSMAE-E    34
Lol p IIIᶜ                  --VDLT---G-DAKT-VLNI--TRPGDTLAE-E     33

Lol p I      IKEKGKDKWIELKESWGAVWRIDTPDKLTGPFTVRYTTEGGTKSEVEDVI  227
Lol p II     L--H-SNE-LA--KNGDG--E-KSDKP-K-ᵣ-NF-FVS-K-MRNVFD--V   84
Lol p III    LRQH-SEE-EPMTKKGN*L-EVKSAKP----MNF-FLSK--M-NVFDE--  83

Lol p I      PEGWKADTSYSAK                                       240
Lol p II     -ADF-VG-T-KPE                                        97
Lol p III    -TAFTVGKT-TPEYN                                      97
```

Figure 7-2 **Homology of *Lol p* III and III.** The amino acid sequences of *Lol p* II and III were aligned with the deduced amino acid sequence of *Lol p* I. A gap represented by an asterisk was introduced in the *Lol p* III sequence to give maximum homology. Residues identical to *Lol p* I are indicated by a dash. Residues considered similar to those of *Lol p* I are shown in **bold** letters. Residues considered to be similar or identical residues between *Lol p* II and III are underlined. Amino acids said to be similar are: A, S, T; D, E; N, Q; R, K; I, L, M, V; F, Y. W. Numbers to the right of the sequence refer to the numbering for each protein.
[a] sequence of *Lol p* I is from Griffith et al., 1991. [b] sequence of *Lol p* II is from Ansari et al. (1989a). [c] sequence of *Lol p* III is from Ansari et al. (1989b).

with *Lol p* I. A number of partial clones that encode different regions of *Cyn d* I were also isolated and these can be grouped into three different families according to similarities in their deduced amino acid sequence. There are 33 differences between the amino acid sequences of the two *Cyn d* I isoforms identified (Smith et al., 1993d). *Cyn d* I is encoded by a multigene family and like *Lol p* I it is expressed specifically in pollen (unpublished results).

Sor h I

Sor h I has been cloned by screening a cDNA expression library with an anti-*Cyn d* I Mab (Avjioglu et al., 1993). The recombinant *Sor h* I (expressed by clone 3S) binds IgE from the sera of some individuals allergic to Johnson grass. Using IgE affinity purified from the fusion protein, the native protein that the gene encodes has been identified as a 35 kDa protein, which also binds to three anti-*Cyn d* I Mabs used in its cloning. Clone 3s has an open reading frame of 783 nucleotides that encodes a protein of 261 amino acids. The mature processed protein apparently consists of 238 amino acids, the remaining 23 amino acids forming a hydrophobic signal peptide.

Other Group I allergens

Several important cereals are members of the family Poaceae, and it is important that the homologous glycoproteins be identified and characterized, as they may well prove to cause occupational allergic diseases in agricultural workers. For example, rice (*Oryza sativa*) pollen is known to be allergenic, and the allergens in rice have been shown to be cross-reactive with those of Kentucky bluegrass (Kimura et al., 1969). *Ory s* I has recently been cloned using Mabs raised against *Cyn d* I (P. Theerakulpisut, unpublished data). *Zea m* I has also been cloned and partially sequenced (P. Bedinger, unpublished data) as part of a study of pollen-specific genes of maize isolated by subtractive hybridization.

Homology between Group I Allergens of Different Grasses

Table 1-4 shows the NH$_2$-terminal sequences of a number of Group I allergens that show strong homology (Singh et al., 1990; Cottam et al., 1986; Singh et al., 1991; Matthiesen et al., 1992). Group I allergens from most temperate grasses have been found to contain hydroxyproline (Matthiesen et al., 1992). There are few amino acid substitutions between these allergens; in the first 15 amino acids there is only one difference between the NH$_2$-terminal sequence of *Lol p* I and those of *Poa p* I and *Phl p* I. *Cyn d* I has more amino acid substitutions in the NH$_2$-terminal amino acid sequence compared to other grasses (eight differences in the first 24 amino acids) and only two of these substitutions are considered

Table 7-4   Comparison of C-terminal Amino Acid Sequences of Group I Allergens

Grass	Allergen	Amino Acid Sequence	IgE Binding
Lolium perenne	*Lol p* I[a,b]	YTTEGGTKSEVEDVIPEGWKADTSYSAK	+
Lolium perenne	*Lol p* I[b]	**********F****************	+
Festuca elatior	*Fes e* I[b]	**********A******	+[g]
Festuca elatior	*Fes e* I[b]	********************	+[g]
Poa pratensis	*Poa p* I[b]	********A*A*********V****E	+[g]
Agrostis alba	*Agr a* I[b]	********A*A************E	+[g]
Anthoxanthum odoratum	*Ant o* I[b]	******K*V*A************E	+[g]
Sorghum halepense	*Sor h* I[c]	I*******TAY************T *T**	?
Cynodon dactylon	*Cyn d* I[d]	L*S*S*GHV*OD*****D**P**V *KS*IQF	−
Cynodon dactylon	*Cyn d* I[e]	L*S***AHLVOD** **AN**P**V*TS*LQFGA	−
Zea mays	*Zea m* I[f]	L*S*S*K*VIAK*I**AN* RP*AV*TSNVQFY	?

Note: Amino acids that do not differ from *Lol p*I are indicated by an asterisk. Underlined amino acids indicate that the amino acid is similar to that of *Lol p*I. Amino acids said to be similar are: A,*S*,T; D,E; N,Q; R,K; I,L,M,V; F,Y,W.

[a] Deduced from cDNA clone 13R (Griffith et al., 1991).
[b] Determined by protein microsequencing (Esch and Klapper, 1989b),
[c] Deduced from cDNA clone 3S (Avjioglu, 1992).
[d] Deduced from cDNA clones 2B and 3B (Smith et al., 1993d).
[e] Deduced from cDNA clone 18B (Smith et al., 1993d).
[f] Deduced from cDNA clone (P. Bedinger, personal communication).
[g] Although the IgE epitope has been localized to this region by inhibition between Mabs and IgE, these peptides have not been shown to bind IgE.

as conservative. *Cyn d* I is also unique in that it does not contain hydroxyproline in its NH$_2$-terminal sequence.

The primary amino acid sequences of four Group I allergens, *Lol p* I, *Cyn d* I, *Zea m* I, and *Sor h* I, have been determined. There is a strong degree of homology between the four proteins (Fig. 7-3). One would expect that this would result in similar responsiveness by individuals to all three proteins. IgE antibodies that bound to Bermuda grass pollen proteins have been identified in sera of Danish individuals allergic to grass pollen even though Bermuda grass is not found in Denmark (Matthiesen and Løwenstein, 1991). This finding supports the theory of cross-sensitization. In contrast, other studies have found that allergens of Bermuda grass show little cross-reactivity with those of other grasses (Marsh et al., 1970; Bernstein et al., 1976; Martin et al., 1985; Schumacher et al., 1985) and separate diagnosis and treatment of individuals allergic to its pollen is required (Bush, 1989). Although all individuals tested in this study who have IgE in their serum that recognizes *Cyn d* I also have IgE specific for *Lol p* I, the reverse is not true. Rabbit monospecific antibodies raised against *Cyn d* I are able to precipitate Group I antigens from other grasses (Matthiesen et al. 1991). However, this reaction is weaker than the analogous reaction using monospecific antibodies against *Lol p* I (Matthiesen and Løwenstein, 1991). It has been suggested that the dominant antigenic epitopes on *Lol p* I are not present on *Cyn d* I and other Group I allergens but that those dominant on *Cyn d* I are present on *Lol p* I but not dominant (Matthiesen et al., 1991). For example, few Mabs raised against *Lol p* I bind to *Cyn d* I but most Mabs raised against *Cyn d* I bind to *Lol p* I (Kahn and Marsh, 1986; Standring et al., 1987; Singh and Knox, 1985; Matthiesen et al., 1991). Anti-*Cyn d* I Mabs have been used in immunoaffinity chromatography to isolate other Group I allergens for NH$_2$-terminal amino acid sequencing (Matthiesen et al., 1992). Thus *Cyn d* I may not share the dominant IgE epitopes of *Lol p* I (which it shares with a number of other grasses; Esch and Klapper, 1989b), but they probably share some of the other IgE epitopes that have been identified on *Lol p* I.

IgE Binding Properties of Recombinant Group I Allergens

The fusion protein of a clone (13R) expressing the C-terminal 144 amino acids (amino acids 96 to 240) of *Lol p* I binds IgE from sera of individuals allergic to ryegrass. This clone encodes a major IgE epitope identified by Esch and Klapper (1989b) which they localized to a 28 amino acid peptide of *Lol p* I, later found to be at the C-terminal end of *Lol p* I (Perez et al., 1990; Griffith et al., 1991). A partial clone expressing amino acids 96 to 221 of *Lol p* I did not bind IgE to the same extent as the recombinant 13R fusion protein (P. Smith unpublished data), suggesting that the last 19 amino acids of *Lol p* I (amino acids 222 to 240) are important for IgE binding. A synthetic peptide consisting

```
Cyn d I.CD1  AIGDKPGPNITATYGSKWLEARATFYGSNPRGAAPDDHGGACGYRNVDKP       50
Cyn d I.3    -X------------D-----K------D--------------------A       50
Lol p Iᵃ     IAKVP-------E--D----D-KS-W--K*-T--G-K-N------KD---A      49
Sor h Iᵇ     PPKVAK-K--------D---RK-ₒW--K*-T--G---N--ₒ----KD-N-A      49


                  60        70        80        90       100
Cyn d I.CD1  PFDGMTACGNEPIFKDGLGCGACYEIKCKEPVECSGEPVLVKITDKNYEH      100
Cyn d I.3    ---S--G--------------S---------A--------I---------      100
Lol p I      --N---G-NGT------R---S-F----TK-ES----A-T-T---D-E-P       99
Sor h I      --NS-G----L----------S-F----DK-A-----A-V-H---M---Q       99
Zea m Iᶜ     LQPV-----V---R-GR---S---VR---KP----N--T-F---M---P        49


                 110       120       130       140       150
Cyn d I.CD1  IAAYHFDLSGKAFGAMAKKGQEDKLRKAGELTLQFRRVKCKYPSGTKITF      150
Cyn d I.3    -------------------E-----------M--------E---D---A-      150
Lol p I      --P--------H---S-----E-Q---S----E-----------D---P--      149
Sor h I      --------A-H-----------E--A---IIDMK--------**-E-V--      147
Zea m I      --P-----------SL--P-LN-ₒ-HC-IMDV-----R----A-E--VE       99


                 160       170       180       190      199
Cyn d I.CD1  HIEKGSNDHYLALLVKYAA*GDGNIVAVDIRPKDSDEFIPMKSSWGAIWR      199
Cyn d I.3    -V-----PN----------*------S----S-G--D-L---Q-------      199
Lol p I      -V--A--PN---I----VD*---DV-----KE-GK-KW-EL-E----V--      198
Sor h I      -V-----PN--------VD*---D--G---KE-GG-AYQ-L-H-------      196
Zea m I      -----C-PN---V---FV-D---D--LME-QD-L-A-TK---E-T---T-      149


                 210       220       230       240
Cyn d I.CD1  IDPKKPLKGPFSIRLTSEGGAHLVQDDVIPANWKPDTVYTSKLQFGA        246
Cyn d I.3    ---P-------T------S-G-VE------ED-------K--I--           244
Lol p I      --TPDK-T---TV-Y-T---TKSEVE----EG--A--S-SA-             240
Sor h I      K-SD--I-F-VTVQI-T---TKTAY-----EG--A--T--A-             238
Zea m I      M-TA-A-R-D--------S-KKVIAK-I----TR--A----NV--Y         192
```

Figure 7-3 **Comparison of deduced amino acid sequences of *Cyn d* I with other Group I allergens.** *Cyn d* I.CD1 is the deduced amino acid sequence of clone CD1. *Cyn d* I.3 is derived from the deduced amino acid sequences of clones KAT-39-1 and 3B (Smith et al., 1993d). Residues identical to *Cyn d* I.CD1 are indicated by a dash. Amino acids similar to *Cyn d* I.CD1 are indicated by **bold** letters. Amino acids said to be similar are: A, S, T; D, E; N, Q; R, K; I, L, M, V; F, Y, W. Underlined amino acids are those different to those found in *Cyn d* I.CD1 but the same as those in *Cyn d* I.3. Asterisks indicate gaps that have been inserted to give maximum homology. Numbers above the sequence refer to the numbering for *Cyn d* I.CD1; numbers to the right of the sequence refer to the numbering for each protein. [a] Sequence of *Lol p* I is from Griffith et al., 1991; [b] sequence of *Sor h* I is from Avjioglu (1992); [c] the sequence of *Zea m* was provided by Dr. P. Bedinger.

of the last 25 amino acids of *Lol p* I has since been shown to bind IgE from the sera of individuals allergic to ryegrass (van Ree et al., 1993). IgE and IgG antibodies that bind to this peptide also cross-react with the C-terminal region of *Lol p* II and III (van Ree et al., 1993).

A number of partial *Cyn d* I clones encoding an analogous region to that expressed by clone 13R do not bind IgE from the serum of individuals allergic to Bermuda grass. These results suggest that *Cyn d* I does not share the C-terminal IgE epitope common to other grasses. Recombinant *Sor h* I also differs in its reactivity to IgE compared to the C-terminal region of *Lol p* I. Of 30 individuals allergic to ryegrass pollen whose sera contained IgE reactive with the fusion protein of clone 13R, only two showed reactivity with recombinant *Sor h* I (E. K. Ong, unpublished data). The fusion protein of a partial clone (2S) expressing amino acid 163 to 238 (corresponding to the C-terminal end of *Sor h* I) does bind IgE suggesting that in *Sor h* I this region is not responsible for IgE binding (Avjioglu, 1992). Table 7-4 shows a comparison of the deduced amino acid sequence of the C-terminal regions of grass Group I allergens. Four nonconservative substitutions occur in both *Cyn d* I isoforms but not in other grasses where the C-terminal epitope occurs. One or more of these substitutions could effect IgE binding. In *Sor h* I there are six amino acid differences compared to *Lol p* I, three of which are nonconservative. It seems likely that these amino acids are important for IgE binding. Experiments involving site directed mutagenesis of *Lol p* I to give the amino acids found in *Cyn d* I or *Sor h* I may identify the amino acids important in the interaction of IgE with this particular epitope.

Although recombinant *Cyn d* I, expressed in the vector pTrc-99A, bound to all Mabs raised against *Cyn d* I, in preliminary testing it did not bind IgE from the serum of individuals allergic to *Cyn d* I. This suggests that recombinant *Cyn d* I does not possess features important for IgE binding and that the IgE epitopes of *Cyn d* I are conformational or require some posttranslational modification in order to be reactive with IgE. The latter is less likely as there is no apparent reduction in IgE binding to deglycosylated *Cyn d* I compared to the nondeglycosylated protein (P. Smith, unpublished data).

Conclusions

The Group I allergens are a family of small glycoproteins (ranging from 27 to 35 kDa) that are specific to the reproductive structures of grasses. Immunochemical studies have shown that Group I allergens from a taxonomically wide range of grasses show cross-reactivity. Group I allergens have been identified in grasses from most subfamilies of the family Poaceae. Those from the subfamily Poodae (including all the most well-studied allergenic grasses) are more closely related, whereas those from other grasses (e.g., Bermuda grass, Johnson grass) are less related.

Molecular characterization has provided the amino acid sequence of a range of Group I allergens and these show strong homology. In the case of *Lol p*I and *Cyn d*I it has also led to the identification of different isoforms of the allergen. Knowledge of the sequences of these allergens will aid in the further identification of allergenic epitopes.

A major allergenic epitope of *Lol p*I has been identified at its C-terminal end and this epitope is expressed by the recombinant protein. Although this epitope is shared by Group I allergens from some closely related grasses, its absence may explain the lack of cross-reactivity between these Group I allergens and those from grasses such as Bermuda and Johnson grass.

References

Ansari, A. A., T. K. Kihara, and D. G. Marsh. 1987. Immunochemical studies of *Lolium perenne* (rye grass) pollen allergens, *Lol p* I, II and III. *J. Immunol.* 139:4034–4041.

Ansari, A. A., P. Shenbagamurthi, and D. G. Marsh. 1989a. Complete amino acid sequence of a *Lolium perenne* (perennial rye grass) pollen allergen, *Lol p* II. *J. Biol. Chem.* 264:11181–11185.

Ansari, A. A., P. Shenbagamurthi, and D. G. Marsh. 1989b. Complete primary sequence of a *Lolium perenne* (perennial rye grass) pollen allergen, *Lol p* III: comparison with known *Lol p* I and II sequences. *Biochem.* 28:8665–8670.

Avjioglu, A. 1992. Molecular analysis of grass pollen allergens. Ph.D. diss., University of Melbourne.

Avjioglu, A., T. Hough, M. B. Singh, and R. B. Knox. 1994. Pollen allergens. In *Genetic control of self incompatibility and reproductive development in flowering plants* (Williams, E. G. Clarke A. E. and Knox R. B. eds.). pp. 336–359. Kluwer Academic Publishers, Dordrecht, The Netherlands.

Bernstein, I. L., M. Perera, J. Gallagher, J. G. Michael, and S.G.O. Johansson. 1976. *In vitro* cross-allergenicity of major aeroallergenic pollens by the radioallergosorbent technique. *J. All. Clin. Immunol.* 57:141–152.

Bond, J. F., D. B. Segal, X-B. Yu, K. A. Theriault, J. Pollock, and H. Yeung. 1993. Human IgE reactivity to purufied recombinant and native grass allergens. *J. Allergy Clin. Immunol.* 91:339.

Bush, R. K. 1989. Aerobiology of pollen and fungal allergens. *J. Allergy Clin. Immunol.* 84:1120–1124.

Chang, Z. N., L. C. Tsai, C. W. Chi, M. C. Wang, H. D. Shen, D. T. Lee, and S. H. Han. 1991. Analysis of allergenic components of Bermuda grass pollen by monoclonal antibodies. *Allergy* 46:520–528.

Cottam, G. P., D. M. Moran, and R. Standring. 1986. Physiochemical and immunochemical characterization of allergenic proteins from rye grass (*Lolium perenne*) pollen prepared by a rapid and efficient purification method. *Biochem J.* 234:305–310.

Ekramoddoullah, A.K.M. 1990. Two-dimensional gel electrophoresis analyses of Kentucky bluegrass and rye-grass pollen allergens. Int. Arch. Allergy Appl. Immunol. 93:371–377.

Esch, R. E. and D. G. Klapper. 1989b. Isolation and characterization of a major cross-reactive grass group I allergenic determinant. *Mol. Immunol.* 26:557–561.

Ford, S. A. and B. A. Baldo. 1987. Identification of Bermuda grass (*Cynodon dactylon*)-pollen allergens by electroblotting. *J. Allergy Clin. Immunol.* 79:711–720.

Freidhoff, L. R., E. Ehrlich-Kautzky, J. H. Grant, D. A. Meyers, and D. G. Marsh. 1986. A study of the human immune response to *Lolium perenne* (Rye) pollen and its components, *Lol p* I and *Lol p* II (Rye I and Rye II). I. Prevalence of reactivity to the allergens and correlations among skin test, IgE antibody, and IgG antibody data. *J. Allergy Clin. Immunol.* 78:1190–1201.

Griffith, I. J., P. M. Smith, J. Pollock, P. Theerakulpisut, A. Avjioglu, S. Davies, T. Hough, M. B. Singh, R. J. Simpson, L. D. Ward, and R. B. Knox. 1991. Cloning and sequencing of *Lol p* I, the major allergenic protein of rye-grass pollen. *FEBS Lett.* 279:210–215.

Han, S., Z. Chang, H. Chang, C. Chi, J. Wang, and C. Lin. 1993. Identification and characterization of epitopes on *Cyn d* I, the major allergen of Bermuda grass pollen. *J. Allergy Clin. Immunol.* 91:1035–1041.

Howlett, B. J. and A. E. Clarke. 1981. Isolation and partial characterization of two antigenic glycoproteins from rye-grass (*Lolium perenne*) pollen. *Biochem. J.* 197:695–706.

Howlett, B. J. and R. B. Knox. 1984. Allergic interactions. In *Cellular Interactions. Encyclopedia of Plant Physiology*. 17, 655–673.

Johnson, P. and D. G. Marsh. 1965a. The isolation and characterisation of allergens from the pollen rye-grass (*Lolium perenne*). *Eur. Polym. J.* 1:63–77.

Johnson, P. and D. G. Marsh. 1965b. 'Isoallergens' from rye grass pollen. *Nature* 206:935–937.

Johnson, P. and D. G. Marsh. 1966a. Allergens from common rye grass pollen (*Lolium perenne*)—I Chemical composition and structure. *Immunochemistry* 3:91–100.

Johnson, P. and Marsh, D. G. 1966b. Allergens from common rye grass pollen (*Lolium perenne*)—II The allergenic determinants and carbohydrate moiety. *Immunochemistry* 3:101–110.

Jones, R. L. and D. G. Robinson. 1989. Protein secretion in plants. *New Phytol.* III: 567–597.

Kahn, C. R. and D. G. Marsh. 1986. Monoclonal antibodies to the major *Lolium perenne* (rye grass) pollen allergen *Lol p* I (Rye I). *Mol. Immunol.* 23:1281–1288.

Kimura, T., M. Todokoro, T. Kuroume, K. Tatemo, and T. Matsumura. 1969. Rice pollen asthma II. Cross antigenicity between rice pollen and other grasses. *Jpn. J. Allergy* 69:1005–1016.

Leiferman, K. M. and G. J. Gleich. 1976. The cross-reactivity of IgE antibodies with pollen allergens. I. Analyses of various species of grass pollens. *J. Allergy Clin. Immunol.* 58:129–139.

Lin, Z., A.K.M. Ekramoddoullah, F. T. Kisil, J. Hebert, and W. Mourad. 1988. Isolation and characterization of *Poa p* I allergens of Kentucky Bluegrass pollen with a murine monoclonal anti-*Lol p* I antibody. *Internat. Arch. Allergy Apl. Immunol.* 87:294–300.

Marsh, D. G., Z. H. Haddad, and P. H. Campbell. 1970. A new method for determining the distribution of allergenic fractions in biological materials: its application to grass pollen extracts. *J. Allergy* 46:107–121.

Marsh, D. G., L. Goodfriend, T. P. King, H. Løwenstein, and T.A.E. Platts-Mills. 1987. Allergen nomenclature. *J. Allergy Clin. Immunol.* 80:639–645.

Martin, B. G., L. E. Mansfield, and H. S. Nelson. 1985. Cross-allergenicity among the grasses. *Ann. Allergy* 54:99–104.

Matthiesen, F., M. J. Schumacher, and H. Løwenstein. 1989. An immunoelectrophoretic analysis of the allergens of *Cynodon dactylon* (Bermuda grass) pollen. *J. Allergy Clin. Immunol.* 83:1124–1134.

Matthiesen, F. and H. Løwenstein. 1991. Group V allergens in grass pollens. II. Investigation of group V allergens in pollens from 10 grasses. *Clin. Exp. Allergy* 21:309–320.

Matthiesen, F., M. J. Schumacher, and H. Løwenstein. 1991. Characterization of the major allergen of *Cynodon dactylon* (Bermuda grass) pollen, *Cyn d* I. *J. Allergy Clin. Immunol.* 88:763–774.

Matthiesen, F., A. K. Nielsen, T. J. Søgaard, S. Klysner, and H. Løwenstein. 1992. NH$_2$-terminal sequences of four immunoaffinity purified grass pollen allergens: *Phl p* I, *Poa p* I, *Sec c* I and *Cyn d* I. Abstract. XVth European Congress of Allergology and Clinical Immunology, Paris, May 10–15.

Mecheri, S., G. Peltre, and B. David. 1985. Purification and characterization of a major allergen from *Dactylis glomerata* pollen: The Ag Dg 1. *Internat. Arch. Allergy Appl. Immunol.* 78:283–289.

Mourad, W., S. Mecheri, G. Peltre, B. David, and J. Hebert. 1988. Study of the epitope specificity of purified *Dac g* I and *Lol p* I, the major allergens of *Dactylis glomerata* and *Lolium perenne* pollens, using monoclonal antibodies. *J. Immunol.* 141:3486–3491.

Mourad, W., D. Bernier, M. Jobin, and J. Hebert. 1989. Mapping of *Lol p* I allergenic epitopes by using murine monoclonal antibodies. *Mol. Immunol.* 26:1051–1057.

Orren, A. and E. B. Dowdle. 1977. Studies on Bermuda grass pollen allergens. *S. Afr. Med. J.* 51:586–591.

Perez, M., G. Y. Ishioka, L. E. Walker, and R. W. Chesnut. 1990. cDNA cloning and immunological characterization of the rye grass allergen *Lol p* I. *J. Biol. Chem.* 265:16210–16215.

Roberts, A. M., R. van Ree, S. M. Cardy, L. J. Bevan, and M. R. Walker. 1992. Recombinant pollen allergens from *Dactylis glomerata:* preliminary evidence that human IgE cross-reactivity between *Dac g* II and *Lol p* I/II is increased following grass pollen immunotherapy. *Immunology* 76:389–396.

Schumacher, M. J., J. Grabowski, and C. M. Wagner. 1985. Anti-Bermuda grass RAST binding is minimally inhibited by pollen extracts from 10 other grasses. *Ann. Allergy* 55:584–587.

Singh, M. B. and R. B. Knox. 1985. Grass pollen allergens. Antigenic relationships detected using monoclonal antibodies and dot blotting immunoassay. *Internat. Arch. Allergy Appl. Immunol.* 78:300–304.

Singh, M. B., P. M. Smith, and R. B. Knox. 1990. Molecular biology of rye-grass pollen allergens. In *Approaches to the Study of Allergens,* Monogr. Allergy, vol. 28, ed. B. A. Baldo, pp. 101–120. Karger Basel.

Singh, M. B., T. Hough, P. Theerakulpisut, A. Avjioglu, S. Davies, P. M. Smith, P. Taylor, R. J. Simpson, L. D. Ward, J. McCluskey, R. Puy, and R. B. Knox. 1991. Isolation of cDNA encoding a newly identified major allergenic protein of rye-grass pollen: Intracellular targeting to the amyloplast. *Proc. Natl. Acad. Sci. USA* 88:1384–1388.

Singh, M. B., P. E. Taylor, and R. B. Knox. 1993. Special preparation methods for immunochemistry of plant cells. In *Immunocytochemistry: A Practical Approach,* ed. J. E. Beesley. IRL Press, England.

Smart, I. J., R. J. Heddle, M. Zola, and J. Bradley. 1983. Development of monoclonal mouse antibodies specific for allergenic components in rye grass (*Lolium perenne*) pollen. *Internat. Arch. Allergy Appl. Immunol.* 72:243–248.

Smith, P. M., A. Avjioglu, L. R. Ward, R. J. Simpson, M. B. Singh, and R. B. Knox. 1993a. Isolation and characterisation of Group I isoallergens from Bermuda grass pollen. *Internat. Arch. Allergy and Immunol.* (in press).

Smith, P. M., M. B. Singh, and R. B. Knox. 1993b. Characterization and cloning of the major allergen of Bermuda grass, *Cyn d* I. In *Molecular Biology and Immunology of Allergens,* ed. D. Kraft. CRC Press, Boca Raton, Louisiana.

Smith, P. M., E. K. Ong, A. Avjioglu, M. B. Singh, R. B. Knox. 1993c. Analysis of ryegrass pollen allergens using two dimensional electrophoresis and immunoblotting. In *Molecular Biology and Immunology of Allergens,* ed. D. Kraft. CRC Press, Boca Raton, Louisiana.

Smith, P. M, C. Suphioglus, I. J. Griffith, K. Theriault, R. B. Knox, and M. B. Singh. 1993d. Cloning and expression of *Cyn d* I, the major allergen of Bermuda grass (*Cynodon dactylon*). *J. Allergy Clin. Immunol.* (in press).

Staff, I. A., P. E. Taylor, P. Smith, M. B. Singh, and R. B. Knox. 1990. Cellular localization of water-soluble allergenic proteins in rye-grass (*Lolium perenne*) pollen using monoclonal and specific IgE antibodies with immunogold probes. *Histochemical J.* 22:276–290.

Standring, R., V. Spackman, and S. J. Porter. 1987. Distribution of a major allergen of rye grass (*Lolium perenne*) pollen between other grass species. *Internat. Arch. Allergy Appl. Immunol.* 83:96–103.

Suphioglu, C., M. B. Singh, P. Taylor, R. Bellomo, P. Holmes, R. Puy, and R. B. Knox. 1992. Mechanism of grass-pollen-induced asthma. *Lancet* 339:569–572.

Suphioglu, C., M. B. Singh, R. J. Simpson, L. D. Ward, and R. B. Knox. 1993. Identification of canary grass (*Phalaris aquatica*) pollen allergens by immunoblotting: IgE and IgG antibody-binding studies. *Allergy* 48:273–281.

Taylor, P. E., I. A. Staff, M. B. Singh, and R. B. Knox. 1994. Localization of the two major allergens in rye-grass pollen using specific monoclonal antibodies and quantitative analysis of immunogold labelling. *Histochem. J.* 26:392–405.

van Ree, R., M.N.B.M. Driessen, W. A. van Leeuwen, S. O. Stapel, and R. C. Aalberse. 1992. Variability of cross-reactivity of IgE antibodies to group I and V allergens in eight grass pollen species. *Clin. Exp. Allergy* 22:611–617.

van Ree, R., W. A. van Leeuwen, and R. C. Aalberse. 1993. Demonstration of human IgE and IgG antibodies against the *Lol p* I C-terminus that crossreact with the C-terminus of *Lol p* II and III. *J. Allergy Clin. Immunol.* 91:187 (Abstract).

Watson, S. H. and C. S. Kibler. 1922. Etiology of hay-fever in Arizona and the Southwest. *JAMA* 78:719–722.

Watson, L. 1990. World grass genera. In *Reproductive Versatility in the Grasses,* ed. G. P. Chapman, pp. 258–265. Cambridge University Press, Cambridge.

8

Structure-Function Relationships of Allergens

Lei Zhang and Shyam S. Mohapatra

Introduction

Allergens are pragmatically defined as a group of antigens able to induce IgE antibody production in atopic patients. In recent years, the application of molecular biology techniques in allergen research has led to rapid advances in our knowledge of structure-function relationships of allergenic molecules. The study of the molecular nature of allergens is not only critical for the understanding of mechanisms of IgE antibody responses but also for development of new immunotherapeutic strategies.

IgE production is controlled by several environmental (including allergens) and host genetic factors. The host factors are deemed to be contributed by at least two sets of genes: (1) MHC loci, which control specific immune response and (2) other non-MHC gene(s) involved in general IgE production. IgE immune responses have been considered a major component of the immune system against helminth and nematode infections (Jarrett and Miller, 1982; Rousseaux-Prevost et al., 1977). However, similar responses lead to allergic diseases when the IgE antibodies are produced against the apparently harmless environmental allergenic molecules. It is now well recognized that specific IgE production is tightly controlled by functionally specialized reciprocal T cells, Th1 and Th2 cells (Mossman and Coffman, 1989; Romagnani, 1991). Recent studies indicate that Th1 and Th2 cells, possibly differentiated from the same precursor Th0 cells (Firestein et al., 1989), are not defective in their capacities of production of one another's cytokines (Yssl et al., 1992). The signals that determine the differentiation of the precursor T cells are currently under investigation. In addition to the molecular structure of allergens *per se,* several other factors have been considered to influence T-cell differentiation and IgE production. These factors include dose and route of immunization; adjuvant-like substances, which may preferentially induce production or secretion of certain cytokines and mediators; and types of APCs, which may differ in their capacities to present different antigens. This chapter reviews the evidence pertaining to the allergen structure with special emphasis on the identification and analysis of the determinants or epitopes that play a critical role in induction and regulation of IgE response.

What Is an Allergen?

For decades the working definition of allergens has been a group of antigens that have the ability to preferentially induce IgE antibodies. Whether this is due to some specific molecular characteristics has been debated. The arguments against the unique nature of allergenic antigens mainly comprise the following: First, the majority of the human population (as well as experimental animals) is not allergic to the ubiquitously distributed antigens, in other words most individuals produce low level of IgE antibodies transiently. Indeed, Th1 cells specific to allergens are found in nonatopic individuals, i.e., the allergen-specific T cells do not provide help to autologous B cells to produce IgE antibodies (Garman et al., 1990; O'Hehir et al., 1989). Second, analysis of the sequence homologies among most of the cloned allergens suggests no common primary sequences that may lead to stimulation of IgE antibody production.

However, a few recent observations argue to the contrary. First, the observation that in the same human individual, or in the experimental animals of the same genetic background, different types of antigens may determine the isotypes of immune responses, as was revealed by the studies of T cells from the atopic patients. The subsets of T cells, Th1 and Th2, were used as markers to examine the effect of different protein antigens on T-cell activation and proliferation. The results obtained from different laboratories are strikingly similar (Yssel et al., 1992; Wierenga et al., 1991; Wierenga et al., 1990; Parronchi et al., 1991), i.e., the antigen-specific T cells from the same individual are found to have two compartments, one is Th1 cells with specificities mainly to bacterial antigens, including tetanus toxoid and antigens from *Candida albicans*, and the other is Th2 cells with specificities primarily to common aeroallergens, including dust mite and grass pollen allergens. These results strongly suggest that under natural exposure, common antigens activate the T cells to differentiate to Th1 type cells, while the allergens preferentially stimulate the T cells to differentiate to Th2 type cells.

Moreover, the components of the antigen in the crude extract from the same source have been reported to induce different types of immune responses (Deuell et al., 1991; Scott et al., 1988; Yamada et al., 1993; Chakrabarty et al., 1981; Matthiesen and Lowenstein, 1991). One study of fungal infection with *Tichophyton tonsurans,* from which most antigens induce delayed-type hypersensitivity (DTH), revealed that a 30kDa protein antigen induced only immediate hypersensitivity but not DTH (Deuell et al., 1991). More convincing results are from the studies of different antigens of *Leishmania major* and *Nippostrongylus brasiliensis*. Two fractions of antigens isolated from *Leishmania major* induced Th1 and Th2-like responses, respectively, when used separately to immunize experimental animals against challenge of the pathogens (Scott et al., 1988). Similarly, natural infection of rats with *Nippostrongylus brasiliensis*, parasites, which are known to induce strong IgE responses, induces also IgG2a antibodies (Yamada et al., 1993). Interestingly, IgG2a and IgE antibodies are induced in fact by different groups of antigens. These two groups of antigens probably have different properties responsible for the isotype of antibody production.

Similarly, studies of grass pollen aqueous extracts indicated that only some of the antigenic components in the grass pollen extracts are recognized by human IgE antibodies (Chakrabarty et al., 1981; Matthiesen and Lowenstein, 1991). Taken together these observations suggest that proteins from the same source differ in their capacities to induce IgE antibodies. Furthermore, since both allergenic and nonallergenic proteins exist in different amounts in the extract, the differences in their capacities to induce IgE antibodies indicate that dose and route of exposure and possible adjuvant-like materials in the sources may not be sufficient to explain why only some proteins induce IgE antibodies.

More recently, we studied a fusion protein comprised of a partial β-galactosidase fragment (69kDa) and an allergenic fragment (20.6kDa) of *Poa p* IX allergen as a recombinant antigen-allergen chimeric (RAAC) protein. Immunization of BDF1 mice with the RAAC protein or the allergenic fragment of the RAAC protein induced only IgE and IgG1 responses whereas the β-galactosidase also induced IgG2a responses (Table 8-1). It is inferred from these results that different antigens induced different types of immune responses in the animals of the same genetic background.

Small peptides with defined sequences and their analogues have also been used to study their capacity to induce IgE antibodies. One is a natural 26 amino acid peptide, melittin, from bee venom, and the other is a synthetic peptide at position 12-26 of bacteriophage λ-cI repressor protein (Soloway et al., 1991; King et al., 1984). There are two striking similarities between these studies: (1) Both peptides induce specific IgE antibodies irrespective of the adjuvant used, and (2) changes in amino acid residues alter their capacity to induce IgE antibodies. Immunization in conjunction with complete Freund's adjuvant or Al(OH)$_3$ led to induction of peptide-specific IgE antibodies. Remarkably, a single amino acid change of the λcI peptide resulted in a total loss of its ability to induce IgE antibody (Soloway et al., 1991). Studies of melittin showed that a peptide of at least 24 amino acids in length with two to four cationic groups at hydrophilic C-terminal region are essential for induction of IgE antibodies (Fehlner et al., 1991a,b).

Taken together these observations indicate the possible role of allergen molecules in the determination of IgE production in atopic individuals. Detailed

Table 8-1 Specific Antibodies Induced After Immunization with rKBG8.3 Allergen and β-gal* in Dextran Sulphate

Ags used in immunization	Antibody titres		
	IgG1	IgG2a	IgE[1]
rKBG8.3[2]	655360	NIL	1280
rKBG8.3 (50 ug)	2621440	320	1280
rKBG8.3 (100ug)	1310720	160	1280
Truncated β-gal[3]	2621440	10240	80
Normal sera to rKBG8.3	NIL	NIL	NIL

*β-galactosidase.

[1]The IgE titers were determined with PCA assay, the other antibody titers were determined with ELISA.

[2]The coating antigen is KBG pollen extract, and each mouse in this group was immunized with 5ug antigen. The other groups immunized with different amount of antigen were indicated.

[3]The coating antigen is native β-gal, and each mouse in this group was immunized with 5ug antigen.

analysis of the molecular structures, including T- and B-cell epitopes, tertiary structures, and possible motifs, of allergen molecules may shed light on such a long-debated issue.

B-Cell Epitopes of Allergens

B-cell epitopes of protein antigens have attracted general interest in the field of immunology because detailed analysis of the structures of the epitopes is expected to help understand the molecular basis of antigenicity with respect to the mechanism of antigen-antibody interaction and its potential use for vaccines.

B-cell epitopes were found to be either sequential or conformational (Benjamin et al., 1984). Recent studies, however, on the crystal structure of antibody and antigen complexes have provided an updated view that most, if not all, of the epitopes are conformational, containing more or less dominant sequential elements (Davis and Padlan, 1990). By comparison of native and denatured allergens, it is possible to show that IgE binding epitopes, like the other antigenic epitopes, are either conformational or sequential nature (King, 1976; Lombardero et al., 1990; Nilsen et al., 1991; Polo et al., 1991). It was also indicated from these studies that carbohydrate groups contribute to IgE binding on some allergens.

To localize the epitopes on protein antigens, different approaches which have been employed are summarized in Table 8-2 with modification (Van Regenmortal, 1989). Only method 1 can be used to visualize the spatial arrangement of the structure of the epitope involved in the antibody binding. However, a notable disadvantage of the method of crystallography is that it provides no information

Table 8-2   Methods used to localize epitopes in protein antigens

Method	Type of epitope recognized	Average number of residues identified in epitopes
1. X-ray crystallography of antigen-Fab complexes	discontinuous epitope reacted with homologous antibody	15
2. Study of cross-reactive binding of peptide fragments with anti-protein antibodies	Continuous epitopes cross-reacting with heterologous antibody	3–8
3. Study of cross-reactive binding of protein with anti-peptide antibodies	Continuous epitopes cross-reacting with heterologous antibody	3–8
4. Determination of critical residues in peptide by systemic replacement with other amino acids	Continuous, cross-reacting epitope containing critical residues interspersed with irrelevant residues	3–5
5. Study of cross-reactivity between homologous proteins or point mutants	Discontinuous epitopes	1–3

on essential amino acids in the epitopes, whereas the other methods are useful to pinpoint the critical residues.

Because of the paucity of purified allergenic molecules and difficulties in obtaining monoclonal IgE antibodies for epitope analysis using techniques of crystallography, the crystal structure of IgE-binding epitopes is so far not available. Thus, there is no report of discontinuous IgE epitopes yet. However, by using other means, continuous IgE binding epitopes are defined in detail for a few allergens. The first allergen characterized with respect to epitope structure is allergen M (*Gad c* I), which is a 12KDa calcium-binding parvalbumin of white muscle tissue from codfish. Employing techniques of trypsin cleavage of the native protein or synthetic peptides, Elsayed and Apold (1983) identified five IgE binding peptides of the 113 amino acid long *Gad c* I. These sequences located at positions 13–32, 33–44, 49–64, 65–74 and 88–96, are responsible for the immunological reactivities of the native protein. These five peptides do not represent the full profile of IgE binding epitopes of the allergen, because the authors showed that peptides 13-32 and 49-64 are at least divalent in nature. The peptide 49-64 encompassed two repetitive sequences D-E-D-K and D-E-L-K separated by a hexapeptide spacer. Interestingly, this unique repetitive structure was found to be cross-reactive with birch pollen allergens (Elsayed et al., 1991).

Another well-characterized allergen, the non-cell-bound hemoglobin of chironomid larvae (*Chi t I*), is from the insect family Chironomidae (Mazur et al., 1990). *Chi t I* is composed of 12 homologous components. Epitope mapping of a component *T* III (136 amino acid in length) using trypsin-digested peptides revealed that three of 11 peptides, at positions 1-15, 91-101 and 110-135, are able to react with human IgE antibodies. These results are slightly different from the previous report of the component *T IV* (Baur et al., 1986), since a peptide at position 32-90 of *T IV* is also reactive with human IgE antibodies. This difference may be caused by the varying lengths of the peptides resulting from the digestion of component *T III*. Within the region of amino acid residues 32-90, six peptides are generated from *T III* and none of them is longer than 22 amino acids (Mazur et al., 1990).

A comprehensive synthetic approach for delineation of continuous epitopes was introduced by Atassi and Atassi (1986; 1990). This approach, using a series of overlapping synthetic peptides, greatly facilitated the identification of IgE binding epitopes of *Amb a* III, an allergen from ragweed pollen. The authors synthesized a series of 15 residue-peptides, with five amino acids overlapping, spanning the entire 101 amino acid residues of the protein. Epitope mapping led to the identification of five peptides, at positions 1-15, 21-35, 31-45, 51-65, and 71-85, which were able to bind to IgE antibodies (54).

Recently, another cost-effective epitope mapping method developed by Geysen et al. (1987) based on solid phase peptide synthesis and ELISA assay, is widely used for analysis of protein antigenic sites. This method has been successfully employed to define a variety of antigenic epitopes. We have also used this method to determine epitopes of the rKBG60, one of the *Poa p* IX allergens. By using murine antibodies raised against Kentucky bluegrass extract, rKBG60, and rKBG8.3, a C-terminal fragment of rKBG60, we demonstrated 13 potential epitopes on this antigen (Fig. 8-1). However, we failed to detect any human IgE binding signals from these peptides. Similarly, the other laboratories have failed to identify IgE binding epitopes by this method (Stewart et al., 1990; Walsh and Howden, 1989). Moreover, the decapeptides numbers 48 to 55 in Figure 8-1 are not reactive to antibodies of any serum in the pin assay, however, this region reacts strongly not only with murine antibodies but also with human IgE and IgG antibodies when 20 residue peptides are used. These peptides were synthesized by bioresin, cleaved and purified by reverse HPLC. The results indicate that the epitopes in this region are more conformational in nature and 10 amino acids may not be long enough to form the stable binding sites for the antibodies. Furthermore, our results using 20 residue peptides revealed also the differences in epitopes recognized by murine antibodies and human IgE and IgG antibodies.

Recently, recombinant DNA techniques were also employed for mapping IgE binding epitopes. An allergen, *Der p* I, from dust mite was cloned and sequenced (Chua et al., 1988). With the recombinant peptides obtained from either a random fragment library or restriction digestion, Thomas and coworkers were able to define the IgE binding sites of this allergen (Greene et al., 1991). Out of 16 overlapping recombinant peptides spanning the entire molecule of 220 amino acids, at least five regions, comprising residues 1-56, 53-99, 98-140, 166-194

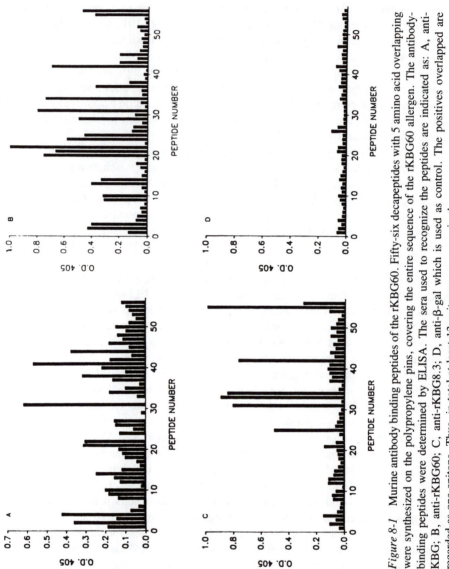

Figure 8-1 Murine antibody binding peptides of the rKBG60. Fifty-six decapeptides with 5 amino acid overlapping were synthesized on the polypropylene pins, covering the entire sequence of the rKBG60 allergen. The antibody-binding peptides were determined by ELISA. The sera used to recognize the peptides are indicated as: A, anti-KBG; B, anti-rKBG60; C, anti-rKBG8.3; D, anti-β-gal which is used as control. The positives overlapped are regarded as one epitope. Thus, in total at least 13 epitopes are recognized.

and 188-222, were found to be reactive with human IgE antibodies. Further, using peptides of various lengths, ranging from 114 residues to 14 residues, they concluded that IgE binding requires peptides of *Der p* I of at least 30 amino acid residues.

The above allergens have been investigated in detail for their profiles of allergenic determinants. Defined IgE-binding peptides, but not systematic analysis, from various sources such as ovalbumin and pollens also have been reported (Elsayed et al, 1991). These studies have led to two major conclusions: (1) although polar or charged amino acids are often found in the epitopes (Elsayed et al., 1991; Mazur et al., 1990; Baur et al., 1986; Atassi and Atassi, 1986, 1990), there exists virtually no consensus sequences among these IgE-binding peptides, and (2) most of the IgE-binding peptides were found also to be reactive with IgG antibodies (Elsayed et al., 1991; Mazur et al., 1990; Baur et al., 1986; Atassi and Atassi, 1986, 1990; Geysen et al., 1987; Stewart et al., 1990; Walsh and Howden, 1989; Greene et al., 1991). Indeed, previous results from application of monoclonal antibodies for inhibition of human IgE to bind to the allergens indicate the same conclusion that IgG and IgE recognize the same or adjacent epitopes (Esch and Klapper, 1989; Moirad et al., 1988).

In addition to general interest, definition of allergenic epitopes enables examination of their potential for immunotherapy. An example of the application of B-cell epitopes is based on the well-established mechanism of allergic reaction, i.e., bridging of two IgE molecules on mast cells or basophils by multivalent allergen or divalent anti-IgE is required for triggering histamine release (Levine and Redmond, 1968; Segal et al., 1977). Indeed, a peptide synthesized from a codfish allergen was shown to block but not elicit allergic reactions (Elsayed et al., 1981). Thus, the well-defined and purified monovalent allergenic peptides are expected to abrogate allergic reactions by interfering with the primary interaction between the native allergens and IgE antibodies.

Another strategy is to induce specific suppression of IgE antibody production to the allergenic epitopes. This attempt is based primarily on the observations that, in various hapten-carrier systems, immunization of carrier-primed mice with a new epitope coupled to the priming carrier induces suppression for high affinity IgG antibody response to the newly introduced epitope without interfering with the response to determinants on the carrier molecules (Herzenberg et al., 1983; Schutze et al., 1989). In this context a recent report demonstrated that the hapten-specific memory B cells had an intrinsic defect that prevented differentiation of the cells into active IgG antibody secreting cells (Galelli and Charlot, 1990).

Other methods include linking the allergenic fragment containing the major epitopes of the allergen to antigenic molecules by genetic engineering. Thus using the RAAC protein we studied the possibility of regulating IgE antibody production to the *Poa p* IX allergens in the context of different adjuvants. After immunization with RAAC protein in CFA, the mice showed no detectable IgE but induced normal IgG1 and IgG2a antibodies to *Poa p* IX allergens and β-galactosidase. Further injection of the RAAC protein in dextran sulfate led to a slight increase in the IgE titer, whereas IgG1 and IgG2a titers to both of the antigens remained unchanged (Fig. 8-2). However, the mice immunized with the rKBG8.3 but not with the RAAC protein, twice in CFA showed high IgE antibody production after further injections with the allergenic polypeptide in dextran sulfate (Fig. 8-2). It is inferred from these results that β-galactosidase influenced the allergenic fragment in terms of IgE induction. Moreover, the allergenic epitopes can be also directly modified as reported previously (Norman et al., 1982) and used to modulate IgE antibody synthesis.

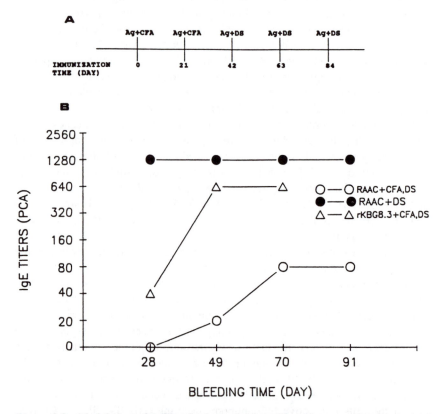

Figure 8-2 Modulation of IgE antibody production. The mice were first immunized twice with either the RAAC protein or the rKBG8.3 in complete Freund's adjuvant, and then challenged with the same protein but in dextran sulfate at three week intervals, as shown immunization schedule 4A. 4B shows the changes of the IgE antibody titers determined by PCA assay, and also the IgE titers of the mice immunized with the RAAC protein and DS alone was used as control.

T-Cell Epitopes of the Allergens

In contrast to the majority of B-cell epitopes, T-cell epitopes are sequential in nature; composed of mostly 8–12 amino acids for $CD8^+$ T cells and relatively diverse lengths for $CD4^+$ T cells (Sette et al., 1989; Nikolic-Zugic and Bevan, 1990; Pala et al., 1988; Maryanski et al., 1990; Romero et al., 1991; Leighton et al., 1991). Because T cells play an essential role in regulating IgE antibody production, T-cell epitopes have recently received more attention than the B-cell epitopes of the allergens.

Again, because of the paucity of information of allergen sequences, T-cell epitopes have been only defined from a limited number of allergens. One well-characterized allergen is *Amb a* III. Kurisaki and coworkers defined T-cell stimulatory peptides at positions, in decreasing levels of responses, 51-65, 1-15, 11-25, 81-95, 71-85 and 91-101 by using ten 15-residue overlapping peptides (Kurisaki et al., 1986). Three of these peptides were found in this study to be coincident with B-cell epitopes. Later on, by induction of antibody production in vivo directly with synthetic peptides, it was shown that all the five antibody-binding peptides also possessed T-cell epitopes, because these peptides induced both the peptide- and native protein-specific antibodies (Atassi and Atassi, 1986).

Lol p I allergen was also analyzed for its T-cell epitopes with peptides synthesized based on the deduced amino acid sequences of a cDNA clone (Perez et al., 1991). Twenty-three 20-residue peptides representing the entire 240 amino acid protein, with 10 amino acid overlapping, were examined with four human T-cell clones specific to *Lol p* I allergen. Interestingly, it was found that all four clones selected from *Lol p* I-stimulated PBL recognized peptide number 20 at position 191-210, indicating that this peptide represented an immunodominant T-cell epitope of the allergen.

T-cell epitopes of a major tree pollen allergen, *Bet v* I, has been mapped recently (Ebner et al., 1993). Eleven T-cell clones from the atopic patients were examined for their reactivities to a series of 75 overlapping peptides. Out of these 11 T-cell clones, nine secreted IL-4 upon allergen stimulation. The other two clones, FS3 and FS5, secreted moderate levels of INF-γ. In total, eight epitopes were defined by using these clones. These epitopes are at positions of 2-16, 11-22, 61-72, 77-88, 85-96, 113-124, 145-156, and 147-158. Although most of these epitopes seem to be able to enhance the T cells to secrete IL-4, they fail to inhibit IgE binding to the native or the recombinant allergen. Thus, the authors suggested that T- and B-cell epitopes may localize differently on these allergens.

Similarly, cloned human T-cell lines have been used to analyze epitopes of *Der p* I allergen of dust mite (Yssel et al., 1992). Three T-cell epitopes of this allergen are identified. Interestingly, p45-67 and p117-143 are recognized by DR-restricted T cells, whereas p94-104 is recognized by all DR2, DRw11, and

DR8 restricted T cells. All three epitopes stimulate the T cells to secrete IL-4 and IL-5. Apparently, this study indicates that some of the T-cell epitopes of Th2 type are not MHC-restricted.

The T-cell epitopes of the other major allergen of mite, *Der p* II, have been studied in various strains of mice (Hoyne et a., 1993). Different strains responded to different epitopes after immunization with the allergen. The peptides comprising epitopes for H-2b mice are peptides 11-35, 78-104; for H-2k mice are peptides 105-129; 36-50 and 78-104; and for H-2d mice is the peptide 36-60. Furthermore, immunization of mice with the synthetic peptides led to definition of two other T-cell epitopes, peptides 1-20 and 22-50. The T cells specific to these two peptides also were found reactive to the allergens, *Der p* II.

Immunization with synthetic peptides is a useful approach to defining T-cell epitopes. This approach not only detects the epitopes accessible on native allergens, but also the cryptic epitopes to which T cells are usually not sensitized by exposure to the native allergens. We examined the coincidence of T- and B-cell epitopes of *Poa p* IX allergen by using twelve 20-residue peptides. Nine of these peptides, #14 (at position 109-128), #16 (129-148), #18 (149-168), #19 (159-178), #21 (179-198), #23 (199-218), #26 (229-248), #27 (239-258), and #28 (249-268), are defined as B-cell epitopes. Three peptides, #5 (19-38), #22 (169-188), and #25 (219-238), did not bind to murine antibodies specific to the allergens. Among these peptides used for immunization, eight induced the allergen-specific antibodies, whereas four peptides including two non-antibody non-binding peptides, #5 and #22, and two antibody-binding peptides, #14 and #21, induced no detectable antibodies (Fig. 8-3).

The role of T-cell epitopes of allergens in activation of different subsets of T cells is unknown. However, some indirect evidence from studies of TCR in cell differentiation and T-cell epitopes from proteins other than allergens suggested that TCRs and their ligands, T-cell epitopes, might be involved in the T-cell subset differentiation (Rocken et al., 1992; Ashbridge et al., 1992; Evavold and Allen, 1991; Evavold et al., 1992).

As stated earlier, Th1 and Th2 cells may originate from the Th0 cells depending on the differential stimulation. One recent report suggested that T-cells differentiation from Th0 cells to either Th1 or Th2 cells does have a relation to the TCR complex (Rocken et al., 1992). In this report, the authors investigated the short-term cultured murine CD4+ T cells. The results show that if the cells activated by staphylococcal enterotoxin are cultured with IL-2 and restimulated with the superantigen, the cells express Th1 phenotype at day 11. However, addition of anti-CD3 monoclonal antibody to the culture led to the expression of Th2 phenotype during the same time. This observation was confirmed by the application of anti-Vβ8 instead of anti-CD3 monoclonal antibody in the same culture. The Vβ8+ but not Vβ8− T cells were induced to differentiate into the Th2 phenotype cells. The strong argument from this study is that the TCR, the receptor of T-cell epitope and MHC molecule, is critical for Th2 cell differentiation.

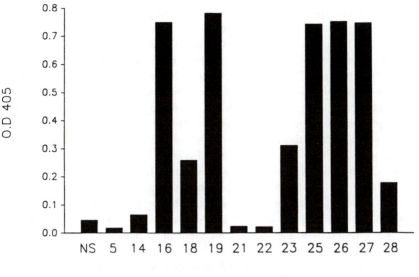

ANTI−PEPTIDE SERA

Figure 8-3 The allergen-specific antibodies induced with the synthetic peptides. BDF1 mice were immunized with the synthetic peptides in dextran sulfate. The antisera from the mice were assayed by ELISA to determine their binding to the recombinant allergen KBG8.3. The nonimmunized mouse serum pool (NS) was used as a negative control.

T-cell epitopes, the natural ligands of TCR, seem also to have effects on the cell differentiation. It is well known from various studies that different antigens in the same crude extract differ in their antigenicity and allergenicity. At the epitope level, recent studies of a 19KDa bacterial protein from *Mycobacterium tuberculosis* showed that different regions of the antigen had preferential effects on activation of different subsets of T cells (Ashbridge et al., 1992). In this study, the overlapping synthetic peptides spanning the antigen of 159 amino acid residues were used to define the T cell epitopes. Five peptides at positions 1-20, 61-80, 76-95, 136-155, and 145-159 were found to be able to stimulate the antigen-specific T cells in proliferation assays. The peptides 76-95 and 136-155 were relatively weak in induction of both Th1- and Th2-like cells. Interestingly, while peptides 1-20 and 61-80 induced preferentially Th2 type responses, peptide 145-159 induced dominant Th1 type response. Together it is suggested that the molecular structure of T-cell epitopes presumably by their varying affinity of interaction with TCR play a role in activation of different subsets of T cells.

Detailed structures of Th1 versus Th2 epitopes were analyzed by Allen and his associates (Evavold and Allen, 1991; Evavold et al., 1992). The report indicates that substitution of a single amino acid of a T-cell epitope resulted in loss of its ability to stimulate T-cell proliferation, while retaining its ability to induce the cells to secrete IL-4 (Evavold and Allen, 1991). From this observation it was concluded that T-cell proliferation and cytokine secretion are two different events and the amino acid residues in the T-cell epitope are critical to these events. Another study from the same group indicates that there may exist fine differences between Th1 and Th2 epitopes (Evavold et al., 1992). A 12-residue peptide at position 64-76 derived from murine hemoglobin was employed to generate a panel of specific Th1 and Th2 clones by immunization of mice with the peptide in complete Freund's adjuvant. With the serial peptide analogues the authors demonstrated that both Th1 and Th2 cell clones required the same essential amino acid residues. However, some differences in fine specificity of the Th1 and Th2 cell clones were noticed, such as phenylalanine at position 71 is critical for the Th1 cell recognition, but substitution of this amino acid had no effect on the Th2 cell stimulation. Thus, although the peptide could induce both Th1 and Th2 subsets and the cells of both subsets shared the gross specificities, the fine differences in their specificities remain an interesting issue for further investigation.

The only allergen of which T cell epitopes have been functionally analyzed is the bee venom melittin. Previous studies demonstrated that its C-terminal amino acid residues 20-26 with cationic residues were essential for specific IgE and IgG1 antibody production, indicating the major T-cell epitope(s) in this region critical for Th2 cell activation. Further studies led to the finding that the region at position 11-19 also possessed a T-cell epitope (Fehlner et al., 1991). This peptide has an α-helical amphiphilic segment and is able to induce Th2 cells in mice. Substitution of the amino acids of the peptide and analysis of its secondary structure indicate that the primary structure of this peptide is more important than its secondary structure for its ability to induce Th2 cells (Fehlner et al., 1991a, b).

Most evidence to date suggests that the B-cell epitopes of allergens are able to react with both IgE and IgG antibodies, although some epitopes reacting with IgE but not IgG and vice versa have been reported. The roles of T-cell epitopes, if any, on T-cell differentiation from Th0 cells to Th1 or Th2 subsets, thereby influencing immunoglobulin class switch, are not clear as yet. Although the results discussed above indicate the different effects of T-cell epitopes on activation of T-cell subsets, more detailed analysis of a variety of such epitopes, in particular the ones from allergens, are required for determination of their roles.

Biological Functions of Allergens

Studies on a few allergens indicate that the biological activities of protein antigens may have implications for their ability to induce IgE antibodies. One major allergen of vespid venom is a hyaluronidase (King et al., 1978); the activity of this enzyme may help the molecule to penetrate the tissue and contact the immune system. Another allergen, phospholipase A2, has been proved to be a potent inducer of prostaglandin synthesis, which in turn elicits IgE antibody responses (Hoffman et al., 1990; Carballido et al., 1992). Unfortunately, the biological functions of most of the protein allergens so far have not been defined.

Amino acid sequences deduced from the nucleotide sequences of the cloned allergens provide valuable information for determining their biological activities by comparison to the known amino acid sequences. Among these allergens, *Der p* I and *Der f* I, the major allergens of dust mite, are homologous to the members of thiol protease family (Chua et al., 1988). Thus, these allergens possess proteolytic activities, which may be essential for them to penetrate the mucous membrane of atopic individuals. Indeed, allergens functioning as protease are not uncommon. *Der p* III and *Der f* III have been recently determined as trypsin by N-terminal sequence comparison with crayfish trypsin and functional analysis (Stewart et al., 1992).

Bet v I, a major birch tree pollen allergen, has significant homologies to the tree pathogenesis-related proteins (Breiteneder et al., 1989). These proteins are induced in responses to stress conditions of the plant. *Bet v* I represents a major group of the tree pollen allergens. This group of allergens has been lately cloned from the tree pollens of alder (*Aln g* I), hazel (*Cor a* I), and hornbeam (*Car b* I) (Breiteneder et al., 1992; Breiteneder et al., in press; Larsen et al., 1992). Comparison of the deduced amino acid sequences revealed above 73% homologies among these allergens. Another tree pollen allergen, *Bet v* II, is found to be profilin (Valenta et al., 1991). Profilins are not only a plant pan-allergen, but also share about 40% homologies with human profilins. Profilins are involved in control of cell shape, movement, and signal transduction by binding to actin. More interestingly, the endogenous profilins released from inflammatory cells may act as booster to maintain the serum specific IgE levels (Valenta et al., 1991).

The biological activities of the major allergens of the bee venom have been well defined. These allergens are hyaluronidase (King et al., 1978), phospholipase A1 (King et al., 1984), and phospholipase A2 (Hoffman et al., 1990). Recently, another group of allergens, including *Dol m* V, *Dol a* V, and *Ves v* V, were cloned and sequenced (Fang et al., 1988; Lu et al., 1993). Similarities were found between these cloned allergens and plant pathogenesis-related proteins. In addition, the allergens have about 30% homologies to mammalian testis proteins. Whether the antibodies induced by the allergens would also recognize the testis and the plant proteins is unclear yet, although cross-reactivities have been shown among these allergens (Lu et al., 1993).

The biological activities of some allergens from fungi have been also determined. *Aspergillus fumigatus* allergen I was found to be a member of the mitogillin family of cytotoxins by analysis of its deduced amino acid sequences (Aruda et al., 1990). Comparison of N-terminal sequences, a major allergen of *Dermatophyte Trichophyton-tonsurans* was shown to share similar sequences as exo

1,3-beta-glucanase (Stewart, 1993). From the above limited data, it appears that the biological functions of some allergens may relate to their abilities to induce IgE antibodies, but others have no apparent relations. Thus, it is deemed that relationships of biological activities and abilities of IgE induction of allergens are coincident rather than a general phenomenon.

References

Aruda, L. K., U. A. Platts-Mills, J. W. Fox, and M. D. Chapman. 1990. *Aspergillus fumigatus* allergen I, a major IgE-binding protein, is a member of the mitogillin family of cytotoxins. *J. Exp. Med.* 172:1529.

Ashbridge, K. R., B. T. Backstrom, H. X. Liu, T. Vikerfors, D. R. Englebretsen, D. R. Harding and J. D. Watson. 1992. Mapping of T helper cell epitopes by using peptides spanning the 19KDa protein of mycobacterium tuberculosis. Evidence for unique and shared epitopes in the stimulation of antibody and delayed-type hypersensitivity responses. *J. Immunol.* 148:2248.

Atassi, H. and M. Z. Atassi. 1986. Antibody recognition of ragweed allergen Ra3: Localization of the full profile of the continuous antigenic sites by synthetic overlapping peptides representing the entire protein chain. *Eur. J. Immunol.* 16:229.

Atassi, H. and M. Z. Atassi. 1990. Mapping of allergen epitopes by antibody and T-cells. In *Epitopes of Atopic Allergens*, eds. A. H. Sehon, D. Kraft, and G. Kunkel, p. 33. The UCB Institute of Allergy. Brussels.

Baur, X., H. Aschauer, G. Mazur, M. Dewair, H. Prelicz and W. Steigemann. 1986. Structure, antigenic determinants of some clinically important insect allergens: chironomid hemoglobulins. *Science* 233:351.

Benjamin, D. C., J. A. Berzofsky, I. J. East, F.R.N. Guard, C. Hannum, S. I. Leach, E. Margoliash, J. Michel, A. Miller, E. M. Prager, M. Reichlin, E. E. Sercarz, S. J. Smith-Gill, P. E. Todd and A. C. Wilson. 1984. The antigenic structure of proteins: a reappraisal. *Ann. Rev. Immunol.* 2:67.

Breiteneder, H., K. Pettenburger, A. Bito, R. Valenta, D. Kraft, H. Rompokl, O. Scheiner, and M. Breiteback. 1989. The gene coding for the major birch pollen allergen *Bet v* 1, is highly homologous to a pea disease resistance gene. *EMBO J.* 8:1935.

Breiteneder, H., F. Ferreira, A. Reikerstorfer, M. Duchene, R. Valenta, K. Hoffmann-Sommergruber, C. Ebner, M. Breiteback, H. Rompold, D. Kraft, and O. Scheiner. 1992. cDNA cloning and expression in E. coli of *Aln g* I, the major allergen in pollen of alder (*Alnus glutinosa*). *J. Allergy Clin. Immunol.* 90:909.

Breitender, H., F. Ferreira, K. Hoffmann-Sommergruber, C. Ebner, M. Breitenback, H. Rumpold, D. Kraft, and O. Scheiner. 1993. Four recombinant isoforms of *Cor a* I, the major allergen of hazel pollen, showed different IgE-binding binding properties. Eur. J. Biochem. 212:355.

Carballido, J. M., N. Carballido-Perres, G. Terres, C. H. Heusser, and K. Blaser. 1992. Bee venom phospholipase A2-specific T cell clones from human allergic and nonallergic individuals: cytokine pattern change in response to antigen concentration. *Eur. J. Immunol.* 22:1357.

Chakrabarty, S., H. Lowenstein, A.K.M. Ekramoddoullah, F. T. Kisil and A. H. Sehon. 1981. Detection of cross-reactive allergens in Kentucky bluegrass pollen and six other grasses by crossed radioimmunoelectrophoresis. *Internat. Arch. Allergy Appl. Immunol.* 66:143.

Chua, K. Y., G. A. Stewart, W. R. Thomas, et al. 1988. Sequence analysis of cDNA coding for a major house dust mite allergen, *Der p* I. Homology with cysteine protease. *J. Exp. Med.* 167:175.

Davis, D. R. and E. A. Padlan. 1990. Antibody-antigen complexes. *Ann. Rev. Biochem.* 59:439.

Deuell, B., L. K. Arruda, M. L. Hayden, M. D. Chapman and T. A. Platts-Mills. 1991. *Trichophyton tonsurans* allergen I. Characterization of a protein that causes immediate but not delayed hypersensitivity. *J. Immunol.* 147:96.

Ebner, C., Z. Szepfalusi, F. Ferreira, A. Jilek, R. Valenta, P. Parronchi, E. Maggi, D. S. Romagnani, O. Scheiner and D. Kraft. 1993. Identification of multiple T cell epitopes on *Bet v* I, the major birch pollen allergen, using specific T cell clones and overlapping peptides. *J. Immunol.* 150:1047.

Elsayed, S. and J. Apold. 1983. Immunochemical analysis of cod fish allergen M: locations of the immunoglobulin binding sites as demonstrated by native and synthetic peptides. *Allergy* 38:449.

Elsayed, S., U. Ragnarsson, J. Apold. 1981. Allergenic synthetic peptide corresponding to the second calcium-binding loop of cod allergen M. *Scand. J. Immunol.* 14:207.

Elsayed, S., J. Apold, E. Holen, H. Vik, E. Florvaag and T. Dybendal. 1991. The structural requirements of epitopes with IgE binding capacity demonstrated by three major allergens from fish, egg and tree pollen. *Scand J. Clin. Lab. Invest.* 204:17, suppl.

Esch, R. E. and D. G. Klapper. 1989. Identification of localization of allergenic determinants on grass group I antigens using monoclonal antibodies. *J. Immunol.* 142:179.

Evavold, B. D. and P. M. Allen. 1991. Separation of IL-4 production from Th cell proliferation by an altered T cell receptor ligand. *Science* 252:1308.

Evavold, B. D., S. G. Williams, B. L. Hsu, S. Buus and P. M. Allen. 1992. Complete dissection of the Hb (64-76) determinant using T helper 1, T helper 2 clones, and T cell hybridomas. *J. Immunol.* 148:347.

Fang, K.S.Y., M. Vitale, P. Felner, and T. P. King. 1988. cDNA cloning and primary structure of a white-face hornet venom allergen, antigen 5. *Proc. Natl. Acad. Sci. USA* 85:895.

Fehlner, P. F., R. Berg, J. P. Tam, and T. P. King. 1991a. Murine T cell responses to melittin and its analogs. *J. Immunol.* 146:799.

Fehlner, P. F., L. Kochoumian, and T. P. King. 1991b. Murine IgE and IgG response to melittin and its analogous. *J. Immunol.* 146:2664.

Firestein, G. S., W. D. Roeder, J. A. Laxer, K. S. Townsend, C. T. Weaver, J. T. Hom, J. Linton, B. E. Torbett, and A. L. Glasbrook. 1989. A new murine CD4$^+$ T cell subset with an unrestricted cytokine profile. *J. Immunol.* 143:518.

Galelli, A. and B. Charlot. 1990. Clonal anergy of memory B cells in epitope-specific regulation. *J. Immunol.* 145:2397.

Garman, R. D., W. Goodwin, A. M. Lussier. 1990. The allergen specific T cells response of atopic and nonatopic individuals. *J. Allergy. Clin. Immunol.* 82:200.

Geysen, H. M., S. J. Rodda, T. J. Mason, G. Tribbick, and P. G. Schools. 1987. Strategies for epitope analysis using peptide synthesis. *J. Immunol. Meth.* 102:259.

Greene, W. K., J. G. Cyster, K. Y. Chua, R. M. O'Brien and W. R. Thomas. 1991. IgE and IgG binding of peptides expressed from fragments of cDNA encoding the major house dust mite allergen *Der p* I. *J. Immunol.* 147:3769.

Herzenberg, L. A., T. Tokuhisa, K. Hayakawa. 1983. Epitope-specific regulation. *Ann. Rev. Immunol.* 1:609.

Hoffman, T., C. Brando, E. F. Lizzio, C. Lee, M. Hanson, K. Ting, Y. J. Kim, T. Abrahamsen, J. Puri, and M. D. Bronviv. 1990. Functional consequences of phospholipase A2 activation in human monocytes. *Adv. Exp. Med. Biol.* 279:125.

Hoyne, G. F., M. G. Callow, M. C. Kuo, and W. R. Thomas. 1993. Characterization of T cell responses to the house dust mite allergen *Der p* II in mice—evidence for major and cryptic epitopes. *Immunology* 78:65.

Jarrett, E.E.E. and H.R.P. Miller. 1982. Production and activities of IgE in helminth infection. *Prog. Allergy* 31:178.

King, T. P., L. Kochoumian, and A. Joslyn. 1984. Melittin-specific monoclonal and polyclonal IgE and IgG antibodies from mice. *J. Immunol.* 133:2668.

King, T. P., L. Kochoumian, and A. Joslyn. 1984. Wasp venom proteins: phospholipase A1 and B. *Arch. Biochem. Biophys.* 230:1.

King, T. P., A. Sobotka, A. Alagon, L. Kochoumian, and M. Lichtenstein. 1978. Protein allergens of white-faced hornet, yellow hornet and yellow jacket venoms. *Biochemistry* 17:5165.

Kurisaki, J. I., H. Atassi, and M. Z. Atassi. 1986. T cell recognition of ragweed allergen Ra3: localization of the full T cell recognition profile by synthetic overlapping peptides representing the entire protein chain. *Eur. J. Immunol.* 16:236.

Larsen, J. N., P. Stroman, and H. Ipsen. 1992. PCR based cloning and sequencing of isogenes encoding the tree pollen major allergen *Car b* I from *Carpinus Betulus,* hornbeam. *Mol. Immunol.* 29:703.

Leighton, J., A. Sette, J. Sidney, et al. 1991. Comparison of structural requirements for interaction of the same peptide with I-E^K and I-E^d molecules in the activation of MHC class II-restricted T cells. *J. Immunol.* 147:198.

Levine, B. B. and A. P. Redmond. 1968. The nature of the antigen-antibody complexes initiating the specific wheal-and-flare reaction in sensitized man. *J. Clin. Invest.* 47:556.

Lombardero, M., P. W. Heymann, T.A.E. Platts-Mills, J. W. Fox, and M. D. Chapman. 1990. Conformational stability of B cell epitopes on group I and group II determatophagoides spp. allergens. Effect of thermal and chemical denaturation on the binding of murine IgG and human IgE antibodies. *J. Immunol.* 144:1353.

Lu, G., M. Villalba, M. R. Coscia, D. R. Hoffman, and T. P. King. 1993. Sequence analysis and antigenic cross-reactivity of a venom allergen, antigen 5, from hornets, wasps, and yellow jackets. *J. Immunol.* 150:2823.

Maryanski, J. L., A. S. Verdini, P. C. Weber, F. R. Salemme and G. Corradin. 1990. Competitor analogs for defined T-cell antigens: peptides incorporating a putative binding motif and polyproline or polyglycine spacers. *Cell* 60:63.

Matthiesen, F. and H. Lowenstein. 1991. Group V allergens in grass pollens. II. Investigation of group V allergens in pollens from 10 grasses. *Clin. Exp. Allergy* 21:309.

Mazur, G., W. Steigemann, and X. Baur. 1990. Chironomid allergens: Localization of B- and T-cell epitopes in three-D structure. In *Epitopes of Atopic Allergens* eds. A. H. Sehon, D. Kraft, and G. Kunkel, p. 48. The UCB Institute of Allergy, Brussels.

Moirad, W., S. Mecheris, G. Peltre. 1988. Studies of the epitope structure of purified *Dac g* I and *Lol p* I, the major allergens of *Dactylis glomerata* and *Lolium perenne* pollens, using monoclonal antibodies. *J. Immunol.* 141:3486.

Mossmann, T. R. and R. T. Coffman. 1989. TH_1 and TH_2 cells: Different patterns of lymphokine secretion lead to different functional properties. *Ann. Rev. Immunol.* 7:145.

Nikolic-Zugic, J. and M. J. Bevan. 1990. Role of self-peptides in positively selecting the T cell repertoire. *Nature* 344:65.

Nilsen, B. M., K. Sletten, B. S. Paulsen, M. O'Neill, and H. V. Halbeek. 1991. Structural analysis of the glycoprotein allergen *Art v* II from the pollen of mugwort (*Artemisia vulgaris* L.). *J. Biol. Chem.* 266:2660.

Norman, P. S., L. M. Lichtenstein, A. Kagey-Sobotka, and D. G. Marsh. 1982. Controlled evaluation of allergoid in the immunotherapy of ragweed hay fever. *J. Allergy Clin. Immunol.* 70:248.

O'Hehir, R. E., V. Bal, D. Quint, J. Lamb. 1989. An in vitro model of allergen dependent IgE synthesis by human B cells: comparison of the response of and atopic and non-atopic individual to *Dermatophagoides spp. Immunology* 66:499.

Pala, P., H. C. Bodmer, R. M. Pemberton. 1988. Competition between unrelated peptides recognized by $H-2K^d$ restricted T cells. *J. Immunol.* 141:2298.

Parronchi, P., D. Macchia, M. Piccinni, P. Biswas, C. Simonelli, E. Maggi, M. Ricci, A. A. Ansari, and S. Romagnani. 1991. Allergen- and bacterial antigen-specific T-cell clones established from atopic donors show a different profile of cytokine production. *Proc. Natl. Acad. Sci. USA* 88:4538.

Perez, M., G. Y. Ishioka, L. E. Walker, and R. W. Chesnut. 1991. cDNA cloning and immunological characterization of the rye grass allergen *Lol p* I. *J. Biol. Chem.* 256:16210.

Polo, F., R. Ayuso, and J. Carreira. 1991. Studies on the relationship between structure and IgE-binding ability of *Parietaria judaica* allergen I. *Mol. Immunol.* 28:169.

Rocken, M., K. M. Muller, J. H. Saurat, and C. Hauser. 1992. Central role for TCR/ CD3 ligation in the differentiation of CD4+ T cells toward a Th1 or Th2 functional phenotype. *J. Immunol.* 148:47.

Romagnani, S. 1991. Human Th1 and Th2 subsets: doubt no more. *Immunology Today* 12:256.

Romero, P., G. Corradin, J. F. Luescher, and J. L. Maryanski. 1991. H-2K$_d$-restricted antigenic peptides share a simple binding motif. *J. Exp. Med.* 174:603.

Rousseaux-Prevost, R., H. Bazin, and A. Capron. 1977. IgE in experimental schistosomiasis. I. Serum IgE levels after infection by *Schistosoma mansoni* in various strains of rats. *Immunology* 33:501.

Schutze, M. P., E. Deriaud, G. Przewlicki, and C. Leclerc. 1989. Carrier-induced epitopic suppression is initiated through clonal dominance. *J. Immunol.* 142:2635.

Scott, P., P. Natovitz, R. L. Coffman, E. Pearce and A. Sher. 1988. Immunoregulation of cutaneous leishmaniasis: T cell lines that transfer protective immunity or exacerbation belong to different T helper subsets and respond to distinct parasitic antigens. *J. Exp. Med.* 168:1675.

Segal, D. M., J. D. Taurog, and H. Metzger. 1977. Dimeric immunoglobulin E serves as a unit signal for mast cell degranulation. *Proc. Natl. Acad. Sci. USA* 74:2993.

Sette, A., L. Andorini, E. Appella, S. M. Colon, C. Miles, S. Tanaka, C. Ehrhardt, G. Doneria, Z. A. Nagy, S. Buus and H. M. Grey. 1989. Structural requirements for the interaction between peptide antigens and I-E^d molecules. *J. Immunol.* 143:3289.

Soloway, P., S. Fish, H. Passmore, M. Gefter, R. Coffee and T. Manser. 1991. Regulation of the immune response to peptide antigens: differential induction of immediate-type hypersensitivity and T cell proliferation due to changes in either peptide structure or major histocompatibility complex haplotype. *J. Exp. Med.* 174:847.

Stewart, G. A. 1993. Sequence similarity between a major allergen from the Dermatophyte Trichophyton-tonsurans and exo 1,3-beta-glucanase. *Clin. Exp. Allergy* 23:154.

Stewart, G. A., L. Armstrong, K. Krska, C. Doyle, P. J. Thompson, K. J. Turner, and H. M. Geysen. 1990. Epitope mapping analysis of the major mite allergens using synthetic peptides. In *Epitope of Atopic Alelrgens* eds. A. H. Sehon, D. Kraft, and G. Kunkel, The UCB Institute of Allergy, Brussels, p. 41.

Stewart, G. A., L. D. Ward, R. J. Simpson, and P. J. Thompson. 1992. The group-III allergen from the house dust mite dermatophagoides-pteronyssinus is a trypsin-like enzyme. *Immunology* 75:29.

Valenta, R., M. Duchene, K. Pettenburger, C. Sillaber, P. Valent, P. Bettelheim, M. Breitenbach, H. Rumpold, D. Kraft, and O. Sheiner. 1991. Identification of profilin as a novel pollen allergen: IgE autoreactivity in sensitized individuals. *Science* 253:557.

Van Regenmortal, M. H. 1989. Structural and functional approaches to the study of protein antigenicity. *Immunol. Today.* 10:266.

Walsh, B. J. and M.E.H. Howden. 1989. A method for detection of IgE binding sequences of allergens based on a modification of epitope mapping. J. Immunol. Meth. 121:275.

Wierenga, E. A., M. Snoek, C. De Groot, I. Chretien, J. D. Bos, H. M. Jansen, and M. L. Kapsenberg. 1990. Evidence for compartmentalization of functional subsets of CD4$^+$ T lymphocytes in atopic patients. *J. Immunol.* 144:4651.

Wierenga, E. A., M. Snoek, H. M. Jansen, J. D. Bos, R. A. Van-Lier and M. L.

Kapsenberg. 1991. Human atopen-specific types 1 and 2 helper cell clones. J. Immunol. 147:2942.

Yamada, M., M. Nakazawa, and N. Arizono. 1993. IgE and IgG2a antibody responses are induced by different antigen groups of the nematode *Nippostrongylus brasiliensis* in rats. Immunology. 78:298.

Yssel, H., K. E. Johnson, P. V. Schneider, J. Wideman, A. Terr, R. Kastelein and J. E. DeVries. 1992. T cell activation-inducing epitopes of the house dust mite allergen *Der p* I. Proliferation and lymphokine production patterns by *Der p* I specific CD4+ T cell clones. J. Immunol. 148:738.

9

Human T-Cell Responses to Grass Pollen Allergens

Sergio Romagnani
Shyam S. Mohapatra

Introduction

Grass pollen allergens constitute a unique class of inhaled antigens that are immunogenic in extremely low doses and are responsible for a number of seasonal allergies, including rhinitis, conjunctivitis, and extrinsic asthma. Grass pollen antigens, as well as other environmental allergens, interact with IgE antibodies specifically and bind to high affinity Fcε receptors (FcεRI) localized on the surface of tissue mast cells and circulating basophils. In response to allergen cross-linkage of their FcεRI, these cells produce and release several potent mediators responsible for allergic symptomatology.

In the last few years the mechanisms responsible for the synthesis of antibodies of the IgE class have been extensively investigated. It has been clearly shown that IgE antibody synthesis in response to aero-allergens is a T-cell-dependent phenomenon (Romagnani, 1990; Wierenga et al., 1990; Parronchi et al., 1991). The reciprocal regulatory role of T cell-derived cytokines, such as IL-4 and IFN-γ, in determining the synthesis of allergen-specific IgE antibody has also been discovered (Del Prete et al., 1988; Pene et al., 1988). This chapter discusses the role and the main features of T cells that initiate and regulate human immune responses to grass pollen allergens.

Allergen Presentation to T Helper (Th) Cells

Once grass pollen allergens are inhaled, the first step of their interaction with the immune system is the uptake of soluble pollen components by major histocompatibility complex (MHC) class II-positive antigen-presenting cells (APC), followed by their presentation to allergen-specific Th cells. Indeed, in contrast to the B-cell antigen receptor that directly recognizes unprocessed native antigens, the T-cell receptor (TCR) only recognizes small stretches of amino acids (epi-topes) on linear peptide fragments of protein antigens (Abbas et al., 1991). These peptides must be presented in a cleft formed by one of the MHC molecules. The TCR simultaneously recognizes the specific antigen epitope and a constant domain of a particular MHC molecule expressed on the surface of autologous APC (Abbas et al., 1991). The molecular basis of allergen recognition has extensively been investigated in the last few years (Marsh et al., 1993).

The nature of APC involved in the grass allergen recognition and presentation is still unknown. Several cell types have been positively identified as expressing MHC class II antigens in healthy or diseased lungs: B cells, activated T cells, fibroblasts, dendritic cells (DC), and macrophages (Schon-Hegrad et al., 1991). Low molecular weight soluble antigens, as those derived from pollen degradation, may be expected to be readily translocated to intraepithelial or submucosal microenvironments via intracellular (pinocytic) or intercellular pathways, and hence gain access to all the potential APC populations stated above. However, the relevant APC require at least two additional properties to act as efficient APC and to induce a primary immune response in a naive host. The first is the capacity to migrate selectively to T-cell zones in regional lymph nodes; the second is the ability to provide Th cells with both the signal represented by the interaction between the peptide complexed with surface MHC class II and the TCR, and a simultaneous "co-stimulatory" signal. Recent data strongly suggest that cells possessing the B7/BB1 molecule, which interacts with the CD28 mole-cule, can efficiently co-stimulate Th cells. Based on these considerations of the repertoire of APCs, DC appears to be more suitable for allergen presentation to Th cells than other APC present in human parenchymal tissue. DC have indeed been estimated to comprise up to 25% of the overall MHC class II-positive cell population resident in the normal lung parenchyma, but constitute the only MHC class II-positive resident cell population within the normal airway epithelium (Schon-Hegrad et al., 1991). DC present in nonlymphoid tissue possess phago-cytic activity and FcγRII are capable of processing exogenous antigens and actively synthesize MHC class II molecules. However, like Langerhans cells present in the skin, they do not constitutively express the B7-BB1 co-stimulatory molecule. This type of DC has been designated as "immature" DC to discriminate them from DC present in draining lymphoid tissues, designated as "mature" DC (Table 9-1). These latter also express surface MHC class II antigens but lack phagocytic activity, FcγRII, as well as the ability to process antigens and to continuously synthesize MHC class II molecules. They exhibit, however, the expression of the B7/BB1 co-stimulatory molecule on their surface (Table 9-1). Thus, it is reasonable to speculate that immature DC present in normal airway epithelium are the professional APC that encounter and process grass pollen allergens in the primary response to aero-allergens. They then probably migrate through afferent lymphatic vessels to draining lymph nodes where they acquire the mature phenotype. Mature DC are able to present the processed peptide to

Table 9-1 Regulatory Factors of Human IgE Synthesis

Positive Regulatory Factors		
Inducing factors		Enhancing factors
1st signal	2nd signal	
Cognate and noncognate T-B cell to cell interaction (CD40 ligand-CD40?)	IL-4	IL-6 TNF-α
Anti-CD40 antibody	IL-13	IL-5
EBV infection		sCD23 (29-37 kD)
Corticosteroids		

Negative Regulatory Factors		
	Acting on T cells	Acting on B cells
	IFN-γ	TGF-β
	IFN-α	IL-8
	PGE$_2$	IL-12
	IL -10	sCD23 (16kD)
		PAF

the specific Th cells and to provide the co-stimulatory signal via the B7/BB1-CD28 interaction. This action results in antigen-specific Th cell activation and proliferation (Fig. 9-1).

Proliferative Responses of Th Cells to Grass Pollen Allergens

The ability of peripheral blood mononuclear cells (PBMC) to proliferate in response to grass pollen extracts or their components was first demonstrated in the late 1960s by a number of investigators (Girard et al., 1967; Brostoff et al., 1969; Romagnani et al., 1973). In 1973, the proliferative response of PBMC from three groups of subjects (untreated grass-sensitive, grass-sensitive treated with specific immunotherapy, and nonatopic) to an aqueous extract of meadow velvet (*Holcus lanatus*-HL-) was examined (Romagnani et al., 1973). PBMC from the great majority of atopic patients exhibited strong proliferative responses to an aqueous HL extract and the response was significantly higher in the atopic than in the nonatopic group. Interestingly, the proliferative response to HL was also significantly higher in untreated grass-sensitive patients in comparison with those that had been treated with specific immunotherapy. On the basis of these data, we suggested that this proliferative response was mainly caused by HL-specific circulating T lymphocytes present in higher concentrations in atopic than

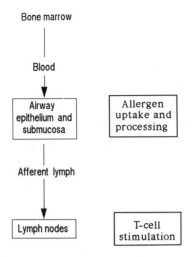

Figure 9-1 Life history of the dendritic cell involved in allergen presentation to the Th cell.

in nonatopic people. Furthermore, a still unknown tolerogenic mechanism was suspected to be responsible for the reduced proliferative response of T cells from both nonatopic subjects and grass-sensitive patients treated with specific immunotherapy (Romagnani et al., 1975).

Substantial progress in the knowledge of T-cell responses to allergens was made in the early 1980s, when in collaboration with A. Lanzavecchia, we established the first T-cell lines and T-cell clones specific for purified grass pollen allergen (*Lolium perenne* Group I or *Lol p* I). These lines and clones were antigen-specific (i.e., responded to the allergen used to raise them and not to other antigens) and were MHC-restricted (i.e., they required that the antigen was presented by autologous APC). Surface marker analysis revealed that *Lol p* I-specific lines comprised mainly cells with a CD4+ phenotype, although a minority of CD8+ cells could also be detected (Lanzavecchia et al., 1983). Subsequently, the introduction of high efficiency T-cell cloning technology played a crucial role in the development and characterization of T cells specific to allergens and furthered the understanding of mechanisms that regulate the allergen-induced IgE antibody synthesis.

Role of Th Cells and of Th Cell-Derived Cytokines in the Induction and Regulation of Human IgE Synthesis In Vitro

In the early 1980s, several attempts to induce IgE production in vitro by stimulating PBMC with polyclonal activators, antigens, T-cell supernatants, or cytokines had been performed, but all proved to be unsuccessful because under these experimental conditions only spontaneously synthesized IgE could be detected (Buckley and Becker, 1978; Romagnani et al., 1980). The unambiguous demonstration that IgE production could be induced by signals delivered in vitro was provided by coculturing B cells with selected alloreactive, autoreactive, or phytohemagglutinin (PHA)-induced T-cell clones (Lanzavecchia and Parodi, 1984; Leung et al., 1986; Romagnani et al., 1987). By assaying the activity of large numbers of phytohemagglutinin (PHA)-induced T-cell clones (TCC) derived from different lymphoid organs, we then showed that human IgE synthesis was strictly regulated by the production of certain cytokines. There was indeed a significant positive correlation between the helper function for IgE synthesis by PHA-induced clones and their ability to produce IL-4 (Del Prete et al., 1988). In contrast, there was significant inverse correlation between the IgE helper activity of PHA-induced T-cell clones [or supernatant (SUP) derived from them] and their ability to produce IFN-γ (Del Prete et al., 1988).

The opposite regulatory role of IL-4 and IFN-γ in the synthesis of human IgE was confirmed by the observations that (1) human recombinant IL-4 can induce the synthesis of IgE in peripheral blood mononuclear cells (PBMNC), and (2) this effect is inhibited by the addition of recombinant IFN-γ (Del Prete et al., 1988; Pene et al., 1988). However, the activity of IL-4 alone is not sufficient for the induction of human IgE synthesis. Both recombinant IL-4 and IL-4-containing SUP were consistently ineffective in inducing IgE synthesis by highly purified B cells (Del Prete et al., 1988; Pene et al., 1988). IL-4-dependent IgE synthesis could be restored by the readdition to highly purified B cells of appropriate concentrations of autologous or allogeneic T cells or of CD4+ TCC (Del Prete et al., 1988). Direct evidence that physical contact between T and B cells is required for IL-4-dependent IgE synthesis was provided by assaying IgE synthesis in a double chamber system in which T and B cells are separated by a millipore membrane that is permeable to molecules but not to cells. In this system, IgE synthesis occurs only when T and B cells are cultured in the same chamber (Del Prete et al., 1988). Kinetics studies indicated that physical interaction with CD4+ T cells is required before the signal provided by IL-4 (Del Prete et al., 1988). This finding suggests that IgE synthesis is dependent on two main signals delivered by CD4+ T cells to B cells: One signal is mediated by physical T-B cell-to-cell contact and another by T-cell-derived IL-4 (Figure 9-2).

Subsequent experiments demonstrated that both noncognate and cognate interaction with CD4+ T cells may render B cells susceptible to the activity of IL-4. This action was shown by assaying the activity of a CD4+ alloreactive TCC (TR46). This clone induced the synthesis of IgE when cultured with B cells possessing the appropriate alloantigen (DR4). When DR4-positive B cells were irradiated, IgE synthesis did not occur, but it was restored by the concomitant addition of DR4-negative B cells. Furthermore, in the presence of exogenously added IL-4, TR46 enabled B cells from all donors tested to produce IgE, irrespec-

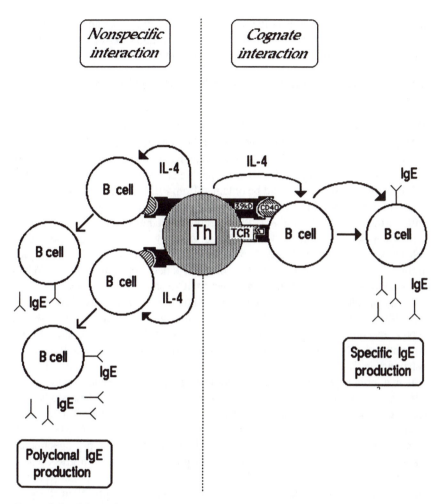

Figure 9-2   Schematic representation of the mechanisms involved in human IgE synthesis.

tive of whether they possessed the DR4 alloantigen (Parronchi et al., 1990). Taken together, these data suggest that both MHC class II-restricted and MHC class II-unrestricted physical contact with helper T cells can lead B cells into cell cycle, rendering them susceptible to the action of IL-4 and/or IFN-γ. This can explain why, in some pathological conditions, production of both specific and polyclonal IgE occurs (Fig. 9-2). More recent data suggest that the crucial physical contact between B cells and activated T cells probably involves the interaction of the CD40 molecule with its 39 kD ligand (Fig. 9-2) (Fanslow et al., 1992). Moreover, besides IL-4 and IFN-γ, other cytokines have also been found to exert positive or negative regulatory effects on the human IgE synthesis. In particular, a novel cytokine (IL-13) has been found to be effective in inducing human IgE synthesis, even if at a lower extent than IL-4 (Punnonen et al., 1993).

Th2-Like CD4+ Subsets Are Involved in the Reactivity to Grass Pollen Allergens

The finding that T-cell clones producing IL-4 but no or low concentrations of IFN-γ are able to induce the synthesis of IgE prompted us to investigate the cytokine secretion profile of T-cell clones reactive with allergens. High numbers of Dermatophagoides group I (*Der p* I)-specific or *Lolium perenne* Group I (*Lol p* I)-specific T-cell clones could be obtained from PBMC of mite or grass pollen-sensitive atopic individuals and compared for their profile of cytokine secretion with purified protein derivative (PPD)-specific or tetanus toxoid (TT)-specific T-cell clones derived from the same donors. Virtually all the allergen-specific T-cell clones produced IL-4 and IL-5 in response to stimulation with PMA plus anti-CD3 antibody, and a proportion of them failed to produce IFN-γ. When assessed with the specific antigen under MHC-restricted conditions, the great majority of allergen-specific clones behaved as murine Th2 cells (Parronchi et al., 1991). Although most TT-specific T-cell clones produced both IL-4 and IFN-γ, resembling Th0 cells, all T-cell clones specific for PPD produced high amounts of IFN-γ, but most of them failed to produce IL-4 and IL-5, thus behaving as Th1 cells. More recently, the prevalent Th2-like profile of grass pollen allergen-specific clones was confirmed by using another grass pollen allergen (the recombinant KBG7.2 or *Poa p* IX) (Table 9.2). This study also led to delineation of the T-cell epitopes of Poa p IX allergen (see Table 9.2).

More importantly, high proportions of grass pollen-specific CD4+ T-cell clones (22% and 14%, respectively) could be isolated from biopsy specimens, taken 48 hr after positive bronchial or nasal provocation test with grass pollen extract, from the bronchial or nasal mucosa of patients with grass pollen-induced asthma or rhinitis. The great majority of these clones reactive with pollen allergens proliferated not only in response to the mixed grass pollen extract but also to the purified *Lol p* I allergen, exhibited a clear-cut Th2 profile, and induced IgE synthesis in autologous PBMC in the presence of *Lol p* I (Del Prete et al., 1993). These findings provide convincing evidence that Th2-like cells specific for grass pollen allergens appear in the mucosa soon after grass pollen inhalation. These cells may be responsible for both the induction of IgE antibody production and the increase of eosinophils via the release of IL-4 and IL-5, respectively, and play a triggering role in the initiation of the allergic cascade.

Cross-Reactivity of Grass Pollen-Reactive Th Cells

To better understand the molecular basis of T-cell responses to grass pollen allergens, the nature of epitopes recognized by different grass pollen-specific

T-cell clones is now being analyzed. T-cell clones against purified *Lol p* I were generated from the PBMC of ryegrass allergic individuals and subsequently screened for their reactivity against other ryegrass allergens, such as *Lol p* II and *Lol p* III, as well as a set of other grass species. Moreover, T-cell clones specific for *Lol p* I or *Poa p* IX were also screened for their reactivity against a set of overlapping peptides, the sequence of which was derived from the cDNA clones and which covered the entire *Lol p* I or *Poa p* IX molecule (Table 9-2). The majority of 24 *Lol p* I-specific T-cell clones derived from four different donors responded to only *Lol p* I, but the others (37%) were stimulated, in addition, by either *Lol p* III, *Lol p* II, or both. In some of the later clones the response to *Lol p* III was relatively higher than that to *Lol p* II (Baskar et al., 1992). Together, these results demonstrate the cross-reactivity among the Lol p proteins at the level of T-cell recognition although, in general, the cross-reactivity to *Lol p* II was relatively less pronounced. Interestingly, seven of the nine *Lol p* I-specific clones reactive to *Lol p* II and *Lol p* III also exhibited cross-recognition of the recombinant *Poa p* IX allergen, rKBG.2 (Mohapatra et al., 1993). One of the above cross-reactive clones was also examined for its proliferative response to eight different pollen extracts and it was found to be reactive to Kentucky bluegrass, orchard, redtop, and smooth brome grasses (Baskar et al., 1992). It is inferred from these studies that major allergens of different grass pollen share cross-reacting T cell epitopes.

The observed cross-reactivity can be explained on the basis of structural analysis of the three Lol p proteins that reveal several homologous segments among them. A recent report identifies one such segment in *Lol p* I (residues 191–210)

Table 9-2   Poa p IX and its peptide specific proliferation of T cell clones

Peptide No.	Sequence	No. of Prolif. clones	% of Poa p IX CPM
*5	GYTPAAPAGAAPKATTDEQK	1	37
*11	AEALSTEPKGAAVDSSKAAL	2	3–16
*12	AAVDSSKAALTSKLDAAYKL	1	18
13	TSKLDAAYKLAYKSAEGATP	1	2
*16	LSEALRIIAGTLEVHGVKPA	2	2–14
*17	TLEVHGVKPAAEEVKATPAG	1	28
*18	AEEVKATPAGELQVIDKVDA	1	15
19	ELQVIDKVDAAFKVAATAAN	2	1–2
*20	AFKVAATAANAAPANDKFTV	1	40
22	FEAAFNDAIKASTGGAYQSY	2	1–2
*23	ASTGGAYQSYKFIPALEAAV	1	7–43
*25	KQSYAATVATAPAVKYTVFE	1	98
*26	APAVKYTVFETALKKAITAM	1	34

*The background proliferation was ¯300 CPM. The peptides capable of inducing proliferation >10% of the *Poa p* IX allergen were considered to represent T cell epitopes of this allergen.

as the immunodominant peptide for T-cell response (Perez et al., 1990). However, our data rather favor the view that different peptides present all along the *Lol p* I molecule are able to stimulate specific T cells. Therefore it is possible that whereas some of these peptides will induce unique T-cell response, others will induce T cells that might cross-react with the other *Lol p*, as well as the other grass species proteins.

Therapeutic Implications of T-Cell Studies

The production of IgE antibodies, which is under T-cell control, constitutes the first step in the allergy cascade. Therefore current efforts for the development of new treatments for allergies are aimed at down-regulating the production of IgE antibodies. It is also known that a successful immune response requires two signals, one involving recognition of the allergen peptide by the T-cell receptor, and the other involving B7/CD28 interaction. Of these only the first signal is responsible for specificity and comprises recognition by T cells of the allergenic peptide, which is presented to the T-cell receptor by the appropriate MHC molecule. A number of different approaches are being examined for the prevention of the allergic disorders that comprise the modulation of the T-cell cytokine profile by the allergen in question or its peptide(s), and the trimolecular (i.e., T-cell receptor, peptide, and MHC) interaction that is expected to interrupt IgE immune response right at the onset. These two approaches are briefly described below.

Modulation of Cytokine Patterns of Allergen-Specific T Cells

A majority of allergen-specific T cells are Th2 type and produce IL-4, which is responsible for the IgE antibody production. However a small frequency of T cells are Th1/Th0 type, which secrete either IFN-γ or IFN-γ plus IL-4. Moreover, IL-4 and IFN-γ are reciprocal in their activity, and the ratio of IL-4 and IFN-γ determines the extent of help to the B cells for the production of IgE antibodies. One possible approach to therapy is to change the cytokine secretion profiles of allergen-specific T cells. Recent studies in our laboratory of *Poa p* IX-specific human T-cell clones derived from PBMC in the presence or absence of IFN-α indicated that IFN-α not only favored the development of Th0/Th1 type of T cells instead of TH2 type of cells, but also selectively amplified Th0/Th1 type of cells with specificity to a single peptide of *Poa p* IX allergen. More recently, IL-12 was also found to be able to favor the in vitro development of allergen-specific T cells into Th0/Th1, instead of Th2, clones (Manetti et al., 1993). Thus, treatment with recombinant allergens and IFN-α or IL-12 may comprise a feasible specific immunotherapeutic approach for allergic disorders.

Modulation of the Trimolecular Interaction and Induction of Specific Tolerance

Allergen specific tolerance may be induced by the functional inactivation of allergen-reactive T cells by treatment of the individual with T-cell epitopic peptides. The peptides of allergens that are recognized by T cells but not B cells may be utilized to make the allergen specific T cells nonresponsive. Since the peptides comprising immunodominant epitopes contain residues for both B- and T-cell recognition, these peptides require appropriate modification for eliminating their immunogenicity without altering their capacity of T-cell recognition. Furthermore, it appears that multiple T-cell epitopes located on several different protein allergens may function as targets for T-cell recognition, which may take this approach impractical. An alternative approach for inhibiting T-cell recognition of specific allergenic determinants may comprise treatment with tailored nonimmunogenic peptides that bind to high-affinity MHC class II molecules of the appropriate specificity. However, again the multiplicity of epitopes each of which are often capable of binding to more than one MHC molecule because of the degenerate nature of MHC-peptide interaction, may restrict the application of this approach. Finally, although the T-cell response in allergic subjects may be directed toward multiple epitopes, the diversity of TCR used may be limited. Therefore, manipulation of T cells by these shared V gene products may constitute a potential means of modulating IgE synthesis. Because the number and nature of the T-cell epitopes may vary from allergen to allergen and epitopes of only a limited number of allergens have been determined, the most efficacious approach for the treatment of allergic disorders remains to be determined.

References

Abbas, A. K., A. H. Lichtman, and J. S. Pober. 1991. *Cellular and Molecular Immunology*. W. B. Saunders Company, Philadelphia, PA.

Baskar, S., P. Parronchi, S. Mohapatra, S. Romagnani, and A. A. Ansari. 1992. Human T cell responses to purified pollen allergens of the grass, *Lolium perenne*. *J. Immunol.* 148:2378–2383.

Brostoff, J., M. W. Greaves, and I. M. Roitt. 1969. IgE in lymphoid cells from pollen-stimulated cultures. *Lancet* i:803–805.

Buckley, R. H. and W. G. Becker. 1978. Abnormalities in the regulation of human IgE synthesis. *Immunol. Rev.* 41:288–298.

Del Prete, G. F., E. Maggi, P. Parronchi, I. Chretien, A. Tiri, D. Machia, M. Ricci, J. Banchereau, J. de Vries, and S. Romagnani. 1988. IL-4 is an essential factor for the IgE synthesis induced in vitro by human T cell clones and their supernatants. *J. Immunol.* 140:4193–4198.

Del Prete, G. F., M. De Carli, M. M. D'Elios, P. Maestrelli, M. Ricci, L. Fabbri, and

S. Romagnani. 1993. Allergen exposure induces the activation of allergen-specific Th2 cells in the airway mucosa of patients with allergic asthma. *Eur. J. Immunol.* 23:1445–1449.

Fanslow, W. C., D. M. Anderson, K. H. Grabstein, E. A. Clark, D. Cosman, and R. J. Armitage. 1992. Soluble forms of CD40 inhibit biologic responses of human B cells. *J. Immunol.* 149:655–660.

Girard, J. P., N. R. Rose, M. L. Kunz, S. Kobayashi, and C. Arbesman. 1967. In vitro lymphocyte transformation in atopic patients; induced by antigens. *J. Allergy* 39:65–81.

Lanzavecchia A., and B. Parodi. 1984. In vitro stimulation of IgE production at a single precursor level by human alloreactive T-helper clones. *Clin. Exp. Immunol.* 55:197–202.

Lanzavecchia, A., P. Santini, E. Maggi, G. F. Del Prete, P. Falagiani, S. Romagnani, M. Ferrarini. 1983. In vitro selective expansion of allergen specific T cells from atopic patients. *Clin Exp. Immunol.* 52:21–28.

Leung, D.Y.M., Y. M. Young, and R. S. Geha. 1986. Induction of IgG and IgE synthesis in normal B cells by autoreactive T cell clones. *J. Immunol.* 136:2851–2856.

Manetti, R., P. Parronchi, M-G. Giudizi, M-P. Piccinni, E. Maggi, G. Trinchieri, and S. Romagnani. 1993. Natural killer cell stimulatory factor (interleukin 12) induces T helper type 1 (Th1)-specific immune responses and inhibits the development of IL-4-producing Th cells. *J. Exp. Med.* 177:1199–1204.

Marsh, D. G., A. Lockhart, and S. T. Holgate. 1993. *The Genetics of Asthma.* Blackwell Scientific Publications, Oxford.

Mohapatra, S. S., S. Mohapatra, M. Yang, A. A. Ansari, P. Parronchi, E. Maggi, and S. Romagnani. 1994. Molecular basis of cross-reactivity among allergen-specific human T cells: T cell receptor Vα gene usage and epitope structure. *Immunology,* 81:15–20.

Parronchi, P., A. Tiri, D. Macchia, M. De Carli, P. Biswas, C. Simonelli, E. Maggi, G. F. Del Prete, M. Ricci, and S. Romagnani. 1990. Noncognate contact-dependent B cell activation can promote IL-4-dependent in vitro human IgE synthesis. *J. Immunol.* 14:2102–2108.

Parronchi, P., D. Macchia, M-P. Piccinni, P. Biswas, C. Simonelli, E. Maggi, M. Ricci, A. A. Ansari, and S. Romagnani. 1991. Allergen- and bacterial antigen-specific T cel clones established from atopic donors show a different profile of cytokine production. *Proc. Natl. Acad. Sci. USA* 88:4538–4542.

Parronchi, P., M. De Carli, M-P. Piccinni, D. Macchia, E. Maggi, G. F. Del Prete, and S. Romagnani. 1992. IL-4 and IFNs exert opposite regulatory effects on the development of cytolytic potential by TH1 or TH2 human T cell clones. *J. Immunol.* 149:2977–2983.

Pene, J., F. Rousset, F. Briere, I. Chretien, J-Y. Bonnefoy, H. Spits, T. Yokota, N. Arai, K-I. Arai, J. Banchereau, and J. de Vries. 1988. IgE production by normal human lymphocytes is induced by interleukin 4 and suppressed by interferons γ and α and prostaglandin E2. *Proc. Natl. Acad. Sci. USA* 85:6880–6884.

Perez, M., G. Y. Ishioka, L. E. Walker, and R. W. Chestnut. 1990. cDNA cloning and immunologic characterization of the rye grass allergen. *Lol p* I. *J. Biol. Chem.* 264:16210–16215.

Punnonen, J., C. G. Aversa, B. G. Cocks, A.N.J. McKenzie, S. Menon, G. Zurawski, R. de Waal Malefyt, and J. de Vries. 1993. Interleukin-13 induces interleukin-4-independent IgG4 and IgE synthesis and CD23 expression by human B cells. *Proc. Natl. Acad. Sci. USA* 90:3730–3734.

Romagnani, S. 1990. Regulation and deregulation of human IgE synthesis. *Immunol. Today* 11:316–321.

Romagnani, S., G. Biliotti, A. Passasleva, M. Ricci. 1973. In vitro lymphocyte response to pollen extract constituents in grass pollen-sensitive individuals. *Int. Arch. Allergy Appl. Immunol.* 44:40–50.

Romagnani, S., G. Biliotti, and M. Ricci. 1975. Depression of grass pollen-induced lymphocyte transformation by serum from hyposensitized patients. *Clin. Exp. Immunol.* 19:83–91.

Romagnani, S., E. Maggi, G. F. Del Prete, R. Troncone, and M. Ricci. 1980. In vitro production of IgE by human peripheral blood mononuclear cells. I. Rate of IgE biosynthesis. *Clin. Exp. Immunol.* 42:167–174.

Romagnani, S., G. F. Del Prete, E. Maggi, and M. Ricci. 1987. Activation through CD3 molecule leads a number of human T cell clones to induce IgE synthesis in vitro by B cells from allergic and nonallergic individuals. *J. Immunol.* 138:1744–1749.

Schon-Hegrad, M. A., J. Oliver, P. G. McMenamin, and P. G. Holt. 1991. Studies on the density, distribution and surface phenotype of intraepithelial class II MHC antigen (Ia)-bearing dendritic cells (DC) in the conducting airways. *J. Exp. Med.* 173:1345–1350.

Wierenga, E. A., M. Snoek, C. de Groot, I. Chretien, J. D. Bos, H. M. Jansen, and M. Kapsenberg. 1990. Evidence for compartmentalization of functional subsets of CD4+ T lymphocytes in atopic patients. *J. Immunol.* 144:4651–4656.

10

Molecular Characterization and Environmental Monitoring of Grass Pollen Allergens

E. K. Ong, R. Bruce Knox, and Mohan B. Singh

Introduction

Grass pollen is the major factor in the outdoor environment triggering hay fever and asthma in spring and early summer. During the past thirty years, the incidence of these allergic diseases in the human population in many parts of the world has doubled (Wuthrich, 1989; Burney et al., 1993), suggesting among many possible explanations either that the amount of allergens in the atmosphere has increased or that environmental air pollution may be exacerbating symptoms. This chapter reviews the molecular biology of certain grass pollen allergens and incidence of these allergens in the air as estimated either by pollen counts or allergen load analysis.

Grass pollen allergens can cause an immediate hypersensitivity (type 1 IgE-mediated) response in susceptible individuals that leads to the symptoms of hay fever and allergic asthma. Hay fever, characterized by inflammation and flow of mucous from eyes and nose, is induced in the mucous membranes of the upper respiratory tract of susceptible individuals (Druce, 1993). If the lower respiratory tract, lungs, or bronchi are affected, the clinical reaction will be that of asthma, where bronchial constriction induces symptoms such as wheezing, cough, and shortness of breath (Hoehne and Reed, 1971). In cool temperate climates Kentucky bluegrass, ryegrass, and timothy grass are the major seasonal cause of these diseases because of the copious production of pollen during flowering (Smart et al., 1979). This chapter focuses on ryegrass.

The major allergens in ryegrass pollen are *Lol p* 1 and *Lol p* 5 (previously referred to as *Lol p* 1b, Singh et al., 1991 and *Lol p* IX, Suphioglu et al., 1992) because of the high frequency of allergic patients responding (>90%) and relative amount of IgE binding to the allergens (Singh et al., 1991). Together, both of these allergens could almost completely inhibit IgE binding to an extract of ryegrass pollen (Bond et al., 1993). cDNA clones encoding these allergens have been isolated (Perez et al., 1990; Griffith et al., 1991; Singh et al., 1991). Both allergens exist in multiple isoforms (Perez et al., 1990; Griffith et al., 1991; Singh et al., 1991; Smith et al., 1993a). These variants exhibit slight physicochemical differences but possess immunologically closely related properties. Comparison between the deduced amino acid sequences of the isoforms may help in elucidating the regions of the molecules that contain allergenic determinants that may be useful for diagnosis and immunotherapy of grass pollen allergy.

Besides identification and characterization of allergens for diagnosis and immunotherapy, allergen avoidance is also an important prevention strategy in the management of hay fever and allergic asthma (Sheffer et al., 1992). The occurrence of airway symptoms in allergic patients is closely associated with the amount of environmental allergens (Sporik et al., 1992; Cogswell, 1992). Allergen avoidance results in an improvement in asthma symptoms and reduction in airway hyperresponsiveness. This has been successfully demonstrated in house dust mite and cat allergens (Colloff et al., 1992). Although it is impossible to avoid airborne pollen completely, knowledge of the occurrence of pollen in the air may help patients to avoid the peak of pollen load. In addition, by comparing the incidence of allergy with known pollen distribution patterns, diagnosis and preventive measures could be taken.

Pollen grains have been shown to be too large to be respirable, for example, grass pollen is ~35 μm in diameter (Hatch and Gross, 1964; Hoehne and Reed, 1971; Wilson et al., 1973). Pollen is deposited in the nose and upper airways of susceptible individuals, causing hay fever. However, some susceptible individuals suffer from asthma. How can pollen trigger an attack of asthma? Recently, allergen-containing micronic particles have been shown to occur naturally in the environment (Stewart and Holt, 1985; Spieksma et al., 1990; Suphioglu et al., 1992). Micronic particles provide a mechanism by which pollen allergens may reach the airways, constricting the bronchi and causing an attack of asthma. Thus pollen counts do not necessarily accurately reflect the total amount of allergen in the air, that is, the allergen load. It is clinically important to estimate the level of allergen load in the air.

Pollen Grains and Allergic Disease

Structure of Pollen Grains

Airborne pollen occurs in a number of shapes, usually spherical to oviodal (Knox, 1984). Sizes vary from 5 to 300 μm in diameter, although most are less than 50 μm in diameter. For example, the diameters of birch, alder, ash, and olive pollen grains range from 19 to 27 μm; cypress, elm, *Eucalyptus*, oak, and poplar from 24 to 38 μm; grass pollen grains average 35 μm; the compound pollen grains of *Acacia* from 38 to 55 μm; *Amaranthus, Rumex,* and ragweed pollen grains from 20 to 32 μm; and plantain from 16 to 40 μm (Smith, 1984).

The wall of the pollen grains consist of an outer layer (the exine) and an inner layer (the intine) (Fig. 10.1). The exine is made of sporopollenin, which is stable and resistant to both physical and enzymatic degradation (Knox, 1984). The outer layer of the exine is sculptured in various group-specific patterns that are an excellent means of identification of the pollen. Pollen grains are considered to be: (1) psilate or smooth and lack sculpturing; (2) scabrate or granular with very fine projections (< 0.5 μm); (3) rugulate with elongate sculptured elements irregularly scattered over the pollen surface; (4) striate with elongated and more or less parallel sculptured elements distributed tangentially on the pollen surface; (5) reticulate with sculptured elements forming an open network or reticulum on the entire pollen surface; (6) verrucate or warty; (7) perforate; (8) foveolate or pitted; and (9) echinate or spiny (Fig. 10-2). Spiny projections are typical of a large number of pollen grains in the family Asteraceae.

Most pollen grains possess apertures, which are a major feature of grain morphology. Apertures are precisely sited exineless areas in the outer wall and may be long furrows (colpi, with long axes more than twice their width), circular pores, or a combination of the two structures (colporate) (Fig. 10-3). Both pores and furrows serve as a mechanism for accommodation to changes in volume and as a place for emergence of the pollen tube. In grass pollen, a small round cap of exine (an operculum) covers the pore area (Fig. 10-4). Some pollen types, such as *Populus,* are inaperture.

A few individual pollen grains adhere together to form tetrads (four-celled) or polyad (multigrain structures). *Acacia* pollen grains are typical of a polyad. Pollen grains in the family Pinaceae possess air bladders (wings) that permit them to travel great distances and make identification easy.

Characters of the intine can also provide identification features. A few pollen grains, such as cypress pollen, have a thick intine. This intine is covered by a thin lightly stained exine, so that the shape is generated by the thick irregular intine whose contact with cytoplasm gives a star-shaped to rectangular outline. Pollen grains in the families Cupressaceae and Taxodiaceae have such a feature.

A pollen grain contains many different type of proteins. There are three major

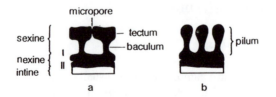

Figure 10-1 Stratification of the pollen walls. Exine is further divided into sexine and nexine. (*a*) Wall with tectate exine. (*b*) Wall with pilate exine. (After Heslop-Harrison, 1968).

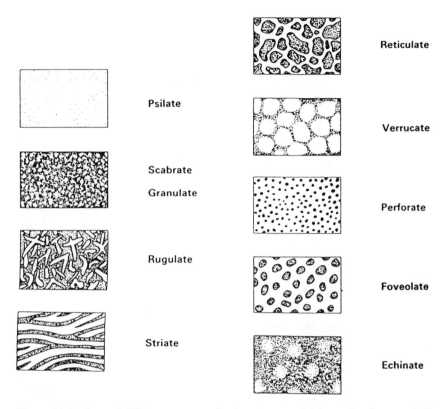

Figure 10-2 Several different patterns of exine structure. In the sculpturing types all raised areas are shown light, all lower areas or holes are shown dark. (From Moore et al., 1991).

protein domains in the pollen grain: the exine, intine, and cytoplasm. Not all proteins act as allergens, sensitizing susceptible individuals. Those proteins that cause an allergic response in susceptible individuals are synthesized in the cytoplasm of the pollen grain and secreted to the pollen wall, where they are released rapidly upon contact with the mucous membranes (Knox et al., 1970; Vithanage et al., 1982; Knox, 1984). For example, *Amb a* I, the major allergen of ragweed pollen, is held mainly in the inner cellulosic intine of the pollen grain wall (Knox and Heslop-Harrison, 1971). In grass pollen grains, the allergens are located in cytoplasmic sites and are rapidly released through channels in the wall when the pollen is moistened (Staff et al., 1990).

Nature of Pollen Allergens

Allergens are those proteins that can cause an immediate hypersensitivity reaction in susceptible individuals following their interaction with cell-bound, specific

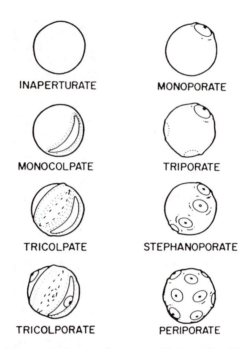

INAPERTURATE   MONOPORATE

MONOCOLPATE   TRIPORATE

TRICOLPATE   STEPHANOPORATE

TRICOLPORATE   PERIPORATE

Figure 10-3   Various apertural types of common airborne pollen (From Solomon, 1984).

IgE antibodies in the airways, gastrointestinal tract, mucous membrane, or skin (Knox et al., 1989). A major allergen can induce an allergic reaction in more than 90% of patients allergic to the protein. A minor allergen is only important for a small number of patients (typically 30 to 70%).

Allergens are generally restricted in size to the M_r of common enzymes, that is, 10 to 40 K (Aas, 1978; Bush, 1989) and are released from pollen grains soon after moistening (Howlett and Knox, 1984). The rate of release of allergens from the grain upon contact with mucous membranes must be rapid, so that interaction occurs before the mucociliary action removes the pollen grain to the glottis (within 6–8 min), where the pollen grain may be swallowed and digested (Bridger and Proctor, 1971). Baraniuk et al. (1988) showed that the nasopharynx was the site of maximum exposure to rapidly released allergens. The ryegrass pollen allergens are mobile molecules, diffusing from the grains within two minutes of moistening (Howlett and Knox, 1984; Staff et al., 1990). The major allergen of ryegrass, *Lol p* 1, which binds IgE in 90% of grass pollen allergic patients, is completely released within 16 minutes (Marsh et al., 1981). Vrtala et al. (1993) showed that most allergens from birch (*Betula verrucosa*) and timothy grass (*Phleum pratense*) pollen after hydration were eluted within the first ten minutes. The major allergens, such as *Bet v* I, *Phl p* I and *Phl p* V, which accounted for IgE binding in 90% of allergic patients were released rapidly within one minute

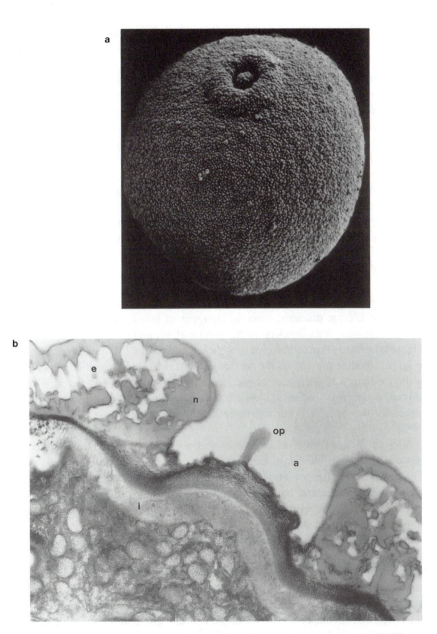

Figure 10-4 Pores of grass pollen grains surrounded by an annulus and covered by an operculum. (*a*) Scanning electron micrograph of pore region of ryegrass (*Lolium perenne*) pollen grain. (*b*) Transmission electron micrograph of pore region of ryegrass (*Lolium perenne*) pollen grain. **a**, aperture, **e**, exine, **i**, intine; **op**, operculum; **n**, annulus. (Courtesy of P. E. Taylor).

and in large quantity. Birch profilin, a minor allergen, accounted for IgE binding in 10% of birch pollen allergic patients, appeared ten minutes later, and appeared in smaller quantity. Marsh et al. (1981) and Baraniuk et al. (1988) demonstrated that allergens from short ragweed pollen were rapidly released within 20 minutes upon hydration. However, the minor allergen, *Amb a* V, (accounting for IgE binding in 13% of allergic patients) was completely released from pollen grains within four minutes. The release of the major allergen, *Amb a* I from the grains was much less rapid. Thus *Amb a* I appears to have much higher allergenic potential than *Amb a* V (Aas, 1978). Although allergenic proteins were released rapidly upon hydration, nonallergenic proteins, such as heat shock protein 70, were preferentially isolated only after using harsh extraction methods, such as boiling in SDS sample buffer and grinding (Vrtala et al., 1993). Thus the allergenic properties of a pollen allergen are linked to the amount and time of release from pollen grains rather than to intrinsic properties (Vrtala et al., 1993).

Prevalence of Allergic Diseases

Airborne pollen allergens of trees, grasses, and weeds are associated with seasonal patterns of allergic diseases, whereas allergens of house dust mite, molds, and animal (cat and dog) danders cause perennial symptoms. Exposure to these allergens leads to (1) production of specific IgE antibodies; (2) binding of allergens to IgE antibodies present on the surface of mast cells and basophils; (3) release of the mediators of allergic response leading to symptoms of hay fever, allergic asthma, and dermatitis. Inhalation of allergens is the primary cause of asthma, especially the asthma occurring in children and young adults (Sporik et al., 1992).

Allergic diseases such as hay fever and allergic asthma have increased in prevalence during the past three decades, particularly in industrial areas (Wuthrich, 1989; Burney et al., 1993) especially in the United States, United Kingdom, New Zealand, and Australia (Sheffer et al., 1992). These increases are alarming at a time when scientific advances have improved our understanding of asthma and provided new therapies.

In London, England, Anderson (1989) demonstrated that children's hospital admissions for asthma rose by 186% and 56% among those aged 0–4 and 5–14 years, respectively, in the seven years between 1978 and 1985. The increase in admissions is because of an increase in the frequency of severe asthma rather than changes in management. The corresponding national trends in England and Wales showed increases of 1300% in the 0–4 age group and 600% in the 5–14 age group in the 27 years between 1958 and 1985. An increase of about 1.8 times in the incidence of both asthma and hay fever in all age groups was found when results were compared of large surveys of morbidity in general practice in England and Wales in 1970–1971 and 1980–1982 (Fleming and Crombie, 1987). This increase was not localized to urban areas and unlikely to be caused by genetic factors over such a short period. The patterns of increase were similar for both asthma and hay fever, suggesting a common denominator. Acute asthma attack rates increased in England from 10.2 in 1976 to 20.2 in 1987 while acute attack rates of bronchitis in the same period rose from 80.3 to 111.8 (Ayers and Fleming, 1989). The rates are expressed as the average attack rate/100,000/week for each year for all ages. In Aberdeen, Scotland, the prevalence of asthma rose from 4.1 to 10.2% and hay fever from 3.2 to 11.9% (Davies, 1993).

In the United States, there was no different in prevalence of asthma among 6 to 11 year-old children from 1963–1965 (5.3%) through 1971–1974 (4.8%) but a significant increase (7.6%) by 1976–1980 (Gergen et al., 1988). The increased incidence of asthma was confirmed by the Rochester Epidemiology Project (Yunginger et al., 1992). The rate of admission of adult patients with asthma more than doubled from 1960 to 1987 (Williams, 1989). The data indicate that patients are being hospitalized for asthma in increasing numbers. This increase does not reflect a tendency to admit patients who are less acutely ill, since a corresponding increase in intubation occurred, reflecting the severity of the disease (Williams, 1989). It is believed that the increased admission rate for asthma is a consequence of a true increase in the prevalence and severity of the disease, rather than changes in management (Gergen and Weiss, 1990).

In New Zealand, a steep rise in hospital admission rates for asthma in children under 15 years of age increased tenfold over a 15-year period (Mitchell, 1985). The prevalence of childhood asthma increased from 7.1% in 1969 to 13.5% in 1982, an increase of 1.9 times over the past decade (Mitchell, 1983). In addition, there has been a significant increase in the prevalence of reported allergic diseases in the families of all children surveyed during the same period.

In New South Wales, Australia, the hospital admission rate for childhood asthma has increased significantly from 3.22 to 4.13 per 1000 population in 1986 (Bauman et al., 1990). Studies of childhood asthma in Melbourne, Australia, have shown more than a doubling of the prevalence of asthma symptoms from 1964 to 1990 (Robertson et al., 1991). The childhood asthma symptoms rose from 19% to 46%, an increase of 141% over the past 26 years. Other places that have reported increases in the prevalence of allergic diseases include Hong Kong (Tseng et al., 1988), Japan (Miyamoto, 1992), Switzerland (Varonier et al., 1984) and Finland (Haahtela et al., 1990).

Molecular Biology of Allergens

Allergen Nomenclature

Table 10-1 shows the International Union of Immunological Societies (IUIS) designation for grass pollen allergens. Under direction from the IUIS allergen nomenclature subcommittee, *Lol p* 1b was renamed *Lol p* IX because of its high homology in nucleotide and deduced amino acid sequences with *Poa p* IX

Table 10-1 IUIS Nomenclature for Some Grass Pollen Allergens with Known Sequence

Allergen source	IUIS nomenclature	M_r (kDa)	Sequence data availability	References
Cynodon dactylon	*Cyn d* I	31/32	C	1
Dactylis glomerata	*Dac g* I	32	P	2
	Dac g V	27/29	P	3, 4
Holcus lanatus	*Hol l* I	34	C	5
	Hol l V	30	P	5
Lolium perenne	*Lol p* 1	35	C	6, 7
	Lol p II	11	C	8
	Lol p III	11	C	9
	Lol p 5/IX	31/33	C	10
Phleum pratense	*Phl p* I	33/35/37	C	11, 12
	Phl p II	10/12	C	13
	Phl p V	32/38	C	14
Poa pratensis	*Poa p* I	33	P	15
	Poa p IX	28/32/34	C	16
Sorghum halepense	*Sor h* I	35	C	17

Notes: P, partial sequence; C, complete sequence.

1. Smith et al. (1993b)
2. Mecheri et al. (1985)
3. Klysner et al. (1992)
4. Roberts et al. (1992)
5. Schramm et al. (1993)
6. Perez et al. (1990)
7. Griffith et al. (1991a)
8. Ansari et al. (1989a)
9. Ansari et al. (1989b)

10. Singh et al. (1991)
11. Petersen et al. (1993a)
12. Petersen et al. (1993b)
13. Dolecek et al. (1993)
14. Bufe et al. (1993)
15. Ekramoddoullah (1990)
16. Silvanovich et al. (1991)
17. Avjioglu et al. (1993b)

(Silvanovich et al., 1991). At the same time that these allergens were cloned, a timothy grass allergen *Phl p* V was isolated by copper chelate and ion exchange chromatography (Matthiesen and Lowenstein, 1991a) and members of the same allergen group were identified in other grasses including ryegrass (Matthiesen and Lowenstein, 1991b). As more data became available, discrepancies occurred in the nomenclature used for group V (*Phl p* V) and group IX (*Lol p* IX and *Poa p* IX) allergens. The following data, including sequence homology as well as antigenic and allergenic cross-reactivities, suggest that there are no differences between group V and group IX allergens:

• The N-terminal region of *Lol p* 5, purified using immunoaffinity chromatography with a mAb raised against *Phl p* V, is identical with that of *Lol
  p* IX (Klysner et al., 1992; Roberts et al., 1992).

• Comparison of N-terminal sequences of *Lol p* IX and other group IX and
  group V allergens revealed high homology (Matthiesen and Lowenstein,
  1991a; Silvanovich et al., 1991).

- There is antigenic and allergenic cross-reactivity between allergens from group V and group IX (Olsen et al., 1991; Ong et al., 1993).
- Comparison of nucleotide and deduced amino acid sequences of several group V and IX allergens (Table 10-2) indicates that the two groups are similar.

Accordingly, we follow IUIS subcommittee for allergen nomenclature in the recommendation that the group IX allergen be designated group V; hence the name *Lol p* 5 is used throughout this chapter.

Grass Pollen Allergens

The best characterized grasses from cool temperate climates include Bermuda grass, canary grass, cocksfoot, Kentucky bluegrass, perennial ryegrass, sweet vernal grass, timothy, and Yorkshire fog. Apart from Bermuda grass, all these grasses belong to a single subfamily Pooideae. The most comprehensive studies have been made of proteins from ryegrass pollen and to a lesser extent Kentucky bluegrass and timothy.

Four groups of allergens, *Lol p* 1, II, III, IV, were originally described in the pollen of ryegrass, *Lolium perenne* by Johnson and Marsh (1965a). *Lol p* 1 is a 35 kDa glycoprotein but the carbohydrate moiety (5% of M_r) is not involved in IgE reactivity (see Chapter 6). *Lol p* II is an acidic protein of 14 kDa (Johnson and Marsh, 1965b) and *Lol p* III has similar M_r as *Lol p* II but has a basic pI (Marsh et al., 1970; Ansari et al., 1987). *Lol p* IV (HMBA) is a 56.8 kDa

Table 10-2 Percentage of nucleotide (in parenthesis) and deduced amino acid sequence comparison between several group V and IX allergens

	Lol p 5A (12R)	*Lol p* 5B (19R)	*Poa p* IX (KBG31)	*Poa p* IX (KBG41)	*Poa p* IX (KBG 60)	*Phl p* VA	*Phl p* VB
Lol p 5A (12R)	—	66.4% [79.4%]	60.7% [74.0%)	73.9% [74.0%]	69.3% [78.6%]	68.9% [76.9%]	66.8% [78.5%]
Lol p 5B (19R)	80.4%	—	62.5% [78.3%]	72.4% [78.0%]	73.6% [83.8%]	76.6% [82.8%]	61.8% [68.7%]
Poa p IX (KBG31)	70.0%	73.9%	—	80.5% [88.2%]	87.6% [91.0%]	80.8% [86.4%]	63.2% [70.4%]
Poa p IX (KBG 41)	79.9%	85.7%	82.0%	—	92.2% [94.4%]	85.5% [89.6%]	68.2% [71.9%]
Poa p IX (KGB 60)	80.8%	81.4%	89.9%	93.5%	—	88.9% [92.1%]	65.4% [75.5%]
Phl p VA	83.2%	84.3%	85.3%	92.3%	91.9%	—	67.1% [75.6%]
phl p VB	74.6%	74.3%	74.6%	76.1%	75.7%	78.6%	—

Note: Homology is the number of identical and functionally similar amino acids (bottom half of the box).

glycoprotein with 17% carbohydrate and a pI of 9.9 (Ekramoddoullah et al., 1983). However, the allergenic spectrum of ryegrass pollen is now known to be more complex (Ford and Baldo, 1986); the international reference preparation for ryegrass contains 17 allergens ranging in M_r from 12 to 89 K (Stewart et al., 1988). These allergenic proteins in pollen have been detected by their ability to bind IgE. Recently, molecular and immunochemical analyses have shown that the allergenic spectrum may not be as complex as it seems. Many of these allergens share cross-reactive epitopes. *Lol p* 1, *Lol p* II, and *Lol p* III have regions showing high levels of homology (see Chapter 6).

Lol p 5 was identified only recently, with M_r ranging from 28–32 K and the gene encoding this allergen has been cloned (Singh et al., 1991). Its amino acid sequence shows no homology with *Lol p* 1, but shows significant homology with an allergen cloned from Kentucky bluegrass pollen (Silvanovich et al., 1991). *Lol p* 5 is not a glycoprotein (Singh et al., 1991). This allergen is equivalent in importance to *Lol p* 1 as judged by both criteria of number of allergic patients responding to it (>90%) and strong IgE binding to the allergen. However, IgE binding appeared to be stronger to *Lol p* 5 than to the more abundant allergen *Lol p* I (Matthiesen and Lowenstein, 1991b). Together, *Lol p* 1 and *Lol p* 5 account for nearly all the IgE binding reactivity of crude ryegrass pollen extract (Bond et al., 1993). Therefore, detailed knowledge of both *Lol p* 1 and *Lol p* 5 characteristics and their B- and T- cell epitopes will be of clinical importance.

Allergens cross-reactive with *Lol p* 1 and *Lol p* 5 have been detected in pollen of other grass genera (Matthiesen and Lowenstein, 1991b; van Ree et al., 1992; Avjioglu, 1992; Avjioglu et al., 1993a, Smith et al., 1993a). While *Lol p* 1 homologues are found in all grass pollens tested, *Lol p* 5 homologues are apparently confined to the Pooideae group, which includes Kentucky bluegrass, orchard grass, timothy, and Yorkshire fog.

Isoforms of Grass Pollen Allergens

When *Lol p* 1 was first isolated from ryegrass, the allergen was found to exist as four proteins that differed slightly in their physical characteristics (mobility on starch gels and amino acid composition) but appeared to be immunologically identical (Johnson and Marsh, 1965a, 1966). These different forms were termed isoallergens. Since then several allergens have been shown to exist as a number of forms differing in physical characteristics (Rohac et al., 1991). *Poa p* I comprises two isoallergens (Lin et al., 1988), one being acidic (35.8 kDa, pI 6.4) and the other basic (33 kDa, pI 9.1). The 35.8 kDa *Poa p* I was further shown to separate into four isoforms differing in charge (Ekramoddoullah, 1990). Five isoforms of *Poa p* V were observed in two-dimensional Western blots of Kentucky bluegrass pollen proteins probed with eluted IgE from recombinant *Poa p* V (KBG 7.2). These differed in M_r but possessed the same basic pI (> 9.5) (Olsen et al., 1991). *Phl p* I consist of seven isoforms that differ in M_r and pI (Petersen et al., 1992a, 1993b): two proteins of 37 kDa with pI 6.4 and 6.6, four of 35 kDa with pI 6.5, 6.8, 7.1, 7.3, and one of 33 kDa with pI 8.5. Eight isoallergens have been demonstrated by their mAb- and IgE-binding in *Phl p* V (Petersen et al., 1992a, 1992b), four isoforms of 32 kDa with pI ranging from 4.8 to 5.9 and four isoforms of 38 kDa with pI ranging from 5.2 to 7.5. Smith et al. (1993a) probed two-dimensional Western blots of ryegrass pollen proteins with IgE antibodies affinity purified from either recombinant *Lol p* 1 or *Lol p* 5. Four isoforms of *Lol p* 1 were detected differing only in pI, and eight isoforms of *Lol p* 5, differing in M_r and pI. *Dac g* V comprises five isoforms of 28 kDa (Walsh et al., 1990). *Cyn d* I consist of two isoforms: 32 kDa, pI 5.8–7.0; 31 kDa pI 7.0–9.0 (Matthiesen et al., 1991; Smith et al., 1993b). The 32 kDa was further separated into ten isoforms with pI ranging from 5.6–7.3 (Han et al., 1993).

Cloning of allergen genes has revealed the amino acid sequences of isoforms of an allergen, e.g., *Lol p* 1, *Lol p* 5, *Poa p* V and *Phl p* V isoforms. Comparison between the deduced amino acid sequences of the isoform revealed high homology (Perez et al., 1990; Griffith et al., 1991; Singh et al., 1991; Ong et al., 1993; Silvanovich et al., 1991; Bufe et al., 1993). Microheterogeneity between isoforms of identical M_r but different pI may be caused by single amino acid substitutions, suggesting that the allergen is encoded by different alleles of the same gene. In contrast, allergens with isoforms of different M_r may have originated by inclusion or exclusion of particular regions of the molecule, suggesting that isoforms of different M_r of an allergen are encoded by a number of genes. Despite an overall homology (>80%), the sequence differences of the isoforms within an allergen may result in differences in reactivity with IgE and IgG antibodies from allergic individuals. For example, the recombinant allergen from ragweed pollen, *Amb a* I.1, was recognized by IgE from the sera of more individuals allergic to ragweed than two other forms, *Amb a* I.2 and *Amb a* I.3 (Bond et al., 1991). These allergens, which are not immunologically identical but highly homologous, should be termed *isoforms* or *polymorphic forms* of an allergen. The question needs to be addressed whether all polymorphic forms are capable of eliciting an allergic response in susceptible individuals.

We have isolated and characterized a cDNA clone, 19R, that encodes an isoform of *Lol p* 5, designated *Lol p* 5B, comprising a 33.8-kDa protein of 339 amino acids with a leader peptide identical to that of *Lol p* 5A (Ong et al., 1993). The deduced amino acid sequence of *Lol p* 5B has 66.4% identity (80.4% similarity) with *Lol p* 5A. However, a *Lol p* 5A-specific monoclonal antibody, FMC A7, does not recognize recombinant proteins encoded by clone 19R, indicating that *Lol p* 5B does not share this epitope. In addition, mAb PpV1 raised against *Phl p* V recognized *Lol p* 5B but not *Lol p* 5A. Cross-reactivity studies showed that both isoforms share similar and unique allergenic epitopes. Immunoblot analysis using sera from a population of 30 patients showed 80% possess IgE antibodies that recognize both *Lol p* 5 isoforms. Variation occurred in the signal intensities of IgE binding. Thus, the differences in sequence between the two polymorphic forms results in differences in their antigenic and allergenic activity (Ong et al., 1993). These differences in antigenic and allergenic activity suggest that polymorphic form-specific IgE epitopes may exist and highlights the need to assess the relative potency of all polymorphic forms.

Expression of Cloned Grass Pollen Allergens

Recombinant *Lol p* 1 and *Lol p* 5 proteins expressed in bacteria are immunoreactive with IgE antibodies from the sera of allergic individuals. IgE antibodies from 90% of patients bound to recombinant *Lol p* 1 and 80% to *Lol p* 5 (Ong et al., 1993). Gene fragmentation has been used (Mehra et al., 1986; Greene et al., 1990) to systematically examine large numbers of peptides of varying length and position within an antigen. Random fragment libraries for these major allergens have enabled expressed peptides to be tested for IgE reactivity. With *Lol p* 5A, a number of peptides ranging in length from 35 to 100 amino acids were found to bind IgE (Suphioglu, 1994). Although different IgE-reactive epitopes were localized throughout the molecule, the highest frequency and intensity of binding was observed in fragments from the C-terminal half of the molecule. These results confirm our original observations that when two "half molecules" of *Lol p* 5A are expressed individually, the majority of the IgE-reactive sites occur on the C-terminal half (Singh et al., 1991).

In addition, 34 overlapping synthetic peptides (12-mers) based on the amino acid sequence of *Lol p* 5A have been tested in a solid phase immunoassay for their reactivity with IgE from the sera of individuals allergic to grass pollen (Suphioglu, 1994). All those with high reactivity with IgE correspond to the C-terminal half of *Lol p* 5A correlating with results obtained from random fragment libraries. This correlation highlights the usefulness of synthetic peptides in mapping of antigenic sites. For example, overlapping synthetic peptides covering the entire *Poa p* V allergen revealed 10 human IgE-reactive peptides (Zhang et al., 1992). The majority of the IgE-reactive peptides were confined to the C-terminal region of the molecule. Further, a recombinant polypeptide fragment that represents the C-terminal amino acid sequence of *Poa p* V has been shown to bind IgE (Olsen et al., 1991). These similar findings with both ryegrass and Kentucky bluegrass suggest that their IgE epitopes are both in the C-terminal region and are expected to show sequence homology. However, with *Lol p* 5B, we have shown that IgE binding epitopes are distributed at the central region of the molecule using gene fragmentation (Ong et al., 1994a). In this region, there are two different IgE binding epitopes that appear to be the major epitopes, as more than 50% of a sampled population have IgE reactive to these epitopes. Comparison of these epitope regions with that of the corresponding regions of *Lol p* 5A shows high similarity between the two isoforms, suggesting that these two epitopes in *Lol p* 5B are conserved in *Lol p* 5A. This hypothesis has been experimentally confirmed since affinity-purified IgE from individual region containing these epitopes bound to *Lol p* 5A.

Short synthetic peptides can also give important information about conformational epitopes, since all conformational epitopes contain linear segments (Gertzoff et al., 1988). Recent studies using peptide based assays have identified the antigenically important regions of discontinuous sites (Redlich et al., 1991; Savoca et al., 1991), even to the resolution of single critical amino acids within the sites (Appel et al., 1990). In future, it will be important to use the synthetic peptide strategy to assess polymorphic form-specific epitopes and relate these epitopes within the context of the folded proteins.

Expression of allergen genes in bacteria has proved difficult in some cases, most notably *Der p* I which when expressed as a fusion protein with glutathione-S-transferase retained only 50% of the IgE binding activity of the natural protein and was recognized by IgE from only 50% of individuals allergic to *Der p* I (Greene et al., 1991). This protein was difficult to express as a nonfusion protein in bacteria but has since been expressed in yeast at high yield and the recombinant protein produced reacted with 9/11 individuals allergic to *Der p* I (Chua et al., 1992). With the cloned ryegrass allergens, the recombinant allergens are recognized by more than 75% of ryegrass allergic individuals, however the percentage of IgE reactivity retained by the recombinant allergen has not been determined. Use of different vectors for expression of the recombinant ryegrass allergens may result in increased IgE reactivity and yield of the recombinant protein.

cDNA clones from cocksfoot (Walsh et al., 1989), Kentucky blue (Mohapatra et al., 1990; Silvanovich et al., 1991) and timothy (Valenta et al., 1992) grass pollens have been isolated and sequenced. The cDNA clone of cocksfoot grass pollen was expressed as a 140 kDa β-galactosidase fusion protein containing 24 kDa of cloned allergen protein. Fusion proteins were recognized by IgE antibodies in 75% (6/8) of the atopic sera, but were not detected by nonatopic sera.

A recombinant peptide of Kentucky bluegrass (KBG) pollen was synthesized as a fusion protein in *E. coli*. (Yang et al., 1991). The fusion protein demonstrated binding to IgE antibodies of >95% of 55 individual sera tested. A positive correlation ($r = 0.90$) was observed between the levels of IgE antibodies corresponding to the fusion protein and natural extracts. Three different protein preparations were compared for IgE-binding, namely total KBG pollen proteins, 27–35 kDa gel-purified pollen proteins, and the recombinant fusion protein. Results indicated that about 50% of the total IgE-binding of KBG pollen proteins was due to the IgE antibodies specific to the fusion protein.

Recombinant fusion proteins of *Phl p* I and *Phl p* V from timothy grass have been used in IgE immunoblots in comparison with natural crude pollen extracts for the diagnosis of grass pollen allergy (Valenta et al., 1992). Out of the 98 tested grass pollen allergic patients, 97 possessed IgE antibodies that recognized both the fusion and natural proteins. These results are in good agreement with

RAST scores and clinical data. Laffer et al. (1993) have demonstrated that up to 70% of IgE in grass pollen allergic patients' sera directed against total timothy pollen extracts could be blocked by preincubation of the sera with the mixture of recombinant *Phl p* I and *Phl p* V.

The use of both recombinant fragments and overlapping synthetic peptides offers considerable scope for the identification and characterization of allergenic epitopes of grass pollen allergens.

Environmental Monitoring of Grass Pollen

Grass Pollen Counts in Relation to Weather Patterns and Seasons

Quantitative Burkard spore trap methods are well established for monitoring the occurrence of grass pollen in the environment. The grass pollen season in areas with temperate climate begins in spring and early summer. In Europe, depending on the countries, the flowering of most grasses occurs from May to July (Nilsson and Palmberg-Gotthard, 1982; Spieksma et al., 1985; Weeke and Spieksma, 1991; Subiza et al., 1992; D'Amato et al., 1992). In the southern hemisphere city of Melbourne, the flowering of most grasses occurs from November to January (Smart and Knox, 1979; Smart et al., 1979).

During the flowering of grasses, the pollen content of the atmosphere depends on the day-to-day prevailing weather. Generally, airborne grass pollen levels increase during warm, dry, and sunny conditions and decrease during cool and rainy periods. Rainfall depletes the pollen in the atmosphere, but later, when the sun shines, rapid increases in pollen concentrations were observed (Davies and Smith, 1973a; Smart et al., 1979; Smart and Knox, 1979; Spieksma, 1980; Knox and Tuohy, 1981; Solomon, 1984; Bush, 1989).

In London, England, Davies and Smith (1973b) analyzed various meteorological factors affecting the daily grass pollen contents in the air based on a 10-year record of grass pollen counts. It was observed that when rain of more than 2 mm falls between 0600 and 1800 hours, the number of grass pollen grains in the air decreases. When rainfall occurs before 0600 hours, provided the weather becomes dry and sunny, there is no effect on the mean daily pollen count. The pollen levels in the air were correlated with wind direction. Winds from the northwest produced higher counts while winds from the northeast and southeast gave lower counts. The pollen count was also appreciably lower on days with most wind and most convection as pollen was dispersed through a deeper layer of air than it was on days with little wind and convection. In addition, the pollen content of the air was positively correlated with day maximum temperature and hours of bright sunshine.

In Galway, Ireland, McDonald (1980) has studied the grass pollen counts with weather parameters over a two-year period. The pollen content of the air in 1977 was three to four times higher than in 1978. This difference in concentration was attributed to the wind direction and velocity, sunshine, and rainfall. In the 1977 season, wind direction varied mainly from west through north to east which was mainly overland; in the 1978 season, the dominant wind direction was from west and south which was mostly over sea. The lower sunshine and more continuous rainfall during the grass pollen season depressed pollen production and dispersal in the 1978 season than in the 1977 season. In addition, the days with high pollen counts followed closely after the higher recorded wind speeds.

In Melbourne, Australia, Smart et al. (1979) observed that grass pollen content of the air increased consistently in days when both maximum and minimum temperatures were high, and on those days with longer sunshine hours and higher wind speeds. Decreases in daily pollen count were associated with rainfall and higher relative humidity. The maximum daily temperatures correlated positively with pollen count ($r = 0.43$, $p = 0.002$) while rainfall and relative humidity were inversely correlated with daily pollen count (for rainfall, $r = -0.21$, $p = 0.03$; for relative humidity, $r = -0.27$, $p = 0.025$). In addition, Smart et al. (1979) have shown that northerly winds carried the highest pollen load.

Besides daily fluctuations, seasonal grass pollen counts vary markedly from one season to the next (Smart and Knox, 1979; Spieksma et al., 1985; Weeke and Spieksma, 1991). For example, in Melbourne, the seasonal grass pollen counts ranged from 2319 in the 1976–77 season to 6798 in the 1979–80 season (the seasonal grass pollen counts during 1990 ranged from 3785 to 5701) (Ong et al., 1994b). These figures are in the same general range, for example, as those for Leiden, The Netherlands (3834 to 5816 pa) (Spieksma et al., 1985; Spieksma, 1986) and for Madrid, Spain (2568 to 6624 pa) (Subiza et al., 1992). The monthly grass pollen count in the air of Melbourne is relatively constant from season to season. November and December are the most important months with an average of 81% (ranging from 74 to 95%) of the total seasonal grass pollen. This season-to-season variation of grass pollen counts may be caused by environmental factors affecting the various grasses (Spieksma et al., 1985; Goldberg et al., 1988).

Grass Pollen Counts, Hay Fever, and Allergic Asthma

Grass pollen is dispersed in the air to effect fertilization and seed setting. During this process, the pollen may impact on the eyes, nose, and upper respiratory tract of sensitive humans, eliciting symptoms of hay fever and allergic asthma (Bush, 1989). The distribution of grass pollen in the air is not homogenous and the concentration varies greatly over a 24-hour period (Smart et al., 1979; Spieksma et al., 1985). Using diurnal pollen count data from 1991 and 1992, we confirmed that in Melbourne, Australia, grass pollen concentration varies markedly over the 24-hour period, although severe counts were only recorded between 1700 and 2100 hours (Ong et al., 1994b).

What is the minimum number of grass pollen grains in the air that can induce the symptoms of hay fever and allergic asthma? So far opinion has varied greatly

(Hyde, 1972; Davies and Smith, 1973a; Spieksma, 1980; Spieksma et al., 1985; Rantio-Lehtimaki et al., 1991). A grass pollen count exceeding 50 grains m^{-3} of air has been shown to provoke allergic reactions in all patients in central London who were clinically sensitive to grass pollen (Davies and Smith, 1973a). Hyde (1972) estimated that 10% of hay fever patients experienced symptoms when the grass pollen concentration exceeded 10 grains m^{-3} air in Cardiff. Rantio-Lehtimaki et al. (1991) showed that early in the season, grass pollen concentration under 30 grains m^{-3} was significantly associated with nasal hyper-sensitivity in Finland. In an epidemiological study in Austria, Zwick et al. (1991) demonstrated that children living in a high pollen count area did not have significantly higher complaints of hay fever and allergic asthma compared to children in a low pollen count area. The grass pollen count in the high area was 1.7 times more than in the low area, and the birch pollen count was 2.8 times higher. Both areas had equal meteorological conditions and high quality air with regard to the air pollutants SO$_2$ and NO$_2$.

Generally, there is a positive correlation between seasonal asthma and the grass pollen count. Hill et al. (1979) showed that childhood asthma increased markedly during the grass pollen season in spring and early summer in Melbourne. The correlation with symptoms was significant only when three-day rolling means of pollen count were used. We have also shown that there is significant correlation between grass pollen counts and hospital presentations for asthma after a lag of two days ($r = 0.33$, $p < 0.05$) (Ong et al., 1994c). However, the relationship was higher after a lag of three days ($r = 0.46$, $p < 0.05$). This is physiologically plausible as late reactions to aeroallergens have been well described.

Lewis et al. (1992) showed that the annual spring and fall peak of airborne grass pollen in Texas was correlated with the frequencies of allergic patients who were skin tested positively to 14 different grasses. Petersen and Sandberg (1981) showed that in Copenhagen, Denmark, there was a significant correlation between the appearance of pollen grains in the air and the occurrence of symptoms and use of medication. The correlations have been made for patients previously diagnosed as allergic to pollen of birch, grasses, and mugwort. In Poland, the occurrence of symptoms of grass pollen allergy was closely correlated with the presence of airborne grass pollen count in two separate studies, 1982–87 and 1988–89 (Obtulowicz, 1990, 1991). Bruce et al. (1977) found that of 29 ragweed-sensitive asthma patients who demonstrated positive skin tests, leucocyte hista-mine release, and inhalation challenge tests, only 13 patients had symptoms that correlated with the atmospheric ragweed pollen counts.

The physical characteristics of a pollen grain strongly influence the site of deposition in the airway. Grass pollen grains (~35 μm in diameter) are too large to be respired (Hatch and Gross, 1964; Hoehne and Reed, 1971; Wilson et al., 1973) and essentially all are deposited in the nose and upper airway. However, a number of hay fever patients develop asthma during the grass pollen season. How can allergen-containing pollen grains become respirable and trigger asthma?

This is achieved by the release of allergen-containing micronic particles, and evidence is presented in the next section.

Allergen Load Analysis

Existence of Micronic Particles in the Atmosphere

Pollen allergens as distinct from the grains, are known to occur naturally in the atmosphere associated with much smaller particles (Solomon et al., 1983; Stewart and Holt, 1985; Schumacher et al., 1988; Spieksma et al., 1990). These micronic particles are within the respirable range (< 5 μm). Micronic particles have been detected in the atmosphere of European and Australian cities during the grass pollen season (Spieksma et al., 1990; Paggiaro et al., 1990; Stewart and Holt, 1985). Similarly, Busse et al. (1972) and Habenicht et al. (1984) have shown that micronic particles containing ragweed allergens exist in aerosol fractions smaller than intact pollen grains. These micronic particles have been detected using different kinds of high volume air sampling and evaluation of particle size distribution (Agarwal et al., 1981; Solomon et al., 1983; Stewart and Holt, 1985; Jensen et al., 1989; Spieksma et al., 1990). During the grass pollen season, the concentrations of micronic particles increase significantly on days following rainfall (Stewart and Holt, 1985; Suphioglu et al., 1992; Knox, 1993).

The origin of these micronic particles was unknown until recently (Spieksma et al., 1990). Solomon (1984, 1986) had suggested several possibilities:

- Pollen fragments produced by physical degradation in the atmosphere.
- The anther lining in ragweed and grasses is coated with orbicules (Ubisch bodies), small spherical particles of 0.02 μm in diameter, that are made of sporopollenin, like the exine.
- Microdroplets containing allergens released from pollen grains in nature.
- Allergen-containing aerosols that bind to physical particulate matter of various submicronic sizes in the environment.

Recently, the principle source of allergen-containing micronic particles from grass pollen has been shown to be intracellular starch granules (Singh et al., 1991). This conclusion is based on two observations. First, a cDNA clone encoding *Lol p* 5 contained a signal peptide with motifs similar to those of chloroplast transit peptides, suggesting that the allergens are synthesized in the cytosol and targeted into chloroplasts. Second, immunogold localization with *Lol p* 5-specific monoclonal antibodies located the allergen within intracellular starch granules (Fig. 10-5), which are formed in amyloplasts, a type of chloroplast (Singh et al., 1991). These granules (< 3 μ) are released from pollen grains during rainfall, when grains osmotically rupture, releasing starch granules into

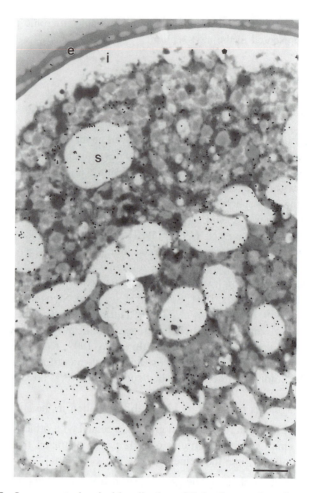

Figure 10-5   Immunocytochemical localization of *Lol p* 5 on section of ryegrass pollen using monoclonal antibody FMC A7 followed by gold probe and silver enhancement. Label is predominantly found on starch granules. Bar = 1 μm, e, exine; i, intine; s, starch.

the atmospheric aerosol as micronic particles (Suphioglu et al., 1992). Starch granules have been detected in the atmosphere during the grass pollen season, on sunny days without rainfall. However, there was a highly significant 50-fold increase in granule number m^{-3} of air on the days following significant rainfall.

   The presence of micronic particles in the air suggests that the true level of allergens in the atmosphere, that is the allergen load, is represented by the sum of intact grass pollen grains and micronic particles. Thus, it is not possible to count these micronic particles with a Burkard volumetric trap as used for pollen counting (Hirst, 1952; Ogden et al., 1974). Special quantitative methods have been developed to estimate allergen load.

Estimation of Allergen Load

Immunochemical analysis of allergen load in the atmosphere has given a more accurate picture of atmospheric allergens. Methods involve use of high-volume air samplers to collect allergens in the air onto fiberglass filters, and the allergens are then eluted and assayed using RAST-inhibition. Three types of methods to elute allergens from fiberglass filters have been described (Agarwal et al., 1981; Jensen et al., 1989); homogenization, descending elution, and recycling methods. Agarwal et al. (1981) reported that with ragweed pollen, descending elution gave a higher yield compared with the homogenization method. However, Jensen et al. (1989) found that with timothy grass pollen, homogenization gave the optimum yield along with greater convenience. In addition, they observed that in the descending elution method carried out for 8 h or longer, different allergens have different elution kinetics and degradation of allergens occurred. These results suggest that different pollen allergens require different methods of elution. This method also generates a high volume of liquid, so that lyophilization is necessary to concentrate the allergens, and with timothy grass allergens, loss of allergenicity occurred after lyophilization (Jensen et al., 1989). These methods of eluting allergens from fiberglass filters for RAST-inhibition assay are laborious and cumbersome.

Recently, Spieksma et al. (1990) reported an in situ method of estimating airborne allergen load without the need for eluting allergens. Allergen load is estimated by applying pooled human sera containing specific immunoglobulin E (IgE) antibodies onto a 20-cm-long particle-loaded impaction strip. Serum is allowed to descend through the strip where specific IgE reacts with allergens and unbound specific IgE is collected and measured using RAST inhibition. This method however, is time consuming as the process of allowing the sera to descend through a strip takes 60 minutes.

To overcome these problems, we have developed an improved immunochemical method for estimating grass pollen environmental allergen load (Ong et al., 1994c). Small circular discs punched from particle impaction filters from a high volume air sampler were directly employed to estimate allergen load (Fig. 10-6). The method utilized IgE RAST-inhibition assay and is rapid, direct and quantitative, using standard RAST discs so that radioactive probes are not required. Fiberglass filters containing grass pollen allergens are used directly without any manipulation, resulting in accurate estimates of grass pollen allergen load.

The allergen load estimates obtained are more sensitive and precise than those previously reported (Agarwal et al., 1981; Jensen et al., 1989; Spieksma et al., 1990). Agarwal et al. (1981) and Jensen et al. (1989) eluted the airborne allergens

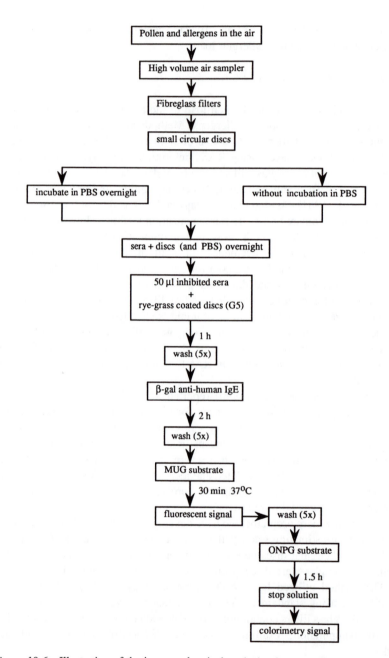

Figure 10-6   Illustration of the immunochemical analysis of grass pollen allergen load.

from fiberglass filters and estimated the eluted allergens with RAST inhibition. In these methods, not all allergens on the fiberglass filters are eluted and some of the allergens are enzymatically degraded or loss of allergenicity occurs during these manipulations. In contrast, Spieksma et al. (1990) did not elute the allergens from the fiberglass filters, rather passing the sera containing IgE onto allergen-impacted filter strips. The IgE in the sera bound to allergens on the strip while descending by gravity. Problems with this method include insufficient time for the IgE-allergen interaction, and allergens on the fiberglass filter may not all contact the IgE. Further, radioactive labeling of secondary antibodies are needed in the RAST-inhibition assay. It seems unlikely that these methods give a precise estimation of total airborne allergen load.

Relationship Between Grass Pollen Allergen Load and Pollen Counts

We have used a combination of high-volume air sampler with immunochemical estimation of grass pollen allergen load and Burkard volumetric trap for the pollen count. We confirm that grass pollen allergen load and pollen counts are correlated (Jensen et al., 1989; Johnsen et al., 1992), but we show that this only applies on dry days during the grass pollen season ($r = 0.85$, $p < 0.001$; Fig. 10-7) (Ong et al., 1994c). On rainy days and on dry days following rainfall, there is no correlation ($r = 0.49$, $p = 0.06$) as pollen grains are washed from the atmosphere by rainfall. The mean grass pollen count on dry days is 3.1 times greater than the mean pollen count on days with rainfall (Table 10-3). The mean grass pollen allergen load showed a different pattern and is only 1.8 times greater. On days following rainfall, the mean pollen count is only half that on dry days, whereas the allergen load is the same as on dry days (Ong et al., 1994c). These differences may be due to the presence of micronic particles in the atmosphere. For example, on December 7, a dry day immediately following significant rainfall, the grass pollen count was 92 m^{-3} and percentage of inhibition was 60% compared to 8, 9, and December 10 where grass pollen count was more than double the value of 7 December but the percentage of inhibition was 42 to 55% (Ong et al., 1994c). At the same time, Suphioglu et al. (1992) have shown previously that allergen-containing starch granules are released into the atmosphere when grass pollen encounters rainfall. These starch granules are a major source of micronic particles and show a 50-fold increase on the day following rainfall in the atmosphere of Melbourne. In our study, we have found significant levels of allergens in the Melbourne atmosphere on rainy days and days following rainfall when the allergen source is likely to be micronic particles.

Relationship Between Micronic Particles and Asthma

Micronic particles have been shown to be allergenic and to contain major allergens. The micronic particles of ryegrass pollen, that is, starch granules, contained

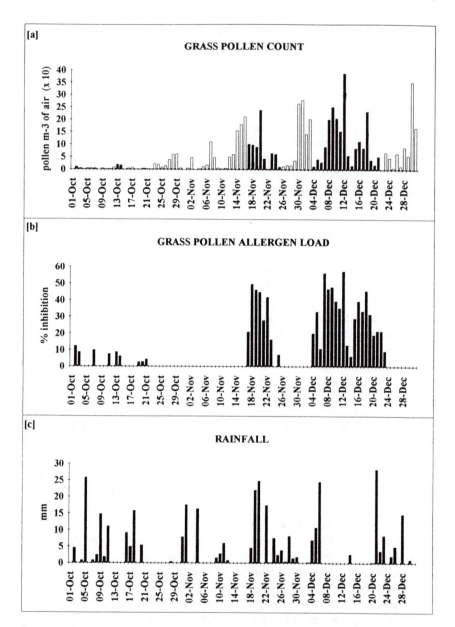

Figure 10-7 Relationship between (a) grass pollen count, (b) grass pollen allergen load, and (c) rainfall in October, November, and December 1992. The solid bars in (a) show the day when particle-loaded impaction filters were collected and analyzed.

Table 10-3   Comparison between Mean Grass Pollen Count and Mean Grass Pollen Allergen Load on Days with and without Rainfall and on Dry Days Following Rainfall

	Mean grass pollen count (m^{-3} of air)	Mean grass pollen allergen load (% inhibition)
Dry days ($n = 11$)	167.0	37.4
Days with rainfall ($n = 15$)	54.5	21.2
Dry days following rainfall ($n = 3$)	74.0	37.7

Lol p 5 allergen (Singh et al., 1991; Suphioglu et al., 1992; Knox, 1993). Western analysis of isolated starch granules with IgE from an allergic serum pool showed at least eight allergens including *Lol p* 5. The presence of *Lol p* 5 in starch granules was further confirmed using mAb FMC A7 which is specific to *Lol p* 5A, an isoform of *Lol p* 5. Skin prick tests with isolated starch granules gave strong positive results in grass pollen allergic and asthmatic patients but no reaction in nonallergic subjects (Suphioglu et al., 1992). Bellomo et al. (1992) showed that isolation starch granules are equally as effective as total pollen extracts in producing cutaneous reactions in grass pollen-sensitive patients and a cohort of thunderstorm asthmatics. In addition, inhalation challenge tests on four thunderstorm asthmatics were positive showing a typical early response with significant bronchial restriction following the administration of the isolated starch granules (Suphioglu et al., 1992). Stewart and Holt (1985) had shown that particles in micronic range (0.5 to 0.8 μm in diameter) of unknown origin were capable of binding to IgE as judged by RAST inhibition, while Spieksma et al. (1990) had found that various particles in the range of 0.6 to 10 μm contained grass pollen allergens.

Ragweed allergens, such as *Amb a* I, are present in particles smaller than 5 μm (Busse et al., 1972; Solomon et al., 1983; Agarwal et al., 1984). Positive skin test results were obtained with particles in the range 0.3 to 6 μm, in short ragweed sensitive patients at dilution of at least 1000-fold higher than on healthy subjects. RAST inhibition assay showed the presence of ragweed pollen allergens including *Amb a* I in these particles.

We have shown that asthma presentations to four major hospitals in Melbourne, Australia, between 1 October and 31 December 1992 correlated more strongly with grass pollen allergen load than with grass pollen counts (Table 10-4) (Ong et al., 1994c). The strongest correlation was between grass pollen allergen load and hospital presentations four days later ($r = 0.52$, $p < 0.001$).

Conclusions

- The prevalence of allergic disease such as hay fever and allergic asthma has increased during the last thirty years.
- *Lol p* 1 and *Lol p* 5 have been shown to be the major allergens in pollen of ryegrass. These allergens exist in multiple isoforms. cDNAs encoding

Table 10-4 Correlations between Emergency Department Presentations for Asthma, Grass Pollen Count and Allergen Load

Lag (days)	Grass pollen counts	Grass pollen allergen load
0	0.116	0.224
1	0.153	0.278
2	0.332*	0.398*
3	0.463*	0.372*
4	0.333*	0.518*
5	0.157	0.399*

*$p < 0.05$

these allergens and some of the isoforms have been isolated and character-ized. Knowledge of the extent of all the polymorphic forms would be of clinical significance.

- Recombinant allergens offer a number of advantages compared with their natural counterparts. First, the technology permits the production of de-fined allergens in large quantities, permitting the development of new diagnostic assays that are of far greater specificity than previously possible and provide the potential for international standards (Yang et al., 1991). Second, determination of the structure and identification of allergenic epitopes is feasible, providing a new understanding of the molecular basis of IgE interactions. Third, there are the exciting possibilities offered by modified allergens or peptides as effective therapeutic modalities (Scheiner, 1992; Bousquet et al., 1993).

- The daily and seasonal total of grass pollen content in the air varied markedly. This form of variation is weather dependent.

- There is a correlation between the incidence of allergic asthma and pollen counts.

- The presence of allergen-containing micronic particles in the air indicates that the true level of allergens in the atmosphere, that is the allergen load, is represented by the sum of intact grass pollen grains and micronic particles.

References

Aas K. 1978. What makes an allergen an allergen. *Allergy* 33:3–14.

Agarwal, M. K., M. C. Swanson, C. E. Reed, and J. W. Yunginger. 1984. Airborne ragweed allergens: Association with various particle sizes and short ragweed plant parts. *J. Allergy Clin. Immunol.* 74:687–693.

Agarwal, M. K., J. W. Yunginger, M. C. Swanson, and C. E. Reed. 1981. An immunochemical method to measure atmospheric allergens. *J. Allergy Clin. Immunol.* 68:194–200.

Anderson, H. R. 1989. Increase in hospital admission for childhood asthma: trends in referral, severity, and readmissions from 1970 to 1985 in a health region of the United Kingdom. *Thorax* 44:614–619.

Ansari, A. A., T. K. Kihara, and D. G. Marsh. 1987. Immunochemical studies of *Lolium perenne* (rye grass) pollen allergens, *Lol p* 1, II and III. *J. Immunol.* 139:4034–4041.

Ansari, A. A., P. Shenbagamurthi, and D. G. Marsh. 1989a. Complete amino acid sequence of a *Lolium perenne* (perennial rye-grass) pollen allergen, *Lol p* II. *J. Biol. Chem.* 264:11181–11185.

Ansari, A. A., P. Shenbagamurthi, and D. G. Marsh. 1989b. Complete primary structure of a *Lolium perenne* (perennial rye-grass) pollen allergen, *Lol p* III: Comparison with known *Lol p* 1 and II sequences. *Biochemistry* 28:8665–8670.

Appel, J. R., C. Pinilla, H. Niman, and R. Houghten. 1990. Elucidation of discontinuous linear determinant in peptides. *J. Immunol.* 144:976–983.

Avjioglu, A. 1992. Molecular analysis of grass pollen allergens. Ph.D. diss. The University of Melbourne, Australia.

Avjioglu, A., M. B. Singh, J. Kenrick, and R. B. Knox. 1993a. Monoclonal antibodies to recombinant allergenic peptide to *Lol p* IX and the identification of corresponding epitopes in different grasses. In *Molecular Biology and Immunology of Allergens,* eds. D. Kraft and A. Sehon. pp. 161–164. CRC Press, Boca Raton, FL.

Avjioglu, A., J. Creaney, P. M. Smith, P. Taylor, M. B. Singh, and R. B. Knox. 1993b. Cloning and characterization of the major allergen of *Sorghum halepense,* a subtropical grass. In *Molecular Biology and Immunology of Allergens,* eds. D. Kraft and A. Sehon, pp. 149–151. CRC Press, Boca Raton, FL.

Ayers, J. G. and D. Fleming. 1989. Underestimation of the rise in asthma over the last decade: evidence from general practice in the UK. *Thorax* 44:365P–366P.

Baraniuk, J. N., R. E. Esch, and C. E. Buckley, III. 1988. Pollen grain column chromatography: Quantitation and biochemical analysis of ragweed-pollen solutes. *J. Allergy Clin. Immunol.* 81:1126–1134.

Bauman, A., D. Lyle, L. Taylor, and P. Thomson. 1990. The use of medical records in epidemiology: A case study using asthma hospitalisations in New South Wales, 1979–86. *Austral. Med. Record J.* 20:101–105.

Bellomo, R., P. Gigliotti, A. Treloar, P. Holmes, C. Suphioglu, M. B. Singh, and R. B. Knox. 1992. Two consecutive thunderstorm associated epidemics of asthma in the city of Melbourne. The possible role of rye-grass pollen. *Med. J. Austral.* 156:834–837.

Bond, J. F., R. D. Garman, K. M. Keating, T. J. Briner, T. Rafnar, D. G. Klapper, and B. L. Rogers. 1991. Multiple *Amb a* I allergens demonstrate specific reactivity with IgE and T cells from ragweed-allergic patients. *J. Immunol.* 146:3380–3385.

Bond, J. F., B. S. Segal, X-B. Yu, K. A. Theriault, M. S. Pollock, and H. Yeung. 1993. Human IgE reactivity to purified recombinant and native grass allergens. *J. Allergy Clin. Immunol.* 91:339.

Bousquet, J., M. Breitenbach, S. Dreborg, J. Kenimer, H. Lowenstein, and P. S. Norman. 1993. Diagnosis of allergy and specific immunotherapy using recombinant allergens and epitopes. In *Molecular Biology and Immunology of Allergens,* eds. D. Kraft and A. Sehon, pp. 311–320. CRC Press, Boca Raton, FL.

Bridger, G. P. and D. F. Proctor. 1971. Laryngeal mucociliary clearance. *Ann. Otol. Rhinol. Laryngol.* 80:445–449.

Bruce, C. A., P. S. Norman, R. R. Rosenthal, and L. M. Lichtenstein. 1977. The role of ragweed pollen in autumnal asthma. *J. Allergy Clin. Immunol.* 59:449–459.

Bufe, A., W. M. Becker, and M. Schlaak. 1993. Recombinant *Phl p* VA (*Phleum pratense*) cDNA clones express at least two general B-cell epitopes of group V grass pollen allergens. In *Annual Meeting of the European Academy of Allergology and Clinical Immunology (abstract),* eds. R. C. Aalberse, M. L. Kapsenberg and P. H. Dieges, p. 127.

Burney, P., J. Bousquet, M. Blumenthal, M. Blumenthal, M. Burr, S. Bryan, D. Charpin, R. A. Kaslow, B. Kay, N.I.M. Kjellman, C. Mapp, T. Miyamoto, P. Paoletti, J. Pollock, R. Ronchetti, E. R. Weeke, V. G. Svendsen, P. Szemer, C. Troise, U. Wahn, K. Weiss, and B. Zweimann. 1993. Evidence for an increase in atopic disease and possible causes. *Clin. Exp. Allergy* 23:484–492.

Bush, R. K. 1989. Aerobiology of pollen and fungal allergens. *J. Allergy Clin. Immunol.* 84:1120–1128.

Busse, W. W., C. E. Reed, and J. H. Hoehne. 1972. Where is the allergic reaction in ragweed asthma. *J. Allergy Clin. Immunol.* 50:289–293.

Chua, K-Y., P. K. Kehal, W. R. Thomas, P. R. Vaughan, and I. G. Macreadie. 1992. High-frequency binding of IgE to the *Der p* allergen expressed in yeast. *J. Allergy Clin. Immunol.* 89:95–102.

Cogswell, J. 1992. How predictive of asthma is atopy? *Clin. Exp. Allergy* 22:597–599.

Colloff, M. J., J. Ayres, F. Carswell, P. H. Howarth, T. G. Merrett, E. B. Mitchell, M. J. Walshaw, J. O. Warner, J. A. Warner, and A. A. Woodcock. 1992. The control of allergens of dust mites and domestic pets: a position paper. *Clin. Exp. Allergy* 22:1–28.

D'Amato, G., S. Dal Bo, and S. Bonini. 1992. Pollen-related allergy in Italy. *Annals Allergy* 68:433–437.

Davies, R. 1993. The impact of pollution on allergic disease. In Annual meeting of the European Academy of Allergology and Clinical Immunology (abstract), eds. R. C. Aalberse, M. L. Kapsenberg and P. H. Dieges, pp. 226.

Davies, R. R. and L. P. Smith. 1973a. Forecasting the start and severity of the hay fever season. *Clin. Allergy* 3:263–267.

Davies, R. R. and L. P. Smith. 1973b. Weather and the grass pollen content of the air. *Clin. Allergy* 4:95–108.

Dolecek, C., S. Vrtala, S. Laffer, D. Kraft, O. Scheiner, and R. Valenta. 1993. Molecular characterization of a cDNA coding for a group II/III allergen from timothy grass (*Phleum pratense*) pollen. In Annual meeting of the European Academy of Allergology and Clinical Immunology (abstract) eds. R. C. Aalberse, M. L. Kapsenberg and P. H. Dieges, pp. 128.

Druce, H. M. 1993. Allergic and nonallergic rhinitis. In *Allergy. Principles and Practice,* eds. E. Middleton, Jr., C. R. Reed, E. F. Ellis, N. F. Adkinson, J. W. Yunginger and W. W. Busse, 4th ed. pp. 1433–1453. Mosby, St. Louis, MO.

Ekramoddoullah, A.K.M. 1990. Two-dimensional gel electrophoresis analyses of Kentucky bluegrass and rye-grass pollen allergens. *Internat. Arch. Allergy Appl. Immunol.* 93:371–377.

Ekramoddoullah, A.K.M., F. T. Kisil, and A. H. Sehon. 1983. Immunological characterization of a high molecular weight basic allergen (HMBA) of rye grass (*Lolium perenne*) pollen. *Mol. Immunol.* 20:465–473.

Fleming, D. M. and D. L. Crombie. 1987. Prevalence of asthma and hayfever in England and Wales. *Brit. Med. J.* 294:279–283.

Ford, D. and B. A. Baldo. 1986. A re-examination of rye-grass (*Lolium perenne*) pollen allergens. *Internat. Arch. Allergy Appl. Immunol.* 81:193–203.

Gergen, P. J. and K. B. Weiss. 1990. Changing patterns of asthma hospitalization among children: 1979 to 1987. *JAMA* 264:1688–1692.

Gergen, P. J., D. I. Mullally, and R. Evans. III. 1988. National survey of prevalence of asthma among children in the United States, 1976 to 1980. *Pediatrics* 81:1–7.

Gertzoff, E. D., J. A. Tainer, R. A. Lerner, and H. M. Geysen. 1988. The chemistry and mechanism of antibody binding to protein antigens. *Adv. Immunol.* 43:1–98.

Goldberg, C., H. Buch, L. Moseholm, and E. R. Weeke. 1988. Airborne pollen records in Denmark, 1977–1986. *Grana* 27:209–217.

Greene, W. K., K. Y. Chua, G. A. Stewart, and W. R. Thomas. 1990. Antigenic analysis of group I house dust mite allergens using random fragments of *Der p* I expressed by recombinant DNA libraries. *Internat. Arch. Allergy Appl. Immunol.* 92:30–38.

Greene, W. K., J. G. Cyster, K-Y. Chua, R. M. O'Brien, and W. R. Thomas. 1991. IgE and IgG binding of peptides expressed from fragments of cDNA encoding the major house dust mites allergen *Der p* I. *J. Immunol.* 147:3768–3773.

Griffith, I. J., P. M. Smith, J. Pollock, P. Theerakulpisut, A. Avjioglu, S. Davies, T. Hough, M. B. Singh, R. J. Simpson, L. D. Ward, and R. B. Knox. 1991a. Cloning and sequencing of *Lol p* 1, the major allergenic protein of rye-grass pollen. *FEBS Letters* 279:210–215.

Haahtela, T., H. Lindholm, F. Bjorksten, K. Koskenvuo, and L. A. Laitinen. 1990. Prevalence of asthma in Finnish young men. *Brit. Med. J.* 301:226–228.

Habenicht, H. A., H. A. Burge, M. L. Muilenberg, and W. R. Solomon. 1984. Allergen carriage by atmospheric aerosol. II. Ragweed-pollen determinants in submicronic atmospheric fractions. *J. Allergy Clin. Immunol.* 74:64–67.

Han, S-H., Z-N. Chang, H. Chang, C-W. Chi, H-J. Perng, C-C. Liu, J-J Tsai, and M. F. Tam. 1993. Use of monoclonal antibodies to isolate and characterize *Cyn d* I, the major allergen of Bermuda grass pollen. *J. Allergy Clin. Immunol.* 92:549–558.

Hatch, T. and P. Gross. 1964. *Pulmonary Deposition and Retention of Inhaled Aerosols,* pp. 27–43. Academic Press, New York.

Heslop-Harrison, J. 1968. The pollen grain wall. *Science* 161:230–237.

Hill, D. J., I. J. Smart, and R. B. Knox. 1979. Childhood asthma and grass pollen aerobiology in Melbourne. *Med. J. Austral.* 1:426–429.

Hirst, J. M. 1952. An automatic volumetric spore trap. *Ann. Appl. Biol.* 39:257–265.

Hoehne, J. H. and C. E. Reed. 1971. Where is the allergic reaction in ragweed asthma? *J. Allergy Clin. Immunol.* 48:36–39.

Howlett, B. J., and R. B. Knox. 1984. Allergic interactions. In *Encyclopaedia of plant physiology,* volume 17, eds. J. Heslop-Harrison and H. F. Linskens, pp. 655–674. Springer, Heidelberg.

Hyde, H. A. 1972. Atmospheric pollen and spores in relation to allergy. *Clin. Allergy* 2:153–179.

Jensen, J., L. K. Poulsen, K. Mygind, E. R. Weeke, and B. Weeke. 1989. Immunochemical estimations of allergenic activities from outdoor aeroallergens, collected by a high-volume air sampler. *Allergy* 44:52–59.

Johnsen, C. R., E. R. Weeke, J. Nielsen, J. Jensen, H. Mosbech, and L. Frolund. 1992. Aeroallergen analyses and their clinical relevance. II. Sampling by high-volume airsampler with immunochemical quantification versus Burkard pollen trap sampling with morphologic quantification. *Allergy* 47:510–516.

Johnson, P. and D. G. Marsh. 1965a. The isolation and characterization of allergens from the pollen of ryegrass (*Lolium perenne*). *Eur. Polym. J.* 1:63–77.

Johnson, P. and D. G. Marsh. 1965b. Isoallergens from rye grass pollen. *Nature* 206:935–937.

Johnson, P. and D. G. Marsh. 1966. Allergens from common ryegrass pollen (*Lolium perenne*)-I Chemical composition and structure. *Immunochem.* 3:91–100.

Klysner, S., K. G. Welinder, H. Lowenstein, and F. Matthiesen. 1992. Group V allergens in grass pollens. IV. Similarities in amino acid compositions and NH_2-terminal sequences of the group V allergens from *Lolium perenne, Poa pratense* and *Dactylis glomerata. Clin. Exp. Allergy* 22:491–497.

Knox, R. B. 1984. The pollen grain. In *Embryology of Angiosperms,* ed. B. M. Johri, pp. 197–271. Springer Verlag, New York.

Knox, R. B. 1993. Grass pollen, thunderstorms and asthma. *Clin. Exp. Allergy* 23:354–359.

Knox, R. B. and J. Heslop-Harrison. 1971. Pollen wall proteins: localisation of antigenic and allergenic proteins in the pollen grain walls of *Ambrosia* spp. (ragweeds). *Cytobios* 4:49–54.

Knox, R. B. and M. Tuohy. 1981. Pollen, plants and people—A review of pollen aerobiology in southern Australia. *Proc. 6th Australian Weeds Conf.* 2:125–141.

Knox, R. B., J. Heslop-Harrison, and C. Reed. 1970. Localisation of antigens associated with pollen grain wall by immunofluorescence. *Nature* 225:1066–1068.

Knox, R. B., M. B. Singh, T. Hough, and P. Theerakulpisut. 1989. The rye-grass pollen allergen, *Lol p* 1. In *Allergy and Molecular Biology,* Advances in Biosciences. eds. A. S. El-Shami and T. G. Merrett. Pergamon Press, Oxford, 74:161–171.

Laffer, S., S. Vrtala, M. Duchene, D. Kraft, O. Scheiner, and R. Valenta. 1993. Recombinant timothy grass (*Phleum pratense*) pollen allergens as tools to determine specific IgE-binding in grass-pollen allergic patients. In *Annual Meeting of the European Academy of Allergology and Clinical Immunology (abstract),* eds. R. C. Aalberse, M. L. Kapsenberg and P. H. Dieges, p. 128.

Lewis, W. H., A. B. Dixir, and H. J. Wedner. 1992. Grass aeropollen of the Western United States Gulf Coast. *Internat. Arch. Allergy Immunol.* 98:80–88.

Lin, Z., A.K.M. Ekramoddoullah, F. T. Kisil, J. Hebert, and W. Mourad. 1988. Isolation and characterization of *Poa p* I allergens of Kentucky bluegrass pollen with a murine monoclonal anti-*Lol p* 1 antibody. *Internat. Arch. Allergy Appl. Immunol.* 87:294–300.

Marsh, D. G., Z. H. Haddad, and D. H. Campbell. 1970. A new method for determining the distribution of allergenic fractions in biological materials: Its application to grass pollen extracts. *J. Allergy* 46:107–121.

Marsh, D. G., L. Belin, C. A. Bruce, L. M. Lichtenstein, and R. Hussain. 1981. Rapidly released allergens from short ragweed pollen. I. Kinetics of release of known allergens in relation to biologic activity. *J. Allergy Clin. Immunol.* 67:206–216.

Matthiesen, F. and H. Lowenstein. 1991a. Group V allergens in grass pollens. Purification and characterization of the group V allergen from *Phleum pratense* pollen, *Phl p* V. *Clin. Exp. Allergy* 21:297–307.

Matthiesen, F. and H. Lowenstein. 1991b. Group V allergens in grass pollens. Investigation of group V allergens in pollens from 10 grasses. *Clin. Exp. Allergy* 21:309–320.

Matthiesen, F., M. J. Schumacher, and H. Lowenstein. 1991. Characterization of the major allergen of *Cynodon dactylon* (Bermuda grass) pollen, *Cyn d* I. *J. Allergy Clin. Immunol.* 88:763–774.

McDonald, M. S. 1980. Correlation of airborne grass pollen levels with meteorological data. *Grana* 19:53–56.

Mecheri, S., G. Peltre, and B. David. 1985. Purification and characterization of a major allergen from *Dactylis glomerata* pollen: The Ag Dg I. *Internat. Arch. Allergy Appl. Immunol.* 78:283–289.

Mehra, V., D. Sweetser, and R. A. Young. 1986. Efficient mapping of protein antigenic determinants. *Proc. Natl. Acad. Sci. USA* 83:7013–7017.

Mitchell, E. A. 1983. Increasing prevalence of asthma in children. *NZ. Med. J.* 96:463–464.

Mitchell, E. A. 1985. International trends in hospital admission rates for asthma. *Arch. Disease in Childhood* 60:376–378.

Miyamoto, T. 1992. Increased prevalence of pollen allergy in Japan. In *Advances in allergology and clinical immunology,* eds. Ph. Godard, J. Bousquet and F. B. Michel, pp. 343–347. The Partheon Publishing Group, UK.

Mohapatra, S. S., R. Hill, J. Astwood, A.K.M. Ekramoddoullah, E. Olsen, A. Silvanovitch, T. Hatton, F. T. Kisil, and A. H. Sehon. 1990. Isolation and characterization of a cDNA clone encoding an IgE-binding protein from Kentucky bluegrass (*Poa pratensis*) pollen. *Internat. Arch. Allergy Appl. Immunol.* 91:362–368.

Moore, P. D., J. A. Webb, and M. E. Collinson. 1991. *Pollen Analysis,* 2nd ed. Blackwell Scientific Publications, London.

Nilsson, S. and J. Palmberg-Gotthard. 1982. Pollen calendar for Huddinge (Sweden), 1977–1981. *Grana* 21:183–185.

Obtulowicz, K., K. Szczepanek, and A. Szczeklik. 1990. The value of pollen count for diagnosis and therapy of pollen allergy in Poland. *Grana* 29:318–320.

Obtulowicz, K., K. Szczepanek, J. Radwan, M. Grzywacz, K. Adamus, and A. Szczeklik. 1991. Correlation between airborne pollen incidence, skin prick tests and serum immunoglobulins in allergic people in Cracowm, Poland. *Grana* 30:136–141.

Ogden, E. C., G. S. Raynor, J. V. Hayes, D. M. Lewis, and J. H. Haines. 1974. *Manual for Sampling Airborne Pollen.* Hafner Press, New York.

Olsen, E., L. Zhang, R. D. Hill, F. T. Kisil, A. H. Sehon, and S. S. Mohapatra. 1991. Identification and characterization of the *Poa p* IX group of basic allergens of Kentucky bluegrass pollen. *J. Immunol.* 147:205–211.

Ong, E. K., I. J. Griffith, R. B. Knox, and M. B. Singh. 1993. Cloning of a cDNA encoding a group-V (group-IX) allergen isoform from rye-grass pollen that demonstrates specific antigenic immunoreactivity. *Gene* 134:235–240.

Ong, E. K., R. B. Knox, and M. B. Singh. 1994a. Mapping of the antigenic and allergenic epitopes of Lol p VB using gene fragmentation. (in preparation).

Ong, E. K., R. B. Knox, S. Farish, and M. B. Singh. 1994b. Seasonal distribution of airborne grass pollen in the atmosphere of Melbourne: Forecasting the onset of the grass pollen season. (in preparation).

Ong, E. K., R. B. Knox, M. Abramson, and M. B. Singh. 1994c. Grass pollen, rainfall and asthma: Relationship between environmental allergen load and pollen counts. (in preparation).

Paggiaro, P. L., F. L. Dente, D. Talini, E. Bacci, B. Vagaggini, and C. Giunini. 1990. Pattern of airway response to allergen extract of *Phleum pratenses* in asthmatic patients during and outside the pollen season. *Respiration* 57:51–56.

Perez, M., G. Y. Ishioka, L. E. Walker, and R. W. Chestnut. 1990. cDNA cloning and immunological characterization of the rye-grass allergen *Lol p* 1. *J. Biol. Chem.* 265:16210–16215.

Petersen, A., W-M. Becker, and M. Schlaak. 1992a. Examination of microheterogeneity in grass pollen allergens. *Electrophoresis* 13:736–739.

Petersen, A., W-M Becker, and M. Schlaak. 1992b. Characterization of isoforms of the major allergen *Phl p* V by two-dimensional immunoblotting and microsequencing. *Internat. Arch. Allergy Immunol.* 98:105–109.

Petersen, A., G. Schramm, A. Bufe, M. Schlaak, and W. M. Becker. 1993a. Investigation of the microheterogeneity in *Phl p* I isoallergens. In *Annual Meeting of the European Academy of Allergology and Clinical Immunology* (abstract). eds. R. C. Aalberse, M. L. Kapsenberg and P. H. Dieges. p. 127.

Petersen, A., W-M. Becker, and M. Schlaak. 1993b. Characterization of grass group I allergens in timothy grass pollen. *J. Allergy Clin. Immunol.* 92:789–796.

Petersen, B. N. and I. Sandberg. 1981. Diagnostics in allergic diseases by correlating pollen/fungal spore counts with patient scores of symptoms. *Grana* 20:219–224.

Rantio-Lehtimaki, A., A. Koivikko, R. Kupias, Y. Makinen, and A. Pohjola. 1991. Significance of sampling height of airborne particles for aerobiological information. *Allergy* 46:68–76.

Redlich, P. N., P. D. Hoeprich, C. B. Colby, and S. E. Grossberg. 1991. Antibodies that neutralize human b interferon biologic activity recognize a linear epitope: Analysis by synthetic peptide mapping. *Proc. Natl. Acad. Sci. USA* 88:4040–4044.

Roberts, A. M., R. van Ree, J. Emly, S. M. Cardy, M.M.A. Rottier, and M. R. Walker. 1992. N-terminal amino acid sequence homologies of group V grass pollen allergens. *Internat. Arch. Allergy Appl. Immunol.* 98:178–180.

Robertson, C. F., E. Heycock, J. Bishop, T. Nolan, A. Olinsky, and P. D. Phelan. 1991. Prevalence of asthma in Melbourne schoolchildren: changes over 26 years. *Brit. Med. J.* 802:1116–1118.

Rohac, M., T. Birkner, I. Reimitzer, B. Bohle, R. Steiner, M. Breitenbach, D. Kraft, O. Scheiner, F. Gabl, and H. Rumpold. 1991. The immunological relationship of epitopes on major tree pollen allergens. *Mol. Immunol.* 28:897–906.

Savoca, R., C. Schwab, and H. R. Bosshard. 1991. Epitope mapping employing immobilized synthetic peptides. How specific is the reactivity of these peptides with antiserum raised against the parent protein? *J. Immunol. Methods* 141:245–252.

Scheiner, O. 1992. Recombinant allergens: Biological, immunological and practical aspects. *Internat. Arch. Allergy Immunol.* 98:93–96.

Schramm, G., A. Bufe, A. Petersen, W. M. Becker, and M. Schlaak. 1993. cDNA cloning and sequencing of *Hol l* I and *Hol l* V, the major allergens of *Holcus lanatus*. In *Annual Meeting of the European Academy of Allergology and Clinical Immunology (abstract)*, eds. R. C. Aalberse, M. L. Kapsenberg and P. H. Dieges, p. 127.

Schumacher, M. J., R. D. Griffith, and M. K. O'Rourke. 1988. Recognition of pollen and other particulate aeroantigens by immunoblot microscopy. *J. Allergy Clin. Immunol.* 82:608–616.

Sheffer, A. L., J. Bousquet, W. W. Busse, T.J.H. Clark, R. Dahl, D. Evans, L. Fabbri, F. E. Hargreave, S. T. Holgate, H. Magnussen, M. R. Partridge, R. Pauwels, R. Rodriguez-Roisin, A. Rubinfeld, M. R. Sotes, M. R. Sears, A. Szczeklik, and J. Warner. 1992. International consensus report on diagnosis and treatment of asthma. *Eur. Respir. J.* 5:601–641.

Silvanovich, A., J. Astwood, L. Zhang, E. Olsen, F. Kisil, A. Sehon, S. Mohapatra, and R. Hill. 1991. Nucleotide sequence analysis of three cDNAs coding for *Poa p* IX isoallergens of Kentucky bluegrass pollen. *J. Biol. Chem.* 266:1204–1210.

Singh, M. B., T. Hough, P. Theerakulpisut, A. Avjioglu, S. Davies, P. M. Smith, P. Taylor, R. J. Simpson, L. D. Ward, J. McCluskey, R. Puy, and R. B. Knox. 1991. Isolation of cDNA encoding a newly identified major allergenic protein of rye-grass pollen: Intracellular targeting to the amyloplast. *Proc. Natl. Acad. Sci. USA* 88:1384–1388.

Smart, I. J. and R. B. Knox. 1979. Aerobiology of grass pollen in the city atmosphere of Melbourne: Quantitative analysis of seasonal and diurnal changes. *Austral. J. Bot.* 27:317–331.

Smart, I. J., W. G. Tuddenham, and R. B. Knox. 1979. Aerobiology of grass pollen in the city atmosphere of Melbourne: Effects of weather parameters and pollen sources. *Austral. J. Bot.* 27:333–342.

Smith, E. G. 1984. *Sampling and Identifying Allergenic Pollens and Molds. An Illustrated Manual for Physicians and Lab Technicians.* Blewstone Press, Texas.

Smith, P. M., E. K. Ong, A. Avjioglu, M. B. Singh, and R. B. Knox. 1993a. Analysis of rye-grass pollen allergens using two dimensional electrophoresis and immunoblotting. In *Molecular Biology and Immunology of Allergens,* eds. D. Kraft and A. Sehon, pp. 141–143. CRC Press, Boca Raton, FL.

Smith, P. M., M. B. Singh, and R. B. Knox. 1993b. Characterization and cloning of the major allergen of Bermuda grass, *Cyn d* I. In *Molecular Biology and Immunology of Allergens,* eds. D. Kraft and A. Sehon, pp. 157–160. CRC Press, Boca Raton, FL.

Solomon, W. R. 1984. Aerobiology of pollinosis, *J. Allergy Clin. Immunol.* 74:449–461.

Solomon, W. R. 1986. Airborne allergens associated with small particle fractions. *Grana* 25:85–87.

Solomon, W. R., H. A. Burge, and M. L. Muilenberg. 1983. Allergen carriage by atmospheric aerosol. I. Ragweed pollen determinants in smaller micronic fractions. *J. Allergy Clin. Immunol.* 72:443–447.

Spieksma, F.Th.M. 1980. Daily hay fever forecast in the Netherlands. Radio broadcasting of the expected influence of the weather on subjective complaints of hay fever sufferers. *Allergy* 35:593–603.

Spieksma, F.Th.M. 1986. Airborne pollen concentrations in Leiden, The Netherlands, 1977–1981. III. Herbs and weeds flowering in the summer. *Grana* 25:47–54.

Spieksma, F.Th.M., J. A. Kramps, A. C. Van Der Linden, B.E.P.H. Nikkels, A. Plomp, H. K. Koerten, and J. H. Dijkman. 1990. Evidence of grass-pollen allergenic activity in the smaller micronic atmospheric aerosol fraction. *Clin. Exp. Allergy* 20:273–280.

Spieksma, F.Th.M., A. D. Van den Assem, and B.J.A. Collette. 1985. Airborne pollen concentration in Leiden, The Netherlands, 1977–1981. II. Poaceae (grasses), variations and relation to hay fever. *Grana* 24:99–108.

Sporik, R., M. D. Chapman, and T.A.E. Platts-Mills. 1992. House dust mite exposure as a cause of asthma. *Clin. Exp. Allergy* 22:897–906.

Staff, I. A., P. E. Taylor, P. Smith, M. B. Singh, and R. B. Knox. 1990. Cellular localization of water soluble, allergenic proteins in rye-grass (*Lolium perenne*) pollen using monoclonal and specific IgE antibodies with immunogold probes. *Histochemical J.* 22:276–290.

Stewart, G. A. and P. G. Holt. 1985. Submicronic airborne allergens. *Med. J. Austral.* 143:426–427.

Stewart, G. A., K. J. Turner, B. A. Baldo, A. W. Cripps, A. Ford, V. Seagroatt, H. Lowenstein, and A.K.M. Ekramoddoullah. 1988. Standardization of rye-grass pollen (*Lolium perenne*) extract. An immunochemical and physicochemical assessment of six candidate international reference preparations. *Internat. Arch. Allergy Immunol.* 86:9–18.

Subiza, J., J. M. Masiello, J. L. Subiza, M. Jerez, M. Hinojosa, and E. Subiza. 1992. Prediction of annual variations in atmospheric concentrations of grass pollen. A method based on meteorological factors and grain crop estimates. *Clin. Exp. Allergy* 22:540–546.

Suphioglu, C. 1994. Molecular biology and immunology of grass pollen allergens. Ph.D. diss. The University of Melbourne, Australia. (submitted).

Suphioglu, C., M. B. Singh, P. Taylor, R. Bellomo, P. Holmes, R. Puy, and R. B. Knox. 1992. Mechanism of grass-pollen-induced asthma. *Lancet* 339:569–572.

Tseng, R.Y.M., Y. M. Lam, and D. P. Davies, 1988. Hospitalisation for childhood asthma in Hong Kong 1976–1985. *Public Health* 102:275–279.

Valenta, R., S. Vrtala, C. Ebner, D. Kraft, and O. Scheiner. 1992. Diagnosis of grass pollen allergy with recombinant timothy grass (*Phleum pratense*) pollen allergens. *Internat. Arch. Allergy Immunol.* 97:287–294.

van Ree, R., M.N.B.M. Driessen, W. A. van Leeuwen, S. O. Stapel and R. C. Aalberse. 1992. Variability of crossreactivity of IgE antibodies to group I and V allergens in eight grass pollen species. *Clin. Exp. Allergy* 22:611–617.

Varonier, H. S., J. de Haller, and C. Schopfer. 1984. Prevalence de l'allergie chez les enfants et les adolescents. *Helv. Paediatr. Acta* 39:129–136.

Vithanage, H.I.M.V., B. J. Howlett, S. Jobson, and R. B. Knox. 1982. Immunocytochemical localization of water-soluble glycoproteins, including Group I allergen, in pollen of ryegrass, *Lolium perenne,* using ferritin-labelled antibodies. *Histochem. J.* 14:949–966.

Vrtala, S., M. Grote, M. Duchene, R. van Ree, D. Kraft, O. Scheiner, and R. Valenta. 1993. Properties of tree and grass pollen allergens: Reinvestigation of the linkage between solubility and allergenicity. *Internat. Arch. Allergy Immunol.* 102:160–169.

Walsh, D. J., J. A. Matthews, R. Denmeade, P. Maxwell, M. Davidson, and M. R. Walker. 1990. Monoclonal antibodies to proteins from Cocksfoot grass (*Dactylis glomerata*) pollen: Isolation and N-terminal sequence of a major allergen. *Internat. Arch. Allergy Appl. Immunol.* 91:419–425.

Walsh, D. J., J. A. Matthews, R. Denmeade, M. R. Walker. 1989. Cloning of cDNA coding for an allergen of cocksfoot grass (*Dactylis glomerata*) pollen. *Internat. Arch. Allergy Appl. Immunol.* 90:78–83.

Weeke, E. R. and F.Th.M. Spieksma. 1991. Allergenic significance of Gramineae pollen. In *Allergenic Pollen and Pollinosis in Europe,* eds. G. D'Amato, F.Th.M. Spieksma and S. Bonini, pp. 109–112. Blackwell Scientific Publications, London.

Williams, M. H. 1989. Increasing severity of asthma from 1960 to 1987. *New Engl. J. Med.* 320:1015–1016.

Wilson, A. F., H. S. Novey, R. A. Berke, and E. L. Surprenant. 1973. Deposition of inhaled pollen and pollen extract in human airways. *N. Engl. J. Med.* 288:1056–1060.

Wuthrich, B. 1989. Epidemiology of the allergic diseases: are they really on the increase? *Internat. Arch. Allergy Appl. Immunol.* 90:3–10.

Yang, M., E. Olsen, J. Dolovich, A. H. Sehon, and S. S. Mohapatra. 1991. Immunologic characterization of a recombinant Kentucky bluegrass (*Poa pratensis*) allergenic peptide. *J. Allergy Clin. Immunol.* 87:1096–1104.

Yunginger, J. W., C. E. Reed, E. J. O'Connell, J. Melton III, W. M. O'Fallon, and M. D. Silverstein. 1992. A community-based study of the epidemiology of asthma. Incidence rates, 1964–1983. *Am. Rev. Respir. Dis.* 146:888–894.

Zhang, L., E. Olsen, F. T. Kisil, R. D. Hill, A. H. Sehon, and S. S. Mohapatra. 1992. Mapping of antibody binding epitopes of a recombinant *Poa p* IX allergen. *Mol. Immunol.* 29:1383–1389.

Zwick, H., W. Popp, S. Jager, C. Wagner, K. Reiser, and F. Horak. 1991. Pollen sensitization and allergy in children depend on the pollen load. *Allergy* 46:362–366.

11

Immunological Characterization of the Major Ragweed Allergens *Amb a* I and *Amb a* II

Bruce L. Rogers, Julian F. Bond, Jay P. Morgenstern,
Catherine M. Counsell, and Irwin J. Griffith

Introduction

Pollen from short ragweed (*Ambrosia artemisiifolia*) is the source of one of North America's most important allergens (King, 1976). *Ambrosia artemisiifolia* is a widespread weed of the Compositae family, which is comprised of several related species (King and Norman, 1986). Although the pollinating season for ragweed varies depending on the geographical location, it is generally from midsummer to late autumn in the eastern and central United States. Approximately 10% of the U.S. population is allergic to ragweed pollen, making this allergen source highly significant in terms of clinical disease.

The predominant allergen from ragweed pollen, *Amb a* I (formerly Antigen E), was first purified and identified almost thirty years ago (King et al., 1964). These early studies established that *Amb a* I constitutes not only a substantial proportion of the protein content of pollen extract (approximately 6%) but also contributes to at least 90% of the allergenic activity found in ragweed pollen as determined by direct skin tests (King et al., 1964; King, 1972). Further biochemical study of this allergen led to the isolation of an immunochemically related allergen, *Amb a* II (formerly Antigen K), that also has high allergenic activity (King et al., 1967; Adolphson et al., 1978). Although both *Amb a* I and *Amb a* II are nonglycosylated proteins of 38-kDa molecular mass, purification often results in the isolation of two noncovalently associated degradation products of 26 and 12 kDa due to the action of a trypsinlike pollen protease (King et al., 1981). These two chains, designated alpha and beta, respectively, can readily be visualized in Coomassie blue stained SDS-PAGE gels of purified *Amb a* I and *Amb a* II preparations (King et al., 1981; Paull et al., 1979).

Ragweed pollen also contains defined allergenic proteins other than *Amb a* I and *Amb a* II since 22 distinct proteins in aqueous pollen extract bind specific human IgE when analyzed by crossed radioimmunoelectrophoresis (Lowenstein and Marsh, 1983). Other defined allergens include *Amb a* III, *Amb a* IV, *Amb a* V and *Amb a* VI (Marsh et al., 1988). The primary structures of *Amb a* III and *Amb a* V have been defined either by protein sequencing, cDNA cloning, or both (Mole et al., 1975; Klapper et al., 1980; Ghosh et al., 1991). RAST analysis has determined that 17–51% of the ragweed allergic patients have IgE that bind to these minor allergens (Adolphson et al., 1978).

Approximately 25 years ago it was established that immunotherapy regimens with ragweed pollen extract were efficacious in relieving symptoms of ragweed hay fever (Lichtenstein et al., 1968). This type of therapy involves administration of subcutaneous injections of soluble ragweed pollen extract in escalating doses (Lichtenstein et al., 1968; Lichtenstein et al., 1966). Significantly, biochemically purified *Amb a* I was also efficacious when used as an immunotherapeutic agent in clinical trials (Norman et al., 1968). However, immunotherapy must be performed with caution since both whole ragweed pollen extract and purified *Amb a* I bind substantial amounts of ragweed-specific IgE and consequently possess considerable anaphylactic activity. As a means to diminish the potential IgE-mediated side effects, some clinical trials have been done using chemically modified (Cockroft et al., 1977; Bacal et al., 1978; Grammer et al., 1982) or peptic fragments of (Litwin et al., 1991) ragweed pollen extract. These chemically treated extracts may be efficacious but the components of the modified preparations are difficult to quantitatively define. Interestingly, in recent years a new focus on the effects of immunotherapy on patients' T cells has emerged. Preliminary studies suggest that after ragweed immunotherapy there is a decrease in patients' T-cell reactivity to *Amb a* I in in vitro proliferation assays (Greenstein et al., 1992). In fact, it has recently been proposed that administration of peptides containing T-cell epitopes may offer a new therapeutic approach to allergy desensitization (Gefter, 1992). It has also been proposed that peptides have a lesser propensity to bind specific IgE and consequently have a lower potential to induce anaphylaxis as well as other IgE-related side effects (Schad et al., 1991; Gefter, 1992). These new approaches are currently undergoing pharmaceutical development and have been a major stimulus to define the complete primary structures of important allergens such as *Amb a* I by cDNA cloning.

The purification of *Amb a* I from pollen (King et al., 1964; King et al., 1967) facilitated the generation of murine monoclonal antibodies directed against the native allergen (Olson and Klapper, 1986). A set of five monoclonal antibodies was used in both inhibition and solid-phase ELISA assays to identify three nonoverlapping antigenic sites on *Amb a* I (Olson and Klapper, 1986). Monoclonal antibodies specific for these three antigenic sites, designated sites A, B, and C, can inhibit the binding of IgE in pooled human sera from ragweed allergic patients. The extent of binding to each site varies in individual human allergic antisera but in many cases the addition of antisite A and antisite B monoclonal antibodies in these assays inhibited up to 80% of the binding of human IgE specific for *Amb a* I (Olson and Klapper, 1986). These data suggest that the native *Amb a* I molecule contains a limited number of major antigenic sites.

Denatured *Amb a* I has a markedly reduced skin test reactivity on sensitive individuals as compared to the native allergen (King et al., 1974). Furthermore, murine monoclonal antibodies directed against the native *Amb a* I react with determinants that are conformationally dependent and do not bind to the denatured antigen (Smith et al., 1988). Therefore, a set of murine monoclonal antibodies was generated against denatured *Amb a* I (Smith et al., 1988). Fourteen out of 15 of these antibodies bind on Western blots to either isolated alpha or beta chains, but not to both. Several antibodies from this panel were used in the isolation and sequencing of two tryptic peptides of *Amb a* I. This was the first report of primary structural information to be derived from *Amb a* I (Smith et al., 1988). Several of these antibodies have been used successfully in the immunochemical characterization and expression cDNA cloning of *Amb a* I (Rafnar et al., 1991).

This chapter reviews the cloning and expression of the *Amb a* I and *Amb a* II multigene family members. Additionally, the immunochemical properties of these highly related proteins in ragweed pollen are described. Ragweed-allergic human IgE binding analysis and human T-cell reactivity to recombinant *Amb a* I and *Amb a* II are also presented. These data are discussed in relation to the potential use of T-cell epitope-containing peptides as potential therapeutic agents for ragweed pollen allergy.

Cloning of *Amb a I* and *Amb a II*

A λgt11 cDNA library was prepared from ragweed pollen mRNA and screened with a pool of mouse monoclonal antibodies raised to denatured *Amb a* I (Smith et al., 1988; Rafnar et al., 1991). Three reactive cDNA clones were isolated and sequenced. All three cDNA clones had open reading frames with extensive homology to the protein sequence derived from biochemically purified *Amb a* I from ragweed pollen (Rafnar et al., 1991). Even though the cDNAs isolated were highly homologous it was evident that each of the translated amino acid sequences had significant differences. Anchor-PCR cloning methods in conjunction with the rescreening of λgt10 and λgt11 cDNA libraries led to the isolation of full-length clones encoding three related proteins, designated *Amb a* 1.1, *Amb a* 1.2 and *Amb a* 1.3 (Rafnar et al., 1991). The translated sequences of cDNAs encoding these proteins are shown in Figure 11-1. In the course of an extensive study of the sequence polymorphism within each of these groups, a fourth isoform was isolated by PCR cloning and was accordingly designated, *Amb a* 1.4 (Griffith et al., 1991a). However, *Amb a* 1.4 cDNAs represent approximately 3% of the total *Amb a* I cDNA clones isolated by PCR amplification and therefore *Amb a* 1.4 protein is likely to be present only in small amounts in ragweed pollen (Griffith et al., 1991a).

```
          10        20        30        40        50
MGIKHCCYILYFTLALVTLLQPVRSAEDLQEILP*VNETRR*LTTSGAYN          I.1
- - - - - - - - - - - - - - - - - - - - - - - - - V E - F - - S A - - - - - S - K A C E - H -   I.2
- - - - Q - - - - - - - - - - - A - - - - - - - - G V G - - - S - - - - - * S - Q A C E - Ḻ -   I.3
- - - - - - - - - - - - - - - - - - - - - - Q̱ - - - S A - - - - * S - - - C - T - -             I.4
- - - - - - - - - - - - - - - - V - A G - L G - E V * D - - - S P - D - - - S - Q G C E - H -     II

          60        70        80        90        100
IIDGCWRGKADWAENRKALADCAQGFGKGTVGGKDGDIYTVTSᴱ_LDDDVA          I.1
- - - K - - - C - - - - - N - - Q - - - - - - - - - - A - - - Y - - - H - - V - - - - - D K - - - - -   I.2
- - - K - - - - - - - - E N - - Q - - - - - - - - - - A - - - Y - - - W - - V - - - - - N - - - - - -   I.3
- - - - - - - - - - - - - - - - - - - - - - A - - - I - - - - - - - - - - - - - - - - - - - - -       I.4
- - - K - - - C - P - - - - - - Q - - G ₙ̲ - - - - - - - - A - H - - - W - - - - M - - - D Q - - - - V   II

          110       120       130       140       150
NPKEGTLRFGAAQNRPLWIIFERDMVIRLDKEMVVNSDKTIDGRGAKVEI          I.1
- - - - - - - - - A - - - - - - - - - - - K - N - - - H - N Q - L - - - - - - - - - - - - V - - N -   I.2
- - - - - - - - - A - - - - - - - - - - K N - - - N - N Q - L - - - - - - - - - - - - V - - - -     I.3
- - - - - - - - - - - - - - - - - - - A̲ - - - - - - - - R - L A I - N - - - - - - - - - - - - -     I.4
- - - - - - - - - - T - D - - - - - - - Q - - - I - Y - Q Q - - - - T - - T̲ - - - - - - - - - - L   II

          160       170       180       190       200
INAGFTLNGVKNVIIHNINMHDVKVNPGGLIKSNDGPAAPRAGSDGDAIS          I.1
- - - L - - M N - - - I - - - - - - I - - I - - C - - - M - - - - - - - P I L - Q Q - - - - - N     I.2
- - G - L - - M N - - - I - - - - - - I - - - - - L - - - M - - - - - - P I L - Q A - - - - T - N   I.3
- - - - - A I Y N - - - I - - - - - I - - - I V - - - - - - - - C - - - - P - - - K - - - - - - - - G   I.4
                                                  - H - - - P V
VYG - I - - M N - - - - - - - - - D I - - - R - L - - - R - - - - G - - - I - - H Q - - - - - - H     II

          210       220       230       240       250
ISGSSQIWIDHCSLSKSVDGLVDAKLGTTRLTVSNSLFTQHQFVLLFGAG          I.1
V A - - - - - - - - - - - - - A S - - - L - I T - - S S H V - - - - C K - - - - - - - - - L - - D   I.2
V A - - - - - - - - - - - - - F - - - - - V T - - S - H V - I - - C K - - - Q S K A I - L R̲ - D   I.3
- - - G - - - - - - - - - A - - - - I - - - H - S - H F T - - - C - - - - - - Y L - - - W D F       I.4
V T - - - D - - - - - - T - - - F - - - - - V N W - S - G V - I - - C K - - H - E K A V - L - - S   II

          260       270       280       290       300
DENIEDRGMLATVAFNTFTDNVDQRMPRCRHGFFQVVNNNYDKWGSYAIG          I.1
- T H Y Q - K - - - - - - - - M - - - H - - - - - - - - - F - - - - - - - - - R - - T - - - -       I.2
- T H V Q - K - - - - - - - - M - - - - - - - - - - Q̲ F - - - - - - - - - - R - - T - - - -         I.3
- - - * * * * - - - - C - - - - - K - - - - - - - - - - N L - - - - V - - - - - - - E R - - - - - L -   I.4
- T H F Q - L K - H V - L - Y - I - - N T - H E - - - - - - F - - - - I - - - F - - R - D K - - - -   II

          310       320       330       340       350
GSASPTILSQGNRFCAPDERSKKNVLGRHGE*AAAESMKWNWRTNKDVLEN          I.1
- - S A - - - - - - - - - - F - - - D I I - Ē̲ - - - A - T - T * G N - - - - S - - - - - D R̲ - L - - -   I.2
- - S A - - - - C - - - - - L - - - D Q I - - - - - A - T - T * G - - - - - A - - - - S D - - L - - -   I.3
- - - G - - - - - - - - - - L - S - * I - - E - V - - Y - - S - M S - - I N - - - - S Y M - - F - -   I.4
- - S N - - - - - - - - K - V - - - F I Y R̲ - - - C L - T - A * Q E P - W - T - - - - - Q N - - - - -   II

          360       370       380       390
GAIFVASGVDPVLTPEQSAGMIPAEPGESALSLTSSAGVLSCQPGAPC          I.1
- - - - L P - - S - - - - - - - - K - - - - - - - - - - A V ᵢ̲ R - - - - - - - - - H Q - - - -       I.2
- - - - - T - S - - - - - V - - - - - - - - - - A - I K - - - - - - - - - R - - - - -               I.3
- - - - P - - - - - - - N - - - - - - A V - R - - - - - - - - - - [ - - - - ]                       I.4
- - - - - - - S - - - - - A - - N - - - M Q - - - - D M V P Q - - M N - - - - T - S - - - -           II
```

Figure 11-1 Composite of *Amb a* I and *Amb a* II sequences relative to *Amb a* I.1. The amino acid sequence of *Amb a* I.1 is given in standard one letter code. Sequences for the other *Amb a* I family members are given relative to *Amb a* I.1 with only differences being shown. A dash indicates identity and an asterisk indicates a spacing introduced to maintain maximal alignment. Amino acid numbering is based on the *Amb a* 1.2 sequence, the longest *Amb a* I family member. Wherever sequence polymorphism has been observed in a given family member, the dominant sequence is given in superscript and the minor sequence is given in subscript at the site of polymorphism. Polymorphisms in a given family member occur as independent events, except for amino acids 183–189 of *Amb a* I.4 where three amino acid differences occur as a block. Bracketed region in *Amb a* I.4 indicates the oligomer sequence used in cloning that has not been independently verified. (After Griffith et al., 1991 (with permission) S. Karger AG, Basel.)

A partial cDNA encoding a portion of *Amb a* II was isolated by hybridization screening with radiolabeled *Amb a* 1.2 cDNA of a λgt10 library made from ragweed flowerhead mRNA (Rogers et al., 1991). The translated sequence of the partial *Amb a* II cDNA clone had detectable homology to the translated *Amb a* I cDNAs and was in complete agreement with amino acid sequence data derived from biochemically purified *Amb a* II (Rogers et al., 1991). A full-length *Amb a* II clone was isolated using the Anchor PCR (Rogers et al., 1991). A comparison of the sequences presented in Figure 11-1 allows the calculation of the level of sequence identity shared by members of this multigene family (Table 11-1). It is clear that *Amb a* II is more divergent from the *Amb a* I family members, having 65% amino acid sequence identity (73% homology) with *Amb a* 1.1 as compared with greater than 73% amino acid identity (81% homology) among the *Amb a* I isoforms. *Amb a* II has other properties that distinguish it from the related *Amb a* I gene products. For example, *Amb a* II mRNA is present in the flowerhead but not detectable in pollen, whereas, *Amb a* I mRNAs are present in both (Griffith et al., 1991a). Also the calculated and observed pl of *Amb a* II differs significantly from that of *Amb a* I (King, 1972; Kuo et al., 1992; see below).

Immunochemical Characterization of *Amb a I* and *Amb a II* in Pollen

Short Ragweed

The definition of the primary structures of the *Amb a* I multigene family of proteins enables a more precise immunochemical definition of these pollen proteins. Select computer-based programs can readily calculate the predicted pIs of the numerous

Table 11-1 Comparison of Identity Between Different *Amb a* I and *Amb a* II Members

	Homology Matrix				
	I.1	I.2	I.3	I.4	II
I.1	—	83.4	83.8	88.5	73.3
I.2	76.8	—	90.7	82.7	78.6
I.3	77.0	86.9	—	81.2	78.6
I.4	82.4	74.5	73.0	—	72.0
II	65.7	69.5	70.2	61.4	—

Source: Griffith et al., 1991 reprinted with permission of S. Karger, A. G., Basel.

Note: Comparisons were made between the amino acid sequences reported for *Amb a* I.1, I.2, and I.3 (Rafnar et al., 1991), *Amb a* I.4, (Griffith et al., 1991a), and *Amb a* II (Rogers et al., 1991). The lower left half of the table gives percentage identity and the upper right half gives percentage homology for each pair of sequences. For example, the sequence of *Amb a* I.4 has 82.4% identity and 88.5% homology with the sequence of *Amb a* I.1. Comparison of a sequence with itself would give 100% identity (shown as —). Similar amino acids used in the calculation of homology: A,S, T; D,E; N,Q; R,K; I,L, M,V; and F,Y,W.

isoforms encoded by the cDNA sequences (Fig. 11-1). A tabulation of these predicted pI as shown in Table 11-2. The presence of multiple naturally occurring forms of *Amb a* I in pollen was previously demonstrated by two-dimensional electrophoresis and Western blotting of aqueous pollen extract (Rafnar et al., 1991). This method, using goat anti-*Amb a* I polyclonal antiserum as a probe, revealed several dominant reactive spots with pIs in the range of 4.5–5.2 (Rafnar et al., 1991). Interestingly, the pI's for the *Amb a* 1.1 isoforms predicted from the primary structure defined by cDNA cloning are within this experimentally observed range (Table 11-2).

The isolation of multiple related *Amb a* I cDNAs, valuable in itself, did not provide any direct information concerning the relative importance of each of the translated products as allergens. Support for the hypothesis that several of these related proteins may be important contributors to ragweed pollen allergy is provided by a two-dimensional Western blot of pollen extract using ragweed allergic patient IgE as a probe (Fig. 11-2). A number of IgE-binding protein spots are observed by this method, as one would expect given the 22 individual ragweed pollen allergens detected previously by crossed immunoelectrophoresis (Lowenstein and Marsh, 1983). A dominant set of proteins with molecular weights of approximately 38 kDa and pI in the range of 4.5–5.85 were detected (Fig. 11-2). These 38-kDa protein spots are believed to correspond to various isoforms of *Amb a* I and *Amb a* II for the following reasons: (1) the molecular weight is appropriate (King et al., 1981; Smith et al., 1988), (2) the observed pI are within the range predicted (Table 11-2), (3) goat anti-*Amb a* I antiserum recognizes proteins of the same pI and molecular weight (Rafnar et al., 1991), (4) antipeptide antisera directed against synthetic peptides based on *Amb a* I and *Amb a* II primary structure recognize several of these species (Fig. 11-3; data not shown), and (5) *Amb a* I and *Amb a* II bind the majority of allergic patient IgE (King et al., 1964; King et al., 1967). Without highly specific antibody reagents it is extremely difficult to unequivocally identify allergens visualized on an IgE-probed 2D Western blot. Analysis is also complicated by the possibility that one spot may represent several unrelated proteins. Nonetheless, the predominant protein spots in Figure 11-2 most probably correspond to different isoforms of *Amb a* I and *Amb a* II, including their previously described proteolyzed alpha (25-kDa) and beta (12-kDa) fragments (King et al., 1981). In a recent publication, rabbit anti-*Amb a* I antiserum very clearly recognized a 35-kDa protein in pollen extract, suggesting the existence of a major proteolytic degradation product of *Amb a* I that had not previously been detected (Bond et al., 1991; see below). This 35-kDa species seems to be present also as several isoforms with pI in the range of *Amb a* I (Fig. 11-2).

The elucidation of the complete primary structure of the *Amb a* I family of proteins permits the generation of antipeptide antisera to specific regions of these molecules. Three synthetic peptides comprising divergent regions of *Amb a* I.1 and *Amb a* II were synthesized, coupled to KLH and used to raise antipeptide

Table 11-2 Predicted pI Values for *Amb a I* and *Amb a II* Isoforms

Family Member	Polymorphic Amino Acids			Predicted pI
Amb a I.1	E_{94}			5.44
	D_{94}			5.44
Amb a I.2	E_{323}	K_{345}	L_{381}	6.31
	E_{323}	K_{345}	I_{381}	6.31
	E_{323}	R_{345}	L_{381}	6.31
	E_{323}	R_{345}	I_{381}	6.31
	K_{323}	K_{345}	L_{381}	6.65
	K_{323}	K_{345}	I_{381}	6.65
	K_{323}	R_{345}	L_{381}	6.65
	K_{323}	R_{345}	I_{381}	6.65
Amb a I.3	L_{49}	G_{248}	Q_{280}	5.45
	Y_{49}	G_{248}	Q_{280}	5.45
	L_{49}	R_{248}	Q_{280}	5.69
	L_{49}	G_{248}	$R_{2\,80}$	5.69
	Y_{49}	G_{248}	R_{280}	5.69
	Y_{49}	R_{248}	Q_{280}	5.69
	L_{49}	R_{248}	R_{280}	5.94
	Y_{49}	R_{248}	R_{280}	5.94
Amb a I.4	E_{31}	E_{122}	$CNDGPPA_{183-189}$	5.13
	E_{31}	E_{122}	$SHDGPPV_{183-189}$	5.24
	E_{31}	A_{122}	$CNDGPPA_{183-189}$	5.24
	Q_{31}	E_{122}	$CNDGPPA_{183-189}$	5.24
	Q_{31}	E_{122}	$SHDGPPV_{183-189}$	5.36
	E_{31}	A_{122}	$SHDGPPV_{183-189}$	5.36
	Q_{31}	A_{122}	$CNDGPPA_{183-189}$	5.38
	Q_{31}	A_{122}	$SHDGPPV_{183-189}$	5.50
Amb a II	D_{71}	T_{139}	K_{322}	5.74
	D_{71}	T_{139}	R_{322}	5.74
	D_{71}	K_{139}	K_{322}	5.86
	D_{71}	K_{139}	R_{322}	5.86
	N_{71}	T_{139}	K_{322}	5.86
	N_{71}	T_{139}	R_{322}	5.86
	N_{71}	K_{139}	K_{322}	5.97
	N_{71}	K_{139}	R_{322}	5.97

Note: The potential isoforms for each *Amb a* I and *Amb a* II family member are identified here by the various combinations of the polymorphic residues presented in Fig. 11-1. Each polymorphic residue was detected independently of other polymorphisms in a given family member, except for aa 183–189 in *Amb a* I.4 where three aa differences occur as a block. The predicted pI of each ragweed isoform was calculated using software contained in PCGENE (Intelligenetics, Mountain View, CA, USA).

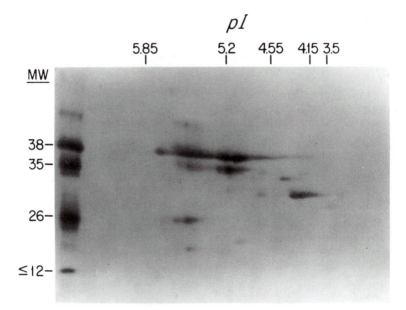

Figure 11-2 Two-dimensional gel electrophoresis and Western blotting of soluble pollen proteins. Crude, soluble pollen proteins were subjected to isoelectric focusing (left to right) followed by 10% SDS-PAGE electrophoresis (top to bottom). The pollen proteins were blotted onto nitrocellulose and probed with a pool of human ragweed-allergic serum. The left hand lane represents blotted pollen proteins from the second dimension separation only.

antisera in rabbits. These sera, directed against peptide RAE 17.1 (EPGESALSL-TSSAGVLSC; residues 375 to 393 of *Amb a* I.1), RAE 50.K (CLRTGAQEP-EWMT; residues 326 to 338 of *Amb a* II), and RAE 51.K (EPGDMVPQLTM-NAGVLTC; residues 375–393 of *Amb a* II) were then used to distinguish *Amb a* I and *Amb a* II on 2D Western blots. The specificity of these sera was demonstrated on SDS-Western blots using recombinant *Amb a* I.1 and *Amb a* II (Fig. 11-3, panel A; data not shown). The anti-RAE 50.K and anti-51.K antisera were specific for *Amb a* II (Fig. 11-3; panel A), whereas anti-RAE 17.1 is specific for *Amb a* I (data not shown). A 2D Western blot of soluble ragweed pollen proteins, probed with the *Amb a* II specific anti-RAE 50.K antipeptide antiserum, clearly visualized three protein spots with pI values of approximately 5.5, 5.7, and 5.85 and did not identify spots in the pI range expected for *Amb a* I (pI 4.5–5.2) (Fig. 11-3; panel B). On a separate blot section the *Amb a* II-specific anti-RAE 50.K recognized coincident spots whereas the *Amb a* I-specific anti-RAE 17.1 antiserum recognized three spots in the pI range of 4.5–5.2, corresponding to that of *Amb a* I (data not shown). Taken together, these data demonstrate the utility of using antipeptide antisera directed against sequences derived from the most divergent regions of individual members of a multigene family

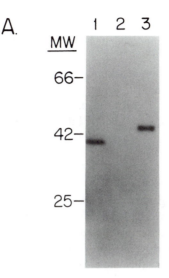

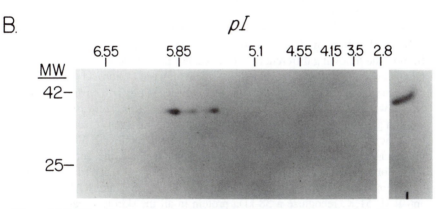

Figure 11-3 Identification of *Amb a* II in pollen using monospecific antipeptide antiserum. (*A*) SDS-PAGE immunoblot of affinity purified recombinant *Amb a* I.1 and *Amb a* II probed with the rabbit antipeptide antiserum, anti-50.K (diluted 1:2000). Lane 1, soluble ragweed pollen extract (12 μg/lane); lane 2; affinity purified recombinant *Amb a* I.1 (5 μg/lane); and lane 3; affinity purified recombinant *Amb a* II (5 μg/lane). (*B*) Two-dimensional immunoblot analyzed for binding of rabbit antipeptide antiserum, anti-50.K (anti-*Amb a* II specific), to ragweed pollen. The tube gel IEF dimension with 30 μg of soluble pollen extract was electrophoresed on 10% acrylamide SDS-PAGE for the second dimension, molecular weight markers, and a lane with 12 μg of pollen extract was included in the SDS-PAGE dimension.

to identify isoforms on a 2D Western blot. Monoclonal antibodies that are even more specific are being generated and their application awaits subsequent screening and selection.

In conclusion, the isolation of several cDNAs encoding unique groups of *Amb a* I multigene products demonstrates the existence of multiple related allergenic proteins in ragweed pollen. Two dimensional Western blot analysis, using specific antipeptide antisera and ragweed allergic IgE as a probe, provides further support for this finding.

Other Ragweed Species

Several suppliers of allergenic extracts provide extracts derived from the pollen of several species of ragweed. It has also been reported that ragweed allergic patients show positive skin tests to pollen extracts derived from several distinct plant species (Leiferman et al., 1976). It would be expected that plants having a close taxonomic relationship would have pollen proteins that have homologous sequences. This is illustrated in the case of *Amb a* V (*A. artemisiifolia;* short ragweed) and *Amb t* V (*A. trifida;* giant ragweed) since they share 49% identity in their amino acid sequences (Ghosh et al., 1991). Similarly, immunochemical analysis of pollens from several trees established the presence of multiple shared epitopes (Rohac et al., 1991).

The immunochemical relatedness between the various ragweed pollens was examined by Western blotting analysis of various commercial skin test reagents. Blots of the following different ragweed skin test reagent samples were prepared: false ragweed (*Franseria acanthicarpa*), giant ragweed (*Ambrosia trifida*), short ragweed (*Ambrosia artemisiifolia*), slender ragweed (*Franseria envifolia*), southern ragweed (*Ambrosia bidentata*), western ragweed (*Ambrosia psilostachya*), western giant ragweed (*Ambrosia aptera*), and wooly ragweed (*Franseria tomentosa*). Several of these blot sections were probed with anti-RAE 50.K (Fig. 11-4, panel A), anti-RAE 51.K (Fig. 11-4, panel B), and the anti-*Amb a* I.1 monoclonal antibody JB1E3-4 (Smith et al., 1988) (Fig. 11-5). The data presented here demonstrate that the anti-*Amb a* II specific antisera, anti-RAE 50.K and anti-RAE 51.K, recognize a 38-kDa protein in all the skin test reagents tested (Fig. 11-4, panels A and B). Since these two antipeptide antisera are specific for *Amb a* II and are directed against separate regions of sequence (residues 326–338 and 375–393, respectively), these data strongly suggest that an *Amb a* II homolog exists in the pollen of all these different ragweed species. An identical blot was probed with the anti-*Amb a* I antipeptide antiserum, anti-RAE 17.1 (data not shown). Another blot with similar samples was probed with a murine monoclonal antibody, JB1E3-4 (Smith et al., 1988), that has been shown using recombinant *Amb a* I.1, I.2, I.3, I.4, and *Amb a* II to react with only *Amb a* I.1 (data not shown). The anti-RAE 17.1 antiserum also recognized a 38-kDa protein in all the skin test reagents tested (data not shown), while the JB1E3-4

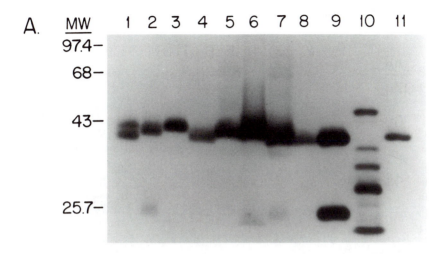

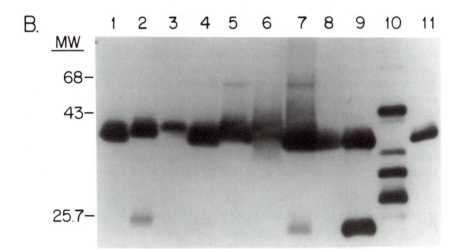

Figure 11-4 Immunoblots of different ragweed pollen extracts analyzed for rabbit anti-*Amb a* II-peptide antiserum binding. SDS-PAGE immunoblot sections were generated with the following antigens: lane 1, Southern ragweed extract (*Ambrosia bidentata*); lane 2, Western ragweed extract (*A. psilostachya*); lane 3, Western giant ragweed extract (*A. aptera*); lane 4, Wooly ragweed extract (*Franseria tomentosa*); lane 5, False ragweed extract (*F. acanthicarpa*); lane 6, Giant ragweed extract (*A. trifida*); lane 7, Short ragweed extract (*A. artemisiifolia*); lane 8, Slender ragweed extract (*F. envifolia*); lane 9, *Amb a* II purified from short ragweed pollen (King et al., 1967); lane 10, recombinant *Amb a* II lysate prepared as described in Bond et al., (1991); lane 11, Short ragweed pollen extract. These blot sections were probed with A, rabbit anti-50.K antipeptide antiserum (anti-*Amb a* II) at a dilution of 1:2000; B, rabbit anti-51.K antipeptide antiserum (anti-*Amb a* II) at a dilution of 1:2000).

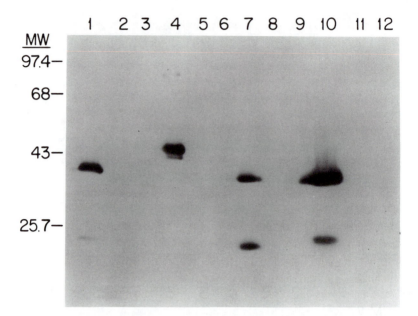

Figure 11-5 Immunoblot of different ragweed pollen extracts analyzed for JB1E3-4 anti-*Amb a* I.1 specific monoclonal antibody binding. The samples loaded were: lane 1, short ragweed pollen extract; lane 2, JM109 host cell lysate; lane 3, affinity purified recombinant *Amb a* II; lane 4, affinity purified recombinant *Amb a* I.1; lane 5, Wooly ragweed extract; lane 6, Western giant ragweed; lane 7, Western ragweed; lane 8, Southern ragweed; lane 9, Slender ragweed; lane 10, Short ragweed; lane 11, Giant ragweed; lane 12, False ragweed. The JB1E3-4 murine monoclonal antibody (Smith et al., 1988) was kindly provided by Dr. David Klapper, University of North Carolina, USA. The scientific names of these different ragweeds are cited in the text and in the legend of Fig. 11-4.

antibody bound to a 38-kDa protein in the pollen from short ragweed (*A. artemisii-folia*) and western ragweed (*A. psilostachya*) but not to any of the others tested (Fig. 11-5). Together, the binding of the rabbit polyclonal antisera suggest that the *Amb a* I and *Amb a* II gene products have not widely diverged throughout evolution of the different ragweed species. However, the binding of the mono-clonal antibody JB1E3-4 suggests that there can be sequence variation at the level of single epitopes. This result may also suggest that the *Amb a* I.1 proteins in short ragweed and western ragweed are more closely related to each other than to *Amb a* I proteins in the other ragweeds.

Now that the complete nucleotide sequences of the *Amb a* I multigene family are known (Rafnar et al., 1991; Rogers et al., 1991; Griffith et al., 1991a), it is possible using standard molecular biology techniques (e.g., λ library screening, PCR) to isolate the genes encoding the homologous proteins in other related ragweed pollens.

Recombinant Expression of *Amb a I* and *Amb a II*: IgE Binding Analysis

The isolation of cDNAs encoding multiple members of the *Amb a* I family of proteins allowed researchers to address questions regarding the roles these proteins play in ragweed allergy. First, in what proportion are each of these proteins represented in ragweed pollen? Second, what is the relative importance of each of these proteins in terms of their contribution to the human allergic response? At present, reagents and purification protocols have not been developed that permit the isolation from pollen of each of the individual *Amb a* I protein family members in native form. As an aid to examining the IgE binding properties and the T-cell reactivity of each of these proteins, the cDNAs of *Amb a* I.1, I.2, I.3, I.4, and *Amb a* II were recombinantly expressed in *E. coli* and purified (Rogers et al., 1991; Morgenstern, unpublished).

The *Amb a* I cDNAs were inserted into the expression vector pTrc99A, B, or C (Amann et al., 1988) and expression was induced as previously described (Bond et al., 1991). Briefly, the pTrc99 vectors contain a strong hybrid *trp/lac* promoter adjacent to the cDNA insertion site that is readily inducible by isopropyl-b-D-thiogalactopyranoside (IPTG). *E. coli* was transformed with the *Amb a* I and *Amb a* II expression vectors, grown to midlog phase, IPTG induced, harvested, and disrupted by sonication. Insoluble material harboring the recombinant product were pelleted and solubilized in 8 M urea (Bond et al., 1991). After dialysis into PBS these lysate preparations were subjected to IgE binding analysis on immunoblots. The IgE binding data demonstrated that ragweed allergic patient IgE specifically recognizes recombinant *Amb a* I.1, I.2, and I.3 (Bond et al., 1991). Approximately 60% of ragweed allergic IgE patients' sera fails to bind to ragweed pollen extract or to the recombinant *Amb a* I forms when examined on Western blots (Bond et al., 1991), probably because of the conformational dependence of many of the IgE epitopes of this molecule (King et al., 1974; Smith et al., 1988).

The same lysate preparations of *Amb a* I.1, I.2, and I.3 were also examined for their capacity to be recognized by T cells derived from 12 ragweed allergic patients. This analysis revealed that these recombinant proteins are all capable of stimulating T-cell proliferation in secondary in vitro assays (Bond et al., 1991). Although these data provided some very interesting information concerning the immunogenicity of these recombinant *Amb a* Is, interpretation of the results in these experiments were complicated by the presence of contaminating *E. coli* proteins as well as different expression efficiencies of each of the *Amb a* I cDNAs. Consequently, expression and purification schemes were applied that allowed a high level of purification of each of these recombinant proteins in a single facile step (Rogers et al., 1991; Morgenstern, unpublished).

Over the last few years, several allergens have been expressed in *E. coli* and the recombinant products characterized (reviewed in Balbo, 1991). In the case of the major dust mite allergens, *Der p* I, *Der f* I, *Der p* II, and *Der f* II, a glutathione-S-transferase fusion-protein based system has been very effective for the expression and purification of the full-length recombinant proteins as well as fragments thereof (Greene et al., 1990). Similarly, an important allergen from Kentucky bluegrass, *Poa p* IX, has been expressed as a fusion protein with β-galactosidase as the fusion partner (Mohapatra et al., 1990). One disadvantage of these expression systems is the presence of substantial nonrelevant sequence contributed by the fusion portion of the recombinant product. To overcome the problems of having large fusion partners, as well as to circumvent the deficiencies of using partially purified lysates (as described previously), a polyhistidine tail of six sequential residues that is capable of chelating Ni^{2+} was introduced at the N-termini of recombinant *Amb a* I.1, I.2, I.3, I.4, and *Amb a* II. The expressed products were solubilized in urea or guanidine HCl and purified by metal-ion affinity chromatography using QIAGEN NTA-agarose, a Ni^{2+} chelating support (Hochuli et al., 1988; Rogers et al., 1991). This method allows the generation of >90% pure protein preparation in a single step, as measured by densitometric analysis of a Coomassie blue stained SDS-PAGE gel (Fig. 11-6).

The binding pattern on Western blots of *Amb a* I.1, I.2, I.3, I.4, and *Amb a* II probed with sera from two individual patients' IgE or pooled sera from 20 patients is shown in Figure 11-7. It is interesting to note that IgE from two ragweed allergic patients binds to *Amb a* I.1, I.3, I.4, but not to *Amb a* I.2 and II (Fig. 11-7; panels A and B). Equal amounts of the recombinant samples were loaded onto the gel and therefore the preferential binding pattern represents the

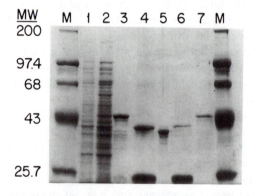

Figure 11-6 SDS-PAGE gel analysis of recombinant *Amb a* Is and *Amb a* II. Samples loaded include: lane M, molecular weight markers; lane 1, uninduced JM109 host cell lysate; lane 2, lysate of JM109 cells harboring the pTrc99-*Amb a* I.1 expression vector after 2 h of IPTG induction; lane 3, affinity purified *Amb a* I.1; lane 4, affinity purified *Amb a* I.2; lane 5, affinity purified *Amb a* I.3; lane 6, affinity purified *Amb a* I.4; and lane 7, affinity purified *Amb a* II. The 10% acrylamide gel was stained with Coomassie blue. The pTrc99 expression vector was a gift kindly provided by Dr. Egon Amann, Behringwerke AG, Germany.

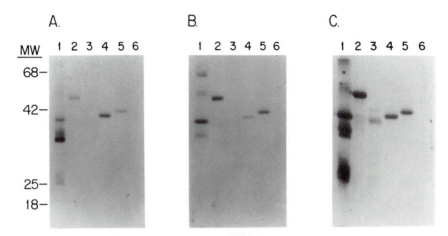

Figure 11-7 Immunoblot analysis of affinity purified recombinant *Amb a* Is and *Amb a* II. SDS-PAGE immunoblot sections were generated with the following antigens: lane 1, Short ragweed pollen extract; lane 2, purified recombinant *Amb a* I.1; lane 3, purified recombinant *Amb a* I.2; lane 4, purified recombinant *Amb a* I.3; lane 5, purified recombinant *Amb a* I.4; and lane 6, purified recombinant *Amb a* II. These blot sections were probed with (A) ragweed allergic patient #143 plasma; (B) ragweed allergic patient #475 plasma; (C) pooled ragweed allergic human plasma. The blotting methods used were essentially as described in Bond et al. (1991).

extent of IgE recognition. This binding pattern is not identical in all patients since the ragweed allergic IgE from the pooled sera binds to *Amb a* I.1, I.2, I.3, I.4, with lesser binding to *Amb a* II (Fig. 11-7, panel C). This hierarchy of binding is not an artifact of Western blotting analysis since a direct binding ELISA assay of IgE from patient #143 illustrates the same binding preference of *Amb a* I.1>I.3>I.4>I.2> II (see Fig. 11-8). It is important to stress that these allergenic proteins are produced in *E. coli* and the purification procedure involves solubilization and chromatography in the presence of denaturing agent. Consequently, the number and nature of the epitopes present on these denatured molecules may be a small subset of those exhibited on the native pollen-derived allergens (Olson and Klapper, 1986). This is supported by the finding that *Amb a* I and *Amb a* II purified from pollen have very similar IgE binding properties (Kuo et al., 1992). Taken together, these data suggest that recombinant *Amb a* I and *Amb a* II, isolated in denatured form, differ significantly in their IgE binding properties, whereas the native molecules isolated from pollen do not (Kuo et al., 1992). It will only be possible to examine the relative IgE binding capacity of the native *Amb a* Is and *Amb a* II after each is purified in their native form from pollen. Studies are underway to develop reagents and purification methods to accomplish this.

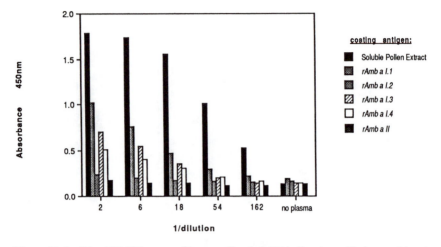

Figure 11-8 Direct ELISA assay of human allergic IgE binding to purified recombinant *Amb a* Is and *Amb a* II. Human plasma from ragweed allergic patient #143 was diluted in PBS and tested in duplicate for specific binding to antigens coated directly onto the surface of the assay plate. The coating antigens used include short ragweed pollen extract (CPE) as well as purified recombinant *Amb a* I.1, I.2, I.3, I.4 and *Amb a* II. The level of IgE binding was measured using a standard peroxidase substrate system (Kirkegaard and Perry Laboratories, Inc.).

Human Ragweed Allergic T-Cell Response to Recombinant *Amb a Is* and *Amb a II*

As shown and discussed in the previous section, ragweed-allergic IgE binds to all the recombinant forms tested, suggesting a high level of cross-reactivity at the B-cell level. The observed high level of sequence homology within the *Amb a* I and *Amb a* II family of proteins also predicts that there could be considerable cross-reactivity at the T-cell level. To address this issue, human T-cell lines were established using soluble ragweed pollen extract as the priming antigen. Subsequent secondary assays with T-cell lines derived from 12 patients measured the proliferation induced in vitro with bacterial lysates containing *Amb a* I.1, I.2, or I.3 as the antigenic stimulus. The results obtained revealed that the pollen extract primed T-cell lines recognized all three recombinant *Amb a* Is (Bond et al., 1991).

The isolation of the individual purified recombinant *Amb a* I.1, I.2, I.3, and I.4 and *Amb a* II proteins allowed the establishment of T-cell lines that had been subjected to serial rounds of stimulation with either *Amb a* I.1 or *Amb a* II. These two related proteins were chosen because they are the most divergent members of this family at the primary sequence level (~66% shared identity; see Table 11-1). As shown in Figure 11-9, a T-cell line established from a ragweed allergic patients' peripheral blood using recombinant *Amb a* I.1 as the priming antigen responds to both *Amb a* I.1 and *Amb a* II even after two rounds of stimulation with *Amb a* I.1 (Rogers et al., 1991). This result suggests that *Amb a* I.1 and *Amb a* II proteins possess common and/or cross-reactive human T-cell epitopes, although confirmatory studies with a greater number of patients needs to be done. Subsequently, the T-cell cross-reactivity of *Amb a* I family members and *Amb a* II was examined in a larger ragweed allergic patient population. Cell lines derived from a set of ragweed allergic patients were established using recombinant *Amb a* I.1 as the priming antigen using previously described methods (Bond et al., 1991; Rogers et al., 1991). These *Amb a* I.1 specific cell lines were then assayed in secondary stimulation assays in vitro with a set of antigens including pollen extract and purified recombinant *Amb a* I.1, I.2, I.3, I.4, and *Amb a* II proteins. Table 11-3 lists the stimulation index, the percent of cultures positive for a particular assay antigen, and the total number of cultures assayed. In addition, *Amb a* I.1 reactive T cells were detected in all cultures stimulated with ragweed pollen extract (12 of 12 patients tested), demonstrating the importance of the *Amb a* I.1 allergen (Bond et al., 1991).

Summary

Over the last three years, the primary structures of several major pollen allergens have been defined by cDNA cloning, including *Amb a* I (Rafnar et al., 1991; Griffith et al., 1991a), *Amb a* II (Rogers et al., 1991), *Lol p* I (Perez et al., 1990; Griffith et al., 1991b), *Lol p* IX (Singh et al., 1991), *Poa p* IX (Silvanovich et al., 1991), and *Bet v* I (Breiteneder et al., 1989). These studies have been extended to include the cloning of various homologs from other related species

Table 11-3 Secondary Human T-cell Proliferation Responses to Purified Recombinant Ragweed Allergens

Assay Antigen	S.I.	Recombinant *Amb a* I.1 Specific Lines	
		% positive	n
Pollen extract	13.5 ± 4.1	88	41
r*Amb a* I.1	25.0 ± 5.1	100	48
r*Amb a* I.2	4.2 ± 0.8	62	8
r*Amb a* I.3	9.1 ± 2.6	52	21
r*Amb a* I.4	8.3 ± 2.5	82	17
r*Amb a* II	14.1 ± 3.4	58	33

Note: Recombinant *Amb a* I.1-primed T cells were tested in secondary T-cell proliferation assays with the antigens listed. Three dilutions of the antigens indicated were tested in duplicate. The mean stimulation index (with T cells, APC, and medium only as the negative control) is shown. The detailed methods used have previously been described in Bond et al., 1991 and Rogers et al., 1991.

such as *Phl p* V from Timothy grass (Bufe et al., 1992), *Cor a* I from hazel (Breiteneder et al., 1992), and *Car b* I from hornbeam (Larsen et al., 1992). One very interesting finding that has emerged is that several of these allergens constitute individual members of multigene families. These studies have been extended to include 2D Western blotting of pollen extracts from Short ragweed (Figs. 11-2, 11-3) and Kentucky bluegrass (Olsen et al., 1991). Due to the presence of minor polymorphisms that contributes to the observed isoforms within each family group (Fig. 11-1; Table 11-2), the immunochemical characterization of these protein isoforms constitutes a formidable undertaking. This is also complicated by the presence of these proteins in the multiple related species of ragweed (Fig. 11-4A, B). Nonetheless, as shown by early studies with the related

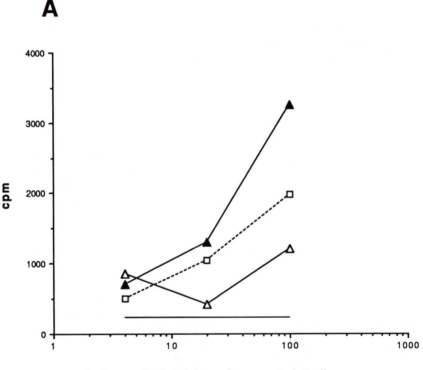

Antigen Preparation (μg protein/ml)

Figure 11-9 Human T-cell responses to affinity purified recombinant *Amb a* I.1 and *Amb a* II. Peripheral blood lymphocytes from ragweed allergic patient #144 were cultured in the presence of recombinant *Amb a* I.1, rested, and restimulated in secondary and tertiary assays with antigens indicated; medium only (−), ragweed pollen extract (□), affinity purified recombinant *Amb a* I.1 (△), and affinity purified recombinant *Amb a* II (▲). *A* represents the results of proliferation of PBMC in a primary stimulation assay.

B

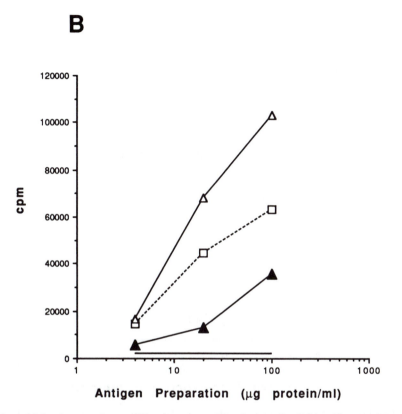

Figure 11-9 *B*, secondary proliferation assay after primary stimulation with recombinant *Amb a* I.1.

proteins *Amb a* I and *Amb a* II, it appears that a substantial proportion of ragweed-allergic human IgE binds to shared B-cell epitopes on these molecules (King et al., 1967; Adolphson et al., 1978). Human allergic patient IgE binds to all the recombinant forms, although there is some variability between patients (Figs. 11-7, 11-8; Bond et al., 1991). This high level of cross-reactive antibody binding to pollen allergens has also been found in the allergenic proteins of closely related grasses (Esch and Klapper, 1989a; Esch and Klapper, 1989b).

There is increasing interest in the allergic-human T-cell recognition of allergenic proteins (O'Hehir et al., 1991) since it is widely acknowledged that CD4+ T cells reactive with protein allergens are required for the production and regulation of allergen-specific IgE (Vercelli and Geha, 1989). It can be expected that the recently published primary structures of the important pollen allergens will lead to the definition of the major human T-cell epitopes of these molecules. Studies have been initiated in this area and have shown that ragweed-allergic human T cells react with all the *Amb a* I and *Amb a* II family members in in

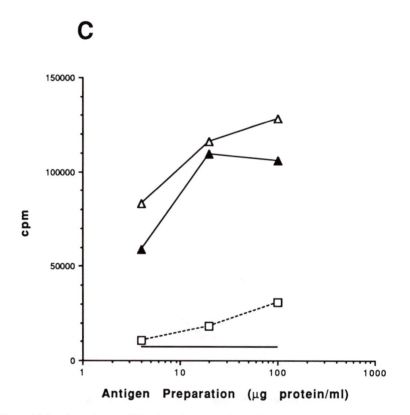

Figure 11-9 *C*, tertiary proliferation after secondary stimulation with recombinant *Amb a* I.1. The proliferation of the T cells was assessed by the uptake of tritiated thymidine in triplicate cultures. (From Rogers et al., 1991, used with permission).

vitro proliferation assays (Fig. 11-9; Table 11-3) (Bond et al., 1991; Rogers et al., 1991).

Acknowledgments

We thank A. Brauer, C. Bizinkauskas, S. Craig, A. K. Nault, J. Pollock, D. Segal and X.-B. Yu for their technical expertise and Joanne Trio for secretarial assistance. We also thank Dr. John Saryan and Dr. Albert Sheffer for supplies of allergic patient blood samples. We acknowledge the careful review of the manuscript by Drs. Richard Garman, Julia Greenstein, Mark Exley, Dean Falb, and Victoria Schad.

References

Adolphson, C., L. Goodfriend, and G. J. Gleich. 1978. Reactivity of ragweed allergens with IgE antibodies. Analyses by leukocyte histamine release and the radioallergosorbent test and determination of cross-reactivity. *J. Allergy Clin. Immunol.* 62:197–210.

Amann, E., B. Ochs, and K.-J. Abel. 1988. Tightly regulated *tac* promoter vectors useful for the expression of unfused and fused proteins in *Escherichia coli*. *Gene* 69:310–315.

Bacal, E., C. R. Zeiss, I. Suszko, D. Levitz, and R. Patterson. 1978. Polymerized whole ragweed: an improved method of immunotherapy. *J. Allergy Clin. Immunol.* 62:289–294.

Balbo, B. A. 1991. Structural features of allergens large and small with emphasis on recombinant allergens. *Current Opinion Immunol.* 3:841–850.

Bond, J. F., R. G. Garman, K. M. Keating, T. J. Briner, T. Rafnar, D. G. Klapper, and B. L. Rogers. 1991. Multiple *Amb a*I allergens demonstrate specific reactivity with IgE and T cells from ragweed allergic patients. *J. Immunol.* 146:3380–3385.

Breiteneder, H., K. Pettenburger, A. Bito, R. Valenta, and D. Kraft. 1989. The gene coding for the major birch pollen allergen *Bet v* I is highly homologous to a pea disease resistance response gene. *EMBO J.* 8:1935–1938.

Breiteneder, H., F. D. Ferreira, K. Hoffmann-Sommergruber, C. Ebner, M. Breitenbach, D. Kraft, and O.Scheiner. 1992. Cloning and expression in *E. coli* of four isoforms of *Cor a* I, the major allergen of hazel pollen. *European J. Allergy Clin. Immunol.* 47:30 (Abstract).

Bufe, A., A. Petersen, C. Schluter, M. Duchrow, and W.-M. Becker. 1992. Cloning and sequencing *Phl p* V isoallergen of 38 kD by use of anchor PCR technique. *J. Allergy Clin. Immunol.* 89:150 (Abstract).

Cockroft, D. W., M. T. Cuff, S. M. Tario, J. Dolovich, and F. E. Hargreave. 1977. Allergen injection therapy with glutaraldehyde-modified-ragweed pollen-tyrosine adsorbate. *J. Allergy Clin. Immunol.* 60:56–62.

Esch, R. E. and D. G. Klapper. 1989a. Identification and localization of allergenic determinants on grass Group I antigens using monoclonal antibodies. *J. Immunol.* 142:179–184.

Esch, R. E. and D. G. Klapper. 1989b. Isolation and characterization of a major cross-reactive grass Group I allergenic epitope. *Mol. Immunol.* 26:557–561.

Gefter, M. L. 1992. A new generation of antigens for immunotherapy, in *Current Therapy in Allergy, Immunology, and Rheumatology,* eds. L. M. Lichtenstein and A. Fauci, pp. 376–378. C. V. Mosby, St. Louis.

Ghosh, B., M. P. Perry, and D. G. Marsh. 1991. Cloning the cDNA encoding the Amb t V allergen from giant ragweed (*Ambrosia trifida*) pollen. *Gene* 101:231–238.

Grammer, L. C., C. R. Zeiss, I. M. Suszko, M. A. Shaughnessy, and R. Patterson. 1982. A double-blind, placebo-controlled trial of polymerized whole ragweed for immunotherapy of ragweed allergy. *J. Allergy Clin. Immunol.* 69:494–499.

Greene, W. K., K. Y. Chua, G. A. Stewart, and W. R. Thomas. 1990. Antigenic analysis of group I house dust mite allergens using random fragments of *Der p* I expressed by recombinant DNA libraries. *Int. Arch. Allergy Appl. Immunol.* 92:30–38.

Greenstein, J. L., J. P. Morgenstern, J. LaRaia, C. M. Counsell, W. H. Goodwin, A. Lussier, P. S. Creticos, P. S. Norman, and R. D. Garman. 1992. Ragweed immunotherapy decreases T cell reactivity to recombinant *Amb a* I.1. *J. Allergy Clin. Immunol.* 89:322 (Abstract).

Griffith, I. J., J. Pollock, D. G. Klapper, B. L. Rogers, and A. Nault. 1991a. Polymorphism of *Amb a* I and *Amb a* II, the major allergens of *Ambrosia artemisiifolia* (short ragweed). *Int. Arch. Allergy Appl. Immunol.* 96:296–304.

Griffith, I. J., P. M. Smith, J. Pollock, P. Theerakulpisut, A. Avjioglu, S. Davies, T. Hough, M. B. Singh, R. J. Simpson, L. D. Ward, and R. B. Knox. 1991b. Cloning and sequencing of *Lol p* I, the major allergenic protein of rye-grass pollen. *FEBS Lett.* 279:210–215.

Hochuli, E., W. Bannwarth, H. Dobeli, R. Gentz, and D. Stuber. 1988. Genetic approach to facilitate purification of recombinant proteins with a novel metal chelate adsorbent. *Bio/Tech.* 6:1321–1325.

King, T. P. 1972. Separation of proteins by ammonium sulfate gradient solubilization. *Biochemistry* 11:367–371.

King, T. P. 1976. Chemical and biological properties of some atopic allergens. *Adv. Immunol.* 23:77–105.

King, T. P. and P. S. Norman. 1986. Standardized extracts, weeds. *Clin. Rev. Allergy* 4:425–433.

King, T. P., P. S. Norman, and J. T. Connell. 1964. Isolation and characterization of allergens from ragweed pollen. II. *Biochemistry* 3:458–468.

King, T. P., P. S. Norman, and L. M. Lichtenstein. 1967. Isolation and characterization of allergens from ragweed pollen. IV. *Biochemistry* 6:1992–2000.

King, T. P., P. S. Norman, and N. Tao. 1974. Chemical modifications of the major allergen of ragweed pollen, Antigen E. *Immunochem.* 11:83–92.

King, T. P., A. Alagon, L. Kochoumian, J. Kuan, A. K. Sobotka, and L. M. Lichtenstein. 1981. Limited proteolysis of Antigen E and K from ragweed. *Arch. Biochem. Biophys.* 212:127–135.

Klapper, D. G., L. Goodfriend, and J. D. Capra. 1980. Amino acid sequence of ragweed allergen Ra3. *Biochemistry* 19:5729–5734.

Kuo, M., X. Zhu, X.-B. Yu, J. Pollock, R. S. Koury, T. P. King, I. J. Griffith, and B. L. Rogers. 1992. Immunochemical characterization of the ragweed allergen *Amb a* II. *J. Allergy Clin. Immunol.* 89:318 (Abstract).

Larsen, J. N., P. Stroman, and H. Ipsen. 1992. PCR based cloning and sequencing of isogenes encoding the tree pollen major allergen *Car b* I from *Carpinus betulus*, hornbeam. *Mol. Immunol.* 29:703–711.

Leiferman, K. M., G. J. Gleich, and R. T. Jones. 1976. The cross-reactivity of IgE antibodies with pollen allergens II. Analyses of various species of ragweed and other fall weed pollens. *J. Allergy Clin. Immunol.* 58:140–148.

Lichtenstein, L. M., P. S. Norman, W. L. Winkenwerder, and A. G. Osler. 1966. *In vitro* studies of human ragweed allergy: changes in cellular and humoral activity associated with specific desensitization. *J. Clin. Invest.* 45:1126–1136.

Lichtenstein, L. M., P. S. Norman, and W. L. Winkenwerder. 1968. Clinical and *in Vitro* studies on the role of immunotherapy in ragweed hay fever. *Amer. J. Medicine*, 44, 514–524.

Litwin, A., A. J. Pesce, T. Fischer, M. Michael, and J. G. Michael. 1991. Regulation of the human immune response to ragweed pollen by immunotherapy. A controlled trial comparing the effect of immunosuppressive peptic fragments of short ragweed with standard treatment. *Clin. Exp. Allergy* 21:457–465.

Lowenstein, H. and D. G. Marsh. 1983. Antigens of *Ambrosia elatior* (short ragweed) pollen. III. Crossed radioimmunoelectrophoresis of ragweed-allergic patients' sera with special attention to quantification of IgE responses. *J. Immunol.* 130:727–731.

Marsh, D. G., L. Goodfriend, T. P. King, H. Lowenstein, and T.A.E. Platts-Mills. 1988. Allergen nomenclature. *Int. Arch. Allergy Appl. Immunol.* 85:194–200.

Mohapatra, S. S., R. Hill, J. Astwood, A. K. Ekramoddoullah, E. Olsen, A. Silvanovitch, T. Hatton, F. T. Kisil, and A. H. Sehon. 1990. Isolation and characterization of a cDNA clone encoding an IgE-binding protein from Kentucky bluegrass (*Poa pratensis*) pollen. *Int. Arch. Allergy Appl. Immunol.* 91:362–368.

Mole, L. E., L.Goodfriend, C. B. Lapkoff, J. M. Kehoe, and J. D. Capra. 1975. The amino acid sequence of ragweed pollen allergen Ra5. *Biochemistry* 14:1216–1220.

Norman, P. S., W. L. Winkenwerder, and L. M. Lichtenstein. 1968. Immunotherapy of hay fever with ragweed antigen E: comparisons with whole pollen extract placebos. *J. Allergy* 42:93–108.

O'Hehir, R. E., R. D. Garman, J. L. Greenstein, and J. R. Lamb. 1991. The specificity and regulation of T-cell responsiveness to allergens. *Ann. Rev. Immunol.* 9:67–95.

Olsen, E., L. Zhang, R. D. Hill, F. T. Kisil, A. H. Sehon, and S. S. Mohapatra. 1991. Identification and characterization of the *Poa p* IX group of basic allergens of Kentucky bluegrass pollen. *J. Immunol.* 147:205–211.

Olson, J. R. and D. G. Klapper. 1986. Two major human allergenic sites on ragweed pollen allergen antigen E identified by using monoclonal antibodies. *J. Immunol.* 136:2109–2115.

Paull, B. R., G. J. Gleich, and M. Z. Atassi. 1979. Structure and activity of ragweed antigen E. II. Allergenic crossreactivity of the subunits. *J. Allergy Clin. Immunol.* 64:539–545.

Perez, M., G. Y. Ishioka, L. E. Walker, and R. W. Chestnut. 1990. cDNA cloning and immunological characterization of the rye grass allergen *Lol p* I. *J. Biol. Chem.* 265:16210–16215.

Rafnar, T., I. J. Griffith, M.-c. Kuo, J. F. Bond, B. L. Rogers, and D. G. Klapper. 1991. Cloning of *Amb a* I (Antigen E), the major allergen family of short ragweed pollen. *J. Biol. Chem.* 266:1229–1236.

Rogers, B. L., J. P. Morgenstern, I. J. Griffith, X.-B. Yu, C. M. Counsell, A. W. Brauer, T. P. King, R. D. Garman, and M.-c. Kuo. 1991. Complete sequence of the allergen *Amb a*II: recombinant expression and reactivity with T cells from ragweed allergic patients. *J. Immunol.* 147:2547–2552.

Rohac, M., T. Birkner, I. Reimitzer, B. Bohle, R. Steiner, M. Breitenbach, D. Kraft, O. Scheiner, F. Gabl, and H. Rumpold. 1991. The immunological relationship of epitopes on major tree pollen allergens. *Mol. Immunol.* 28:897–906.

Schad, V. C., R. D. Garman, J. L. Greenstein. 1991. The potential use of T cell epitopes to alter the immune response. *Seminars Immunol.* 3:217–224.

Singh, M. B., T. Hough, P. Theerakulpisut, A. Avjioglu, S. Davies, P. M. Smith, P. Taylor, R. J. Simpson, L. D. Ward, J. McCluskey, R. Puy, and R. B. Knox. 1991. Isolation of cDNA encoding a newly identified major allergenic protein of rye-grass pollen: intracellular targeting to the amyloplast. *Proc. Natl. Acad. Sci. USA* 88:1384–1388.

Silvanovich, A., J. Astwood, L. Zhang, E. Olsen, F. Kisil, A. Sehon, S. Mohapatra, and R. Hill. 1991. Nucleotide sequence analysis of three cDNAs coding for *Poa p* IX isoallergens of Kentucky bluegrass pollen. *J. Biol. Chem.* 266:1204–1210.

Smith, J. J., J. R. Olson, and D. G. Klapper. 1988. Monoclonal antibodies to denatured ragweed pollen allergen *Amb a* I: Characterization, specificity for the denatured allergen, and utilization for the isolation of immunogenic peptides of *Amb a* I. *Mol. Immunol.* 25:355–365.

Vercelli, D., and R. S. Geha. 1989. Regulation of IgE Synthesis in Humans. *J. Clinical Immunol.* 9:75–83.

12

The *Amb* V Allergens from Ragweed

Thorunn Rafnar, William J. Metzler, and
David G. Marsh

Introduction

Ragweed pollen contains multiple allergenic components that differ in their ability to elicit an allergic response in humans. There are the "major" allergens, such as *Amb a* I, to which the great majority of ragweed-allergic individuals develop antibodies. Conversely, there are proteins that are reactive in a small percentage (10–20%) of ragweed-allergic subjects, even though they can be highly immunogenic in susceptible individuals. These are referred to as "minor" allergens. A group of minor allergens, the *Amb* Vs, has been used as a model to study the genetics of human immune response. The *Amb* Vs were selected for two primary reasons. First, the immunizing doses are extremely low; for *Amb a* V the dosage is probably less than 60 ng per year in the Baltimore area (Marsh, 1975). Under these conditions approximately 10% of the ragweed-allergic population develop IgE antibody and a somewhat greater portion develop IgG antibody to *Amb a* V (Marsh et al., 1982a). This condition mimics the responder/nonresponder delineation observed in experimental animals. Second, the *Amb* Vs' small size and simple structure suggested that immune recognition might be limited to a single, or very few, immunodominant epitopes. Thus, in the polymorphic human population, only individuals with a particular genetic makeup would be responsive to these allergens. Recently, studies of the *Amb* Vs have been expanded to include analysis of the molecular interactions that occur during immune response to inhaled allergens. Particular attention has been paid to the mechanisms of antigen processing and T-cell activation.

Structural and Biochemical Properties of the *Amb* Vs

The allergens from the two most abundant North American species of ragweed, *Amb a* V from short ragweed (*Ambrosia artemisiifolia*, formerly Ra5) and *Amb t* V from giant ragweed (*Ambrosia trifida*) are the most extensively studied *Amb* V allergens (Figure 12-1). Studies have recently included a third homologue, *Amb p* V from western ragweed (*Ambrosia psilostachya*). The *Amb* Vs are among the smallest pollen allergens defined to date, consisting of a single polypeptide chain of only 40–45 amino acids with no detectable carbohydrate or lipid. *Amb a* V and *Amb t* V are basic proteins of 5,000 and 4,400 molecular weight, respectively (Lapkoff and Goodfriend, 1974; Roebber et al., 1985). *Amb a* V and *Amb t* V have amino acid sequence identity of about 45%, with each protein containing 8 cysteine residues. All of the cysteines participate in disulfide-bond

```
                                                     -22
Amb a V                        attactttgttaattttttatat ATG AAT AAT
                                                     Met Asn Asn
Amb t V                        acatttcatagtttaaagaaattatc ATG AAG AAC
                                                     Met Lys Asn

                                    -19              -15
Amb a V  --- --- --- --- --- --- --- --- --- -GA AAA AAT GTC TCG TTT
         --- --- --- --- --- --- --- --- --- --- Glu Lys Asn Val Ser Phe
Amb t V  ATA TTT ATG CTT ACA CTT TTT ATT CTT ATT ATT ACT TCG ACC ATT
         Ile Phe Met Leu Thr Leu Phe Ile Leu Ile Ile Thr Ser Thr Ile

              -10                    -5               -1
Amb a V  GAA TTT ATA GGA TCC ACA GAT GAA GTC GAT GAA ATA AAA --- ---
         Val Phe Ile Gly Ser Thr Asp Glu Val Asp Glu Ile Lys --- ---
Amb t V  AAG GCT ATA GGA TCC ACA AAT GAA GTC GAT GAA ATA AAA CAA GAA
         Lys Ala Ile Gly Ser Thr Asn Glu Val Asp Glu Ile Lys Gln Glu

          1               5               10              15
Amb a V  --- TTA GTG CCG TGT GCC TGG GCG GGG AAC GTT TGT GGT GAA AAG CGT
         --- Leu Val Pro Cys Ala Try Ala Gly Asn Val Cys Gly Glu Lys Arg
Amb t V  GAC GAT GGA CTT TGT --- TAT GAG GGG ACC AAT TGT GGT AAA GTG GGC
         Asp Asp Gly Leu Cys --- Tyr Glu Gly Thr Asn Cys Gly Lys Val Gly

          16              20              25              30
Amb a V  GCA TAT TGT TGT AGC GAC CCA GGG CGG TAC TGT CCT TGG CAA GTG
         Ala Tyr Cys Cys Ser Asp Pro Gly Arg Tyr Cys Pro Try Gln Val
Amb t V  AAA TAC TGT TGT AGC CCC ATT GGG AAG TAC TGT --- --- --- ---
         Lys Tyr Cys Cys Ser Pro Ile Gly Lys Tyr Cys --- --- --- ---

          31              35              40              45
Amb a V  GTT TGT TAC GAA TCC AGC GAA ATA TGC AGC AAA AAA TGC GGC AAA
         Val Cys Tyr Glu Ser Ser Glu Ile Cys Ser Lys Lys Cys Gly Lys
Amb t V  GTC TGT TAT GAT TCC AAG GCA ATA TGC AAC AAA AAT TGT ACT TAA
         Val Cys Tyr Asp Ser Lys Ala Ile Cys Asn Lys Asn Cys Thr ---

Amb a V  ATG CGG ATG AAC GTC ACT AAG AAC GCA ATA TAA
         Met Arg Met Asn Val Thr Lys Asn Ala Ile
Amb t V  --- --- -tg aat gtc act aag cac aca ata taa

Amb a V  aattaattgcagtttaaatgcttatgcagttatctattaaatgaaataatcgtttatat(a)n
Amb t V  aaataattacagttt(a)n
```

Figure 12-1 Nucleotide sequences of the cDNAs of *Amb a* V and *Amb t* V with translated amino acid sequences. The identical codons and stop codons are shown in bold type. The postulated amino acid sequences of the signal peptides deduced from the nucleotide sequences are shown in italics. The highly homologous regions at the 5′- and 3′-ends are underlined. The *Amb a* V translation initiation codon (ATG) at position minus 22 is putative. (Adapted from Ghosh et al., 1991, 1993a)

pairing, conferring resistance to heat-denaturation that is commonly observed in small highly disulfide-bonded molecules (Baer et al., 1980).

The high degree of sequence homology among the *Amb* V allergens suggests that the molecules adopt similar three-dimensional folding patterns. This has been verified experimentally. The three-dimensional solution structures of *Amb a* V and *Amb t* V have been determined by two-dimensional NMR spectroscopy (Figure 12-2; Metzler et al., 1992a, 1992b). Both molecules have been found to contain a small segment of antiparallel β-sheet, a C-terminal α-helix, and several loops. At the interface of the helix and sheet is a rigid core, comprised mainly of hydrophobic residues and two disulfide bonds. Although the structures of the two molecules are topologically similar, significant differences exist in the packing of side chains in the hydrophobic core of the molecules (Metzler et al., 1992b).

The location and function of the *Amb* Vs in the pollen is, as yet, unknown. *Amb a* V is released rapidly from pollen in an aqueous solution (completed within 4 minutes) (Marsh et al., 1981), suggesting that it may reside in the exine layer of the pollen grain and possibly has a role in the initial steps of fertilization.

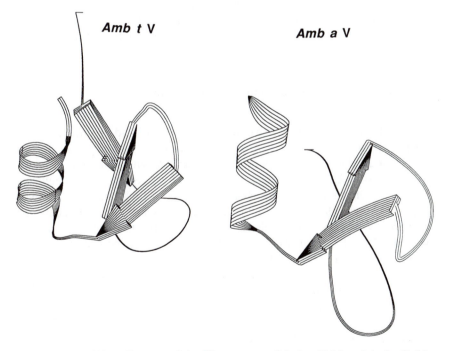

Amb t V   **Amb a V**

Figure 12-2 Ribbon diagrams of the 3D structures of *Amb a* V (a) and *Amb t* V (b), derived by two-dimensional NMR spectroscopy.

Based on their pattern of closely spaced disulfides, it has been suggested that the *Amb* Vs were members of a class of proteins related to wheat germ agglutinin and erabutoxin-b that adopt the "toxin-agglutinin fold". The disulfide pairs of the *Amb* Vs (and overall three-dimensional structures) have recently been determined to be different from those of the "toxin-agglutinin" proteins (Metzler et al., 1992a, 1992b); thus the biological function of the *Amb* Vs is also expected to be unrelated to these proteins. The *Amb* Vs show remarkable three-dimensional similarity to the Kunitz trypsin inhibitors (D. Bassolino-Klimas and W. J. Metzler, unpublished results); however, despite the structural similarity, *Amb* Vs have not been found to show any inhibitory activity for trypsin (T. Rafnar, unpublished observation).

Despite their overall structural homology, *Amb a* V and *Amb t* V are essentially non-cross-reactive antigenically, suggesting that the binding regions for antibodies are dissimilar (Roebber et al., 1985). Immunodominant B-cell epitopes tend to be primarily on the surface of native molecules, in regions subject to the fastest evolutionary change (Benjamin et al., 1984). Examination of the structures of *Amb a* V and *Amb t* V indicates that the less well-conserved residues tend to be located at the exposed surfaces. The greatest homology between *Amb a* V and *Amb t* V is in regions of the molecules that are important for their overall structural integrity; in general, the conserved amino acids are buried within the molecule and are relatively inaccessible to antibody binding. This pattern is commonly seen among structurally related molecules.

HLA Association of Human Immune Response to the *Amb* Vs

In groundbreaking studies, Marsh et al. (1982a, 1982b) found significant associations between responsiveness to *Amb a* V and the MLR specificity Dw2. In their studies of unrelated Caucasians, 95% of 38 subjects with serum IgE to *Amb a* V had HLA-Dw2, whereas only 22% of 139 *Amb a* V-IgE negative individuals had Dw2. Similar results were obtained for IgG. After immunotherapy with ragweed extract, all 22 Dw2+ subjects in this study had made IgG antibodies to *Amb a* V whereas the prevalence of IgG Ab responders in Dw2− subjects was only 28%. Dw2 was also found to be strongly associated with the quantity of IgG produced. Dw2 responders were also positive for the serologically defined specificity DR2 (i.e., were found to be DR2.2+). The association between DR2 and IgE mediated skin-test reactivity to *Amb a* V was confirmed by Goodfriend et al. (1985) and Blumenthal et al. (1992).

To define further this polymorphism of the *Amb a* V response, Zwollo et al. (1991) and Marsh et al. (1990a) used the molecular biology techniques of restriction fragment length polymorphism (RFLP) mapping, sequence-specific oligonucleotide (SSO) typing, and DNA sequencing of the polymorphic second exon regions of DRB1, DRB3/4, DQA, and DQB. These studies suggested that the phenotype DR2.2 is associated with an MHC class II molecule that is the major restriction element for *Amb a* V. This finding has recently been confirmed in an international HLA and allergy study (Marsh et al., 1993).

Measurements of *Amb t* V-specific IgG antibodies in subjects who had received immunotherapy with giant ragweed extract demonstrated that responsiveness to *Amb t* V was also significantly associated with DR2.2 (Roebber et al., 1985). However, association between DR2 and skin test reactivity to *Amb t* V was not found to be statistically significant in a group of Canadian patients, although 7 of 10 *Amb t* V-sensitive individuals typed DR2 (Goodfriend et al., 1985). Strikingly, both studies reported that all individuals that were sensitive to both *Amb a* V and *Amb t* V possessed DR2/Dw2. A systematic analysis of IgG responses to sequential, low doses of *Amb a* V, *Amb t* V and *Amb p* V to matched pairs of 12 DR2+ and 12 DR2- ragweed allergic subjects confirmed the association of all three homologues to DR2 (Marsh et al. 1990b).

T-cell studies have now provided conclusive evidence for a cause-and-effect relationship between DR2 and response to *Amb a* V. Analysis of the MHC class II restriction of a T-cell line and three T-cell clones from a DR2.2+, ragweed-allergic individual showed that the T cells proliferated only when either DR2.2 or DR2.12 was present on the antigen-presenting cells (Huang et al., 1991). Moreover, the responses of the T-cell line and clones were abolished by anti-DR but not anti-DQ or anti-DP antibodies (Huang and Marsh, 1991). Specifically, it was shown that the monoclonal antibody Hu30 (anti-DRα/$\beta_1$2) blocked, in a dose-dependent manner, the T-cell responses to *Amb a* V. Taken together, these data confirm that the heterodimers DR (α,β1*1501) and DR(α,β1*1502) are the restriction elements for *Amb a* V.

Subsequent studies, using the *Amb a* V-specific clones described above, have shown that *Amb a* V, *Amb t* V and *Amb p* V may share similar regions that interact with DR (agretopes) but possess distinct T-cell epitopes. The *Amb a* V homologues were unable to stimulate the T-cell clones directly but were able to inhibit, in a dose-dependent manner, the stimulation of the T cells by *Amb a* V (Huang and Marsh, 1991). To define the agretopes more precisely, Huang and Marsh (1991) tested a series of synthetic peptides for their ability either to stimulate directly the *Amb a* V-T cells or to inhibit stimulation of the *Amb a* V-T cells by *Amb a* V. The results of this study indicated that a peptide spanning amino acids 31–44 in *Amb a* V was able to block presentation of *Amb a* V to the T cells but could not stimulate the *Amb a* V-T cells directly. This result suggests that the peptide can bind to DR(α,β1*1501) and that either the cysteine residues are required for T-cell recognition or that a part of the T-cell epitope lies N-terminally to the peptide. More recently, a second synthetic peptide spanning amino acids 26–39 in *Amb a* V was found to stimulate the *Amb a* V-specific T-cell clones to an extent equal or to greater than the native molecule (Rafnar et al., 1992). The results of these studies indicate that at least one of the *Amb a* V residues between positions 26 and 30 are crucial for T-cell recognition.

Comparison of the amino acid sequence of *Amb a* V with the sequence of *Amb t* V (which cannot stimulate the *Amb a* V-specific T cells) supports this hypothesis (Figure 12-1), in particular, amino acids 27–30 in *Amb a* V have no counterparts in *Amb t* V.

Molecular Biology of the *Amb* Vs

The *Amb* Vs represent minor fractions of the total proteins in ragweed pollen extracts (about 0.1%). The isolation and purification of any particular *Amb* V in a quantity sufficient for immunological studies requires large amounts of pollen and multiple purification steps. This tedious process has been described elsewhere (Roebber et al., 1985). To circumvent this problem, Ghosh et al. (1991, 1993a) have cloned the cDNAs for both *Amb t* V and *Amb a* V from their respective pollens. This cloning has facilitated the production of recombinant *Amb* V proteins in large amounts, free from contamination with other pollen proteins (Rafnar et al., 1992). The cloning and expression of the *Amb* Vs have presented considerable technical difficulties, but have provided significant advances; one is now able to produce recombinant mutant *Amb* Vs in vitro and analyze antigen processing and presentation, as well as the interactions between *Amb a* V and the specific T-cell receptor.

Conventional cDNA cloning involves the screening of cDNA libraries with either antibodies against the native protein, oligonucleotides corresponding to a known part of the protein sequence, or both. Unfortunately, small RNAs are particularly vulnerable to being lost or degraded during conventional cDNA cloning. Moreover, the structure of the *Amb* Vs is dependent on the formation of correct disulfide bonds, which are unlikely to form in the *E. coli* cytoplasm. Antibodies made to native *Amb* Vs may be unable to recognize the recombinant products with an affinity high enough for library screening. Not surprisingly therefore, extensive screening of pollen cDNA libraries with *Amb* V antibodies and degenerate oligonucleotide mixtures failed to detect any *Amb* V-specific sequences.

In an alternate method for cloning the *Amb* V genes, Ghosh et al., (1991, 1993a) used an anchor-PCR method to amplify *Amb a* V and *Amb t* V sequences directly from both pollen RNA and genomic DNA. Analysis of the *Amb a* V and *Amb t* V sequences obtained in this manner revealed a number of interesting features (Figure 12-1). Both molecules contain hydrophobic signal sequences required for secretion of the proteins. However, a single nucleotide deletion at position −56 in *Amb a* V moves the prospective ATG start codon at position

---

*These peptides were synthesized by substituting Ala residues for Cys residues. This was done to reduce the aggregation caused by intra- and interpeptide disulfide linkages.

−64 (and used by *Amb t* V) out-of-frame with the *Amb a* V protein sequence. This mutation was consistently found in many separate cloning experiments, both when *Amb a* V was amplified from pollen RNA and genomic DNA. It has been speculated that either an RNA-editing mechanism or translation frameshift mechanism may be involved in the synthesis of *Amb a* V or, alternatively, another initiation codon may be used (Ghosh et al., 1993a). A second unexpected finding was that the *Amb a* V DNA sequence extends 30 nucleotides beyond the prospective stop codon inferred from the protein sequences, i.e., instead of a stop codon at position 136 there is a Met codon. Thus the expressed *Amb a* V sequence would contain extra 10 amino acids at its C-terminus. It is possible that a trypsinlike enzyme may remove these 10 amino acids posttranslationally to yield the native 45 amino acid form. It can also be speculated that the 44 amino acid *Amb a* V, formerly isolated by traditional protein purification procedures (Roebber et al., 1982), may have arisen by a mutation of the Lys codon AAA at position 133 to the TAA stop codon.

Considerable sequence homology existed in the regions flanking the *Amb a* V and *Amb t* V coding sequences. This finding was of some surprise because the noncoding portions of a gene are generally under less evolutionary pressure to remain unchanged and therefore tend to be more polymorphic than the protein coding regions. It was speculated that these regions were important in regulating the expression of the *Amb* Vs and might be conserved among other closely related ragweed species. Oligonucleotide primers based on sequences in these regions were used to amplify *Amb* V sequences from western ragweed pollen (Ghosh et al., 1993b). These experiments have yielded cDNA clones encoding two major sequence forms. The first group of sequences, termed *Amb p* VA, is highly homologous to *Amb a* V (one or two amino acid differences) and contains the same features in their 3′ and 5′ ends observed in *Amb a* V, i.e., the out-of-frame start codon and 10 extra C-terminal amino acids. The second *Amb p* V form, *Amb p* VB, shares about 50% homology with *Amb a* V and is missing one of the four disulfide pairs observed in *Amb a* V. At present it is not clear why such variable sequences are present in western ragweed pollen.

Taken together, the information gathered from the cloning of the *Amb* Vs has indicated that these homologues contain different types of structural elements that are either highly conserved or are polymorphic. Also, *Amb p* V contains multiple isoforms that are expressed at the RNA level and quite probably at the protein level as well.

To facilitate structure-function studies, recombinant *Amb a* V (r*Amb a* V) and r*Amb t* V were produced in *E. coli* (Rafnar et al., 1992). r*Amb t* V was shown by inhibition studies and NMR spectroscopy to be indistinguishable from native *Amb t* V (n*Amb t* V). In contrast, although all of the allergic subjects tested were able to recognize some *Amb a* V determinants on the recombinant molecule, r*Amb a* V lacked at least one immunodominant B cell epitope present on n*Amb a* V. This suggested that the r*Amb a* V molecule was not completely correctly folded. The results of these experiments indicated that, although both r*Amb t* V and r*Amb a* V are good candidates for MCH/T cell epitope studies and diagnostic testing, only r*Amb t* V is useful for complete B-cell epitope analysis when expressed in this system. This was an unexpected, but very significant result, for it accentuates the need for extreme caution when using recombinant proteins for immunogenic analyses. It is crucial for the determination of the B-cell epitopes that the structure of the recombinant protein is identical to that of the native protein. This is particularly true for highly disulfide-bonded molecules which often form aggregates in the *E. coli* cytoplasm. It is possible that other eukaryotic expression systems that secrete the recombinant proteins, such as baculovirus or yeast, may be able to produce fully active r*Amb a* V.

Conclusions

Because of their small size and defined three-dimensional structure, the *Amb* Vs present an excellent model to study the relationship between structure and function of an airborne allergen, especially for B-cell determinants that are generally dependent on the three-dimensional structural integrity of the molecule. Similarly, the *Amb* Vs provide a simple model to study antigen processing and presentation of peptides to T cells by MHC class II molecules. The observation that *Amb a* V, *Amb t* V and *Amb p* V are all restricted by the same DR molecule suggests that intracellular processing of the molecules by human antigen presenting cells may be similar and that only a very limited number of the peptides generated during antigen processing are able to bind with high affinity to MHC class II molecules.

References

Baer, H., M. C. Anderson, R. Hale, and G. J. Gleich. 1980. The heat stability of short ragweed pollen extract and the importance of individual allergens in skin reactivity. *J. Allergy Clin. Immunol.* 66:281.

Benjamin, D. C., J. A. Berzofsky, I. J. East, F.R.N. Gurd, C. Hannum, S. J. Leach, E. Margoliash, J. G. Michael, A. Miller, E. M. Prager, M. Reichlin, E. E. Sercarz, S. J. Smith-Gill, P. E. Todd, and A. C. Wilson. 1984. The antigenic structure of proteins: a reappraisal. *Ann. Rev. Immunol.* 2:67.

Blumenthal, M., D. Marcus-Bagley, Z. Awdeh, B. Johnson, E. J. Yunis, and C. A. Alper. 1992. HLA-DR2, [HLA-B7, SC31, DR2], and [HLA-B8, SC01, DR3] haplotypes distinguish subjects with asthma from those with rhinitis only in ragweed pollen allergy. *J. Immunol.* 148:411.

Ghosh, B., M. P. Perry, and D. G. Marsh. 1991. Cloning the cDNA encoding the Amb t V allergen from giant ragweed (Ambrosia trifida) pollen. *Gene* 101:231.

Ghosh, B., M. P. Perry, T. Rafnar, and D. G. Marsh. 1993a. Cloning and expression of immunologically active recombinant Amb aV allergen of short ragweed (Ambrosia artemisiifolia) pollen. 150:5391.

Ghosh, B., M. P. Perry, T. Rafnar, D. G. Klapper and D. G. Marsh. 1993b. *J. Allergy Clin. Immunol.* (Abstract).

Goodfriend, L., A. M. Choudhury, D. G. Klapper, K. M. Coulter, G. Dorval, J. DelCarpio, and C. K. Osterland. 1985. Ra5G, a homologue of Ra5 in giant ragweed pollen: isolation, HLA-DR-associated activity and amino acid sequence. *Mol. Immunol.* 22:899.

Huang, S.-K. and D. G. Marsh. 1991. Human T-cell responses to ragweed allergens: Amb V homologues. *Immunology* 73:363.

Huang, S.-K., P. Zwollo, and D. G. Marsh. 1991. Class II major histocompatibility complex restriction of human T cell responses to short ragweed allergen, Amb a V. *Eur. J. Immunol.* 21:1469.

Lapkoff, C. B. and L. Goodfriend. 1974. Isolation of a low molecular weight ragweed pollen allergen: Ra5, *Int. Arch. Allergy Appl. Immunol.* 46:215.

Marsh, D. G. 1975. In *The Antigens,* Vol. 3 ed. M. Sela, pp. 271. Academic Press, New York.

Marsh, D. G., L. Belin, C. A. Bruce, L. M. Lichtenstein, and R. Hussain. 1981. Rapidly released allergens from short ragweed pollen. *J. Allergy Clin. Immunol.* 67:206.

Marsh, D. G., S. H. Hsu, M. Roebber, E. Ehrlich-Kautzky, L. R. Freidhoff, D. A. Meyers, M. K. Pollard, and W. B. Bias. 1982a. HLA-Dw2: A genetic marker for human immune response to short ragweed pollen allergen Ra5. I. Response resulting primarily from natural antigenic exposure. *J. Exp. Med.* 155:1439.

Marsh, D. G., D. A. Meyers, L. R. Freidhoff, E. Ehrlich-Kautzky, M. Roebber, P. S. Norman, S. H. Hsu, and W. B. Bias. 1982b. HLA-Dw2: A genetic marker for human immune response to short ragweed pollen allergen Ra5. II. Respoonse after ragweed immunotherapy. *J. Exp. Med.* 155:1452.

Marsh, D. G., P. Zwollo, S.-K. Huang, B. Ghosh, and A. A. Ansari. 1990a. *Cold Spring Harbor Symposium Quant. Biol.* 54:459.

Marsh, D. G., P. Zwollo, L. Freidhoff, D.B.K. Golden, A. A. Ansari, E. Ehrlich-Kautzky, D. A. Meyers, and C. L. Holland. 1990b. Studies of human immune response to the Amb V (Ra5) homologues. *J. Allergy Clin. Immunol.* 85:201.

Marsh, D. G., M. N. Blumenthal, T. Ishikawa, A. Ruffili, S. Sparholt, and L. R. Freidhoff. 1993. In *Proc. 11th International Histocompatibility Workshop and Conference,* eds. K. Tsuji, M. Aizawa, and T. Sasazuki, Oxford Univ. Press, Oxford.

Metzler, W. J., K. Valentine, M. Roebber, M. S. Friedrichs, D. G. Marsh, and L. M. Mueller. 1992a. Determination of the three-dimensional solution structure of the ragweed allergen Amb a V by nuclear magnetic resonance spectroscopy. *Biochemistry* 31:5117.

Metzler, W. J., K. Valentine, R. Roebber, D. G. Marsh, and L. M. Mueller. 1992b. Proton resonance assignments and three-dimensional solution structure of the ragweed allergen Amb a V by nuclear magnetic resonance spectroscopy. *Biochemistry* 31:8697.

Rafnar, T., B. Ghosh, W. J. Metzler, S.-K. Huang, M. P. Perry, L. M. Mueller, and D. G. Marsh. 1992. Expression and analysis of recombinant Amb a V and Amb t V allergens. Comparison with native proteins by immunological assays and NMR spectroscopy. *J. Biol. Chem.* 267:21119.

Roebber, M., D. G. Klapper, and D. G. Marsh. 1982. Two isoallergens of short ragweed component Ra5. *J. Immunol.* 129:120.

Roebber, M., D. G. Klapper, L. Goodfriend, W. B. Bias, S. H. Hsu, and D. G. Marsh. 1985. Immunochemical and genetic studies of Amb. t. V (Ra5G), an Ra5 homologue from giant ragweed pollen. *J. Immunol.* 134:3062.

Zwollo, P., E. Ehrlich-Kautzky, A. A. Ansari, S. J. Scharf, H. A. Erlich, and D. G. Marsh. 1991. Sequencing of HLA-D in responders and nonresponders to short ragweed allergen, Amb a V. *Immunogenetics* 33:141.

13

Characterization of a Dominant Epitope of the Major Allergens of *Parietaria*

Anna Ruffilli, Anna Scotto d'Ambusco, and C. De Santo

"Je suis la mauvaise herbe, braves gens, brave gens, qui pousse en liberté dans les jardins mal frequentés"

G. Brassens

Introduction

Recently physicochemical characterization of allergens isolated from native sources has been accompanied by molecular cloning of allergen genes. Several epidemiologically and clinically relevant pollen allergens, as well as those that have been used as models for classical studies, are now being studied at the level of recombinant DNA products (Bond et al., 1991; Breiteneder et al., 1989, 1992; Perez et al., 1990; Griffith et al., 1991a; Gosh et al., 1991; Valenta et al., 1991a, 1992a; Silvanovich et al., 1991; Rafnar et al., 1991; Rogers et al., 1991; Singh et al., 1991). This approach promises to facilitate structure/function analyses on the immune recognition of allergens. In addition, the wealth of information gathered on structure and genomic organization of allergens may prompt studies aiming at clarifying their function in the organisms of origin. By placing these proteins in their natural perspective, it is possible to achieve a more profound understanding of the basic mechanisms of allergy.

Recombinant DNA technology has been extensively used to delineate B-cell epitopes of protein antigens: The antigenicity of linear segments of natural proteins is often conserved in the chimeric fusion polypeptides. Until now this approach, when applied to the dissection of the antibody response to allergens, has met with a limited success. Yet, the definition of linear segments of allergens bearing epitopes for B and T cells, would provide precious tools for exploring the mechanisms of immune recognition.

Table 13-1 lists fragments of allergens defining IgE-binding epitopes. Allergens from which IgE binding peptidic fragments have been derived by synthesis or enzymatic cleavage include the major allergen of *Betula verrucosa, Bet v I,* the parvalbumin allergen of codfish, *Gad c I,* and ovalbumins (Elsayed et al., 1989; Elsayed 1980, 1981; Johnsen and Elsayed, 1990). Overlapping peptides encom-

Table 13-1 IgE Binding Fragments of Allergens

Enzymatic or Chemical Cleavage Fragments

Bet v I (17 kDa). A 10 res. peptide, inhibiting 76% of the IgE binding to the allergen (human pool, RAST inhib.) (Elsayed et al., 1989)

Gad c I (12.3 kDa). Two peptides (16 aa res), having partial aa homology, inhibiting 58% and 38.6% of the IgE binding to codfish extract (human pool, RAST inhib.). (Elsayed et al., 1980, 1981)

Gal d II (ovalbumin: 42.7 kDa). A peptide (17 res.), inhibiting 25% of the IgE binding to the allergen (human pool and two individual sera) (Johnsen and Elsayed, 1990).

Overlapping Peptides

Amb a III (12.3 kDa). Four synthetic peptides, binding IgE from a pool of allergic patients' sera (Atassi and Atassi, 1986)

Der p II (14.1 kDa). A peptide (14 res), binding (weakly) IgE from few individual sera of allergic patients (van't Hof et al., 1991)

Chi t I (comp.III) (16 kDa). Four IgE binding peptides, including an 11 aa peptide (serum pool). Mazur et al., 1990)

Recombinant DNA Products

Der p II (14.1 kDa) glutathione transferase fusion proteins expressing cDNA fragments, screened with sera of 58 allergic patients. Only some large fragments had a weak binding activity with IgE of some sera (Chua et al., 1991)

Der p I (25.4 kDa). Glutathione transferase fusion proteins expressing 16 peptides (40–50 aa), screened with individual sera. Peptides of less than 30 res. had low reactivity. (Greene et al., 1990, 1991; Greene and Thomas, 1992)

Lol p Ib (31 kDa). An IgE binding fragment of appr. 30 kDa, a breakdown product of a glutathione-S transferase fusion protein (Singh et al., 1991)

Par o I (14 kDa). A 26 bp cDNA expressing an IgE binding betagalactosidase fusion protein (Scotto d'Abusco et al., 1992)

passing the polypeptide sequences have been used to map B-cell epitopes of a minor pollen allergen of *Ambrosia, Amb a III,* (Atassi and Atassi, 1986), of the major mite allergens *Der p I* (Greene et al., 1990, 1991; Greene and Thomas, 1992) and *Der pII* (Chua et al., 1991), and of the chironomid allergen *Chi t I* (Mazur et al., 1990).

Recombinant DNA technology has been used to map dominant epitopes of the major mite allergens *Der p I* and *Der p II*. The recombinant polypeptides had been shown to maintain most (*Der p II*) or part (*Der p I*) of the IgE binding activity of the natural allergens (Greene and Thomas, 1992; Chua et al., 1990), which are known to possess conformation-dependent epitopes (Lombardero et al., 1990). Glutathione transferase fusion proteins, obtained by cloning *Der p I* cDNA fragments in the vector pGEX, had IgE binding activity only if the expressed fragment consisted of more than 30 amino acid residues (Greene et al., 1990; Greene et al., 1991; Greene and Thomas, 1992): There was evidence that the five IgE binding sites identified formed a larger discontinuous epitope. A similar analysis of *Der p II* peptides expressed by random cDNA fragments cloned into λgt11 or by large restriction endonuclease fragments expressed by pGEX vectors did not reveal substantial IgE binding (Chua et al., 1991). Finally, a study on *Lol p Ib*, a recently described allergen of *Lolium perenne,* identified initially as a recombinant DNA product, reported that a breakdown product of a glutathione-S transferase fusion protein of approximately 30 kDa had IgE binding activity (Singh et al., 1991).

Among described allergens, the major allergens of *Parietaria* have unique properties in that they possess a linear dominant antibody binding site. This chapter reports the characterization of this epitope encoded by 26-bp cDNA sequence.

Allergy to Parietaria

Parietaria is a plant of the family *Urticaceae* (nettle). In the Mediterranean area, its presence characterizes abandoned fields and gardens. The vulgar name, in English (pellitory-of-the-wall), Italian, French and Spanish, refers to its preference for acidic soil. *Parietaria* has been a constant companion of human settlements, appreciated once for its pharmacological properties and now by archaeologists as a marker to locate ancient town walls. Its popularity as a specific agent of pollinosis is relatively recent, the first case of *Parietaria*-specific respiratory allergy having been described in 1927.

In the Mediterranean regions, the species *P. judaica* L. (1763), *P. officinalis* L. (1753), *P. lusitanica* L. (1753), *P. cretica* L. (1753) and *P. mauritanica* Dur., occupy partially overlapping areas. Recently, the presence of this weed has been reported in regions with a temperate climate such as Australia and California (Raffaelli, 1977; D'Amato et al., 1991).

In the areas of its diffusion, *Parietaria* is usually a major cause of respiratory allergy. Table 13-2 reports the prevalence of skin sensitization to the extracts of *Parietaria,* Gramineae and mite for 94 unrelated Italian patients, along with that of the serum IgE antibodies reactivity to the major allergens *Par o I, Der p I,* and *Lol p V*. Allergy to *Parietaria* is characterized by a comparatively high proportion of monosensitized subjects.

Reported prevalence of *Parietaria*-specific allergy in northern Italy is 10–40%; in southern Italy and some areas of northern Italy such as the region of Liguria and the town of Brescia 70–80% in Spain 10–60%; in Mediterranean France 25% (D'Amato et al. 1991). The long duration of the flowering season, in some regions lasting from March to October; the preference for areas near human settlements, the small size of its pollen grains, and the high allergenicity of some of the pollen proteins are among the factors that account for the high allergenicity of this weed.

Allergy to *Parietaria* is an important cause of asthma. Table 13-3 reports the prevalence of asthma among the patients described in Table 13-2, which are here grouped according to skin sensitivity to *Parietaria,* mite, and grass.

Table 13-2 Prevalence of Skin Test Positivity,
Monosensitization and IgE Antibody Reactivity to
Major Allergens

Parietaria (67 patients)	
sk+	71.3%
monos	24.5%
Anti *Par o I* IgE ab	44.7%
Grass (43 patients)	
sk+	45.7%
monos	3.2%
anti *Lol p V* IgE ab	15.6%
Mite (35 patients)	
sk+	37.3%
monos	8.5%
anti *Der p I* IgE ab	13.8%

Note: Sk+, skin test positivity to extracts; monos, monosensitization. These data are a part of the HLA and Allergy study of the XI Histocompatibility Workshop, directed by D. G. Marsh.

Table 13-3 Prevalence of Asthma Among Subjects
Skin Test Positive to *Parietaria,* Mite, and Grass
Pollen Extracts in 94 Italian Allergic Patients

	Total	Asthma (%)
Parietaria sk+	67	44.8
monos.	23	39.1
Par o I IgE+	42	48
Grass sk+	43	53.5
monos	3	67
Lol p V IgE+	15	40
Mite sk+	35	60
monos	8	87.5
Der p I IgE ab+	13	76.9

Note: Sk+, skin test positivity to extracts; monos, monosensitization. These data are a part of the HLA and Allergy study of the XI Histocompatibility Workshop, directed by D. G. Marsh.

The major allergens of *P. judaica* and *P. officinalis* have been purified from the pollen extracts (Corbi et al., 1985; Ayuso et al., 1988; Giuliani et al., 1987; Oreste et al., 1991). Figure 13-1 shows the immunoblot analysis of the pollen extract of *P. officinalis* fractionated by gel filtration: IgE antibodies of a pool of sera of allergic patients bound with high affinity to the 14 kDa major allergen.

Purified *Par j I* and *Par o I* preparations contain closely related molecular

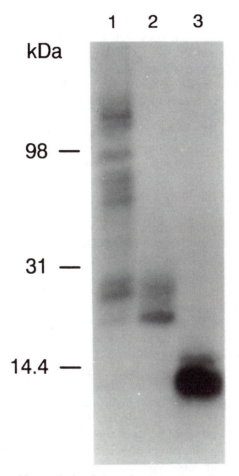

Figure 13-1 Immunoblot analysis of three fractions eluted from Sephadex G75 gel filtration of *P. officinalis* pollen extract using a serum pool of patients allergic to *Parietaria*.

species different in charge and electrophoretic mobility (Coscia et al., 1992; Polo et al., 1990). N-terminal sequences of *Par o I* and of three isoforms of *Par j I* have been determined and show high homology (Oreste et al., 1991; Polo et al., 1990) in agreement with previous findings documenting the extensive cross-reactivity of these allergens (Corbi et al., 1985).

Par o I and *Par j I* are highly allergenic. Table 13-2 shows that 61% of *Parietaria*-sensitive individuals had serum IgE antibodies specific for *Par o I*. This prevalence was higher than that observed for *Der p II*- and *Lol p V*-specific IgE among mite and grass allergic individuals respectively. The high allergenicity of these molecules was documented also by the observation that for approximately one-third of *Parietaria*-allergic patients they appeared to be the only or major target of serum antibodies, as documented by immunoblotting analyses (Ford et al., 1986; Coscia et al., 1989).

The genetic background of the populations exposed, and more specifically the frequency of particular HLA-class II alleles, could influence the predisposition to develop antibody responses specific for the major allergens (Scotto d'Abusco et al., 1989; Marsh, 1990).

Nothing is known of the physiological (in the plant) function of *Par o I* and *Par j I*. As for tissue distribution, mRNA encoding *Par o I* has been shown to be present in *P. judaica* inflorescences (Scotto d'Abusco, 1990).

A cDNA Sequence Encoding a Dominant Epitope

cDNA clones expressing a major epitope of *Par o I/Par j I* were obtained by immunoscreening a *P. judaica* pollen cDNA library. Poly (A)+ RNA was transcribed into double-stranded cDNA using Moloney murine leukemia virus reverse transcriptase and oligo (dT) primers. cDNA was ligated with EcoRI/Not I adaptors and inserted into dephosphorylated λgt11 arms. The recombinant phage library was immunoscreened using IgGs of rabbit anti-*Par j I* antiserum: Out of a total of 3×10^5 plaques screened, 35 were found to be positive.

Immunoblot analysis of the lysates confirmed that the recombinant polypeptides bound with high affinity Igs of rabbit anti-*Par j I* antiserum and IgEs of a pool of sera of allergic patients (human pool). Figure 13-2 represents the analysis of several clones with a rabbit antiserum specific for *Par o I* and shows that the expressed epitope was shared between the major allergens of *P. judaica* and *P. officinalis*. In agarose gel electrophoresis, EcoRI digested DNA of the clones shown in Figure 13-2 formed bands of approx. 100 bp.

Inserts DNA was purified from three immunopositive clones (Manfioletti and Schneider, 1988), subcloned into the plasmid vector pUC18, and sequenced (Sanger et al., 1977). Each clone yielded the same 26-bp nucleotide sequence (6a sequence) represented in Figure 13-3.

The 6a sequence contains an inverted (palindromic) and a direct repeat of an eight-bp segment. Elements consisting of short repeated nucleotide sequences, frequently found in the genomes of eukaryotes and prokaryotes, are known to promote instability. Direct repeats are a feature of a number of models of mutagenesis based on recombination, replication, or repair whereas inverted repeats, facilitating the formation of secondary DNA structures, are thought to be involved in insertion/deletion events creating polymorphisms.

The identity of the insert was validated using two independent approaches: PCR amplification of pollen cDNA primed by a degenerate oligonucleotide deduced from a known segment of the amino acid sequence of *Par o I* and analysis of the specificity of an antiserum raised by immunizing a rabbit with the fusion protein of one of the clones containing the 6a insert (6a FP).

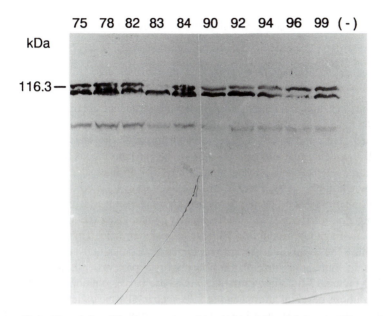

Figure 13-2 Reactivity of fusion proteins with rabbit anti *Par o I*. Immunoblot analysis of the lysates of recombinant clones.

Single-stranded cDNA obtained from total *P. judaica* or *P. officinalis* pollen RNA was PCR amplified using the degenerate oligonuclotide represented in Figure 13-4, deduced from the amino acid sequence: Val gln gly lys gly lys pro (res. 17–24 of *Par o I*) as primer for the (−) strand. A 23-bp portion of the 26-bp sequence was the primer for the (+) strand (Figure 13-4).

After 40 amplification cycles (annealing at 38°C, 4 min extension at 72°C and 2 min denaturation at 94°C) amplified DNA bands of approximately 120 bp were obtained. The nucleotide sequence of amplified DNA is shown in Figure 13-5.

The 126 bp sequence is not a PCR artifact as it was amplified in several independent experiments from *P. judaica* and *P. officinalis* pollen cDNA. It is apparently related to genes expressing *Par o I-Par j I;* its reading frame is interrupted by a stop codon (TAA). Possibly it represents a segment of pseudogene, deriving from recent duplication and conserved among species. Computer analysis (GenEMBL databank) did not reveal significant homology with known plant DNAs. The sequence was significantly similar (60.2% homology

GGCACGAGCTCGTGCCCGACTCGTGCC

overline: direct repeat.
underline: inverted repeat

Figure 13-3 The 6a sequence.

(+) strand   GTNCAA/GGGNAAA/GGAA/GAAA/GCC
(−) strand   ACGAGTCGGCACGAGCCTCGTGCC

Figure 13-4   Primers for PCR amplification.

in 123 bp) to a human telomere-associated repeat sequence obtained from human placental male DNA (Brown et al., 1990). Experiments are in progress to identify expressed genes and to define their relationships with the sequence described. Recently, evidence that allergens belong to polygene families has been accumulating (Rafnar et al., 1991; Bond et al., 1991; Griffith et al., 1991a, 1991b; Silvanovitch et al., 1991; Perez et al., 1990).

Additional evidence that the 6a nucleotide sequence is a segment of gene(s) encoding *Par o I/Par j I,* has been obtained by analyzing the specificity of the antiserum of a rabbit immunized with the 6a fusion protein. This antiserum (anti 6a), unlike the preimmune serum, contained antibodies that specifically bond *Par o I* in immunoblotting of the pollen extract proteins (not shown) and in ELISA. Figure 13-6 compares the dose response curves obtained by incubating serial dilutions of rabbit anti-6a and of a *Parietaria*-allergic patient serum known to have a high level of anti-*Par o I* IgG antibodies with solid phase *purified Par o I.* The binding affinity of purified IgGs of rabbit anti-6a and of rabbit anti-*P. officinalis* pollen extract with increasing amounts of solid phase *P. officinalis* and *P. judaica* pollen extract are compared in Figure 13-7. Preincubation of anti-6a with *P. judaica* pollen extract, but not with *E. coli* lysate, extensively inhibited the binding.

Characterization of 6a Epitope

Reactivity with Human Antibodies

The 6a sequence, in frame with β-galactosidase, encodes the following peptide: gly thr ser ser cys arg leu val pro. That a segment of nine amino acids is sufficient to form a sequential epitope is amply documented (Lerner, 1984). More recently, antibodies to peptides of five to seven amino acids have been shown to be able to provide sequence specific recognition of linear epitopes (Holley et al., 1991).

1
  GTCCAGGGAA AGGAGAAGCC AAGAGTGAAG AACCAACCCC AAGGAAAAGG TAAACCTACA
61
  GAAGTGTAA GCTAAGCTGC CCCATACcGT GCTCGACGAC GGGCACGAGC TCGTGCCGAC
121
  TCGTGCC

Figure 13-5   Amplified DNA sequence.

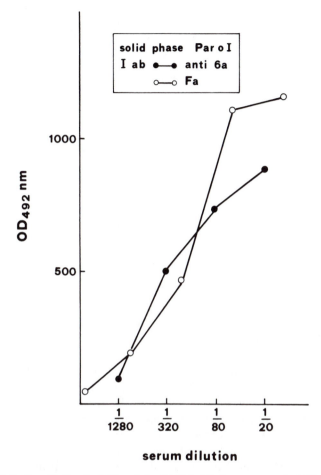

Figure 13-6 Specificity of rabbit anti-6a. ELISA. Binding curves of rabbit anti-6a and of a human serum (Fa) to solid phase purified *Par o I* (5 ng/well). II ab.; goat antirabbit IgG, rabbit antihuman IgG. The assay was performed as described by Voller et al. (1976).

The ability of the 6a fusion protein to absorb pollen allergen-specific antibodies from sera of *Parietaria*-allergic patients was assessed by ELISA and by double antibody radioimmunoassay (DARIA). Preincubation with appropriate amounts of the fusion protein extensively inhibited IgE- and IgG-antibody binding. Inhibition of the binding of IgE antibodies to solid phase *P. officinalis* pollen extract proteins (ELISA) and of IgG antibodies to purified radiolabeled *Par o I* in the DARIA are represented respectively in Figures 13-8 and 13-9.

Antibodies of the rabbit antiserum anti-6a competed with human IgG and IgE antibodies for the binding to *Par o I*. This is shown by the experiment summarized in Table 13-4. Anti-6a extensively inhibited the binding of antibodies of the

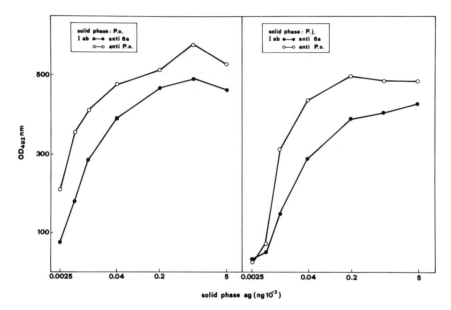

Figure 13-7 Specificity of rabbit anti-6a determined by ELISA. Binding curves of IgGs of rabbit anti-6a and *P. officinalis* pollen extract to solid phase *P. judaica* (right) and *P. officinalis* (left) pollen extracts (5 ng/well). II ab.: goat antirabbit IgG.

human pool to purified radiolabeled *Par o I* in DARIA. Reciprocally the human pool inhibited the binding of rabbit IgG antibodies (Table 13-4).

Previously several studies have addressed the issue of reactivity of other allergens with human antibodies. Antibody response to the mite allergen *Der p I* have been compared by Greene et al. (1992): antibodies of either isotypes bound significantly to several recombinant peptides of this allergen, delineating epitope maps which differed among individuals, but for each individual response were largely overlapping for the two isotypes. Mazur et al. (1990) have analyzed the binding affinity of antibodies from a human pool to a panel of peptidic fragments of the chironomid allegens *Chi t 1,* showing that an epitope, defined by an eleven amino acid res. peptide, was the main target of IgE and IgG antibodies.

An approach that has been used to investigate the structural relationship between the variable regions of immunoglobulins of different isotype sharing antigen specificity, is the analysis of idiotypic determinants. IgE and IgG antibodies in the human response to tetanus toxoid, rye pollen proteins, and shrimp allergens have been shown to share idiotypes (Geha, 1982; Bose et al., 1984; Nagpal et al., 1989). An analysis of the idiotypes of human IgE and IgG specific for house dust mite extract proteins has revealed both shared (Saint-Remy et al., 1988) and isotype specific idiotypes (Saint-Remy et al., 1986).

Studies on BALB/c mice include the analysis of the responses to synthetic

Table 13-4   Par o I DARIA Inhibition

Competing Serum	I ab	II ab	Inhibition (%)
anti 6a	H. pool	a H IgE	63
anti 6a	H. pool	a H IgG	83
H. pool	anti 6a	a R IgG	82

Note: The competing serum was added to I ab in the proportion 2/1, 6/1 and 8/1, for the determination of human IgE, human IgG, and rabbit IgG, respectively (final vol. of 100 ul). II ab.: rabbit antihuman IgE (a H IgE) or IgG (a H IgG), or goat antirabbit IgG (a R IgG). Radiolabeled *Par o I:* 3 ng (20.000 cpm.).

polypetide antigens (Dessain et al., 1980; Lowy et al., 1980), documenting idiotypic cross-reactivity of IgE and IgG antibodies and a study on the response to phosphorylcholine reporting sharing of idiotypes in the late but not in the early immune response (Weber et al. 1986).

Dominance

Although it is generally accepted that any portion of a protein can evoke an antibody response provided that suitable hosts and ways of immunization are used, the range of specificities of antibodies present in the serum during the individual response to a defined antigen is usually limited: Fine specificity analyses have shown that often the majority of antibodies is directed toward a few of the potential epitopes (Benjamin et al. 1984). The cellular and molecular basis of this phenomenon are not well understood.

The term dominance, describing a quality of epitopes of an antigen, in the scientific literature is used in various connotations. In the analysis of individual responses, dominance has been attributed to epitopes binding a relatively high proportion of the antigen-specific antibodies. This definition reflects the existence of mechanisms controlling the selection of B-cell epitopes, elimination or anergy of self-reactive clones, immune response genes control, idiotypic networks, and in the case of parasite antigens, coevolution with the hosts are among those postulated.

Dominance, in population studies, is attributed to epitopes recognized by antibodies of the majority of the individuals responding to a given antigen. Such epitopes are indicated here as epidemiologically dominant or highly allergenic/antigenic. In this context, dominance reflects the existence of mechanisms controlling the diversity of epitope-specific responsiveness among members of a population.

Dominance in individual response and dominance at the population level are two entirely different concepts, although some of the controlling mechanisms may be common. These include the influence of MHC class II genes, which in the response to some antigens have been shown to control epitope-specific antibody responsiveness (review Berzofsky 1984). The diversity of the population frequen-

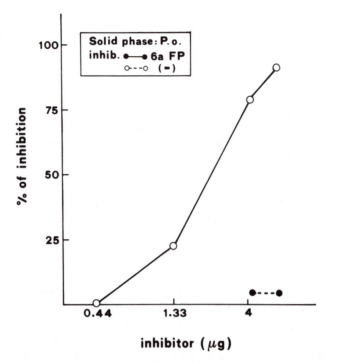

Figure 13-8 Inhibition of human IgE antibody binding of 6a fusion protein by ELISA. The patients serum pool (1/3 dilution) was preincubated with the amount of the 6a fusion protein indicated in the abscissae. II ab.: radiolabeled goat antihuman IgE.

cies of distinct alleles at these loci may influence the antigenicity of specific epitopes.

The ability of the 6a fusion protein to absorb a large proportion of Parietaria allergen-specific antibodies from a serum pool (Figs. 13-8 and 13-9) suggested that the majority of the sera forming the pool possessed antibodies specific for this epitope. Concordantly, competition with rabbit anti-6a inhibited extensively the binding of IgE and IgG antibodies of the pool to *Par o I* (Table 13-4). The sera of 13 allergic patients were monitored for the presence of anti-6a IgE antibodies by antibody competition ELISA. The results, summarized in Figure 13-10, show that rabbit anti-6a inhibited for each serum a large proportion of the IgE binding to solid phase *P. officinalis* pollen allergens. Taken together, the results of these experiments indicate that the 6a epitope is epidemiologically dominant.

That major allergens have epidemiologically dominant epitopes is not a tautology. Thus, for example, the analysis of the fine specificity of individual antibody responses to *Der p I* performed by Greene et al. (1991) did not reveal epidemiologically dominant epitopes: IgE and IgG antibodies bound to several members of

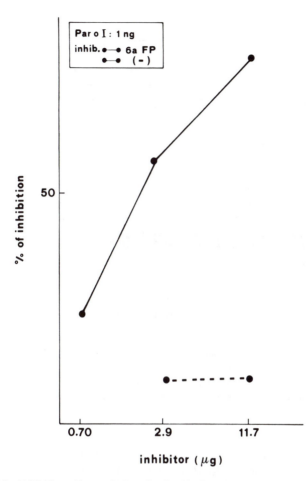

Figure 13-9   Inhibition of human IgG antibodies binding of 6a fusion protein by DARIA. Inhibition of the binding to radiolabeled *Par o I* (1 ng). The serum of an allergic patient (dil. 1/10) was preincubated with the amount of the 6a fusion protein indicated in the abscissae. II ab.: goat antihuman IgG. The assay was performed essentially according to Platts-Mills et al. (1978).

a panel of overlapping recombinant peptides but epitopes dominant in individual responses differed among the eight sera analyzed. A study by Lin et al. (1990) has shown that four mAbs-defined epitopes of *Poa p I* inhibited 30 to 55% of the IgE binding from three sera: one of these mAbs had previously been shown to inhibit 70% of the binding from a pool of six sera (Lin et al., 1988).

Putative epidemiologically dominant epitopes include the linear epitopes defined by two peptides of *Gad c I, which* inhibited respectively about 58 and 71% of the IgE binding to the solid phase allergen for five sera (Elsayed et al., 1980,

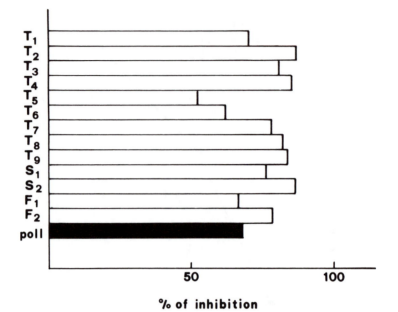

% of inhibition

Figure 13-10 Inhibition of allergenicity of the 6a epitope by ELISA. Competition between rabbit anti-6a and IgE antibodies of 13 allergic patients and of a pool. Wells containing *P. officinalis* pollen extract (5 μg) were incubated with rabbit anti-6a (dil. 1/2) prior to the addition of the human sera (individual sera), or the assay was performed with a 1/1 mixture of the competing sera (1/2 dil.) (serum pool). II ab.: radiolabeled anti human IgE.

1981) and mAbs-defined conformational epitopes. A *Der f I* epitope was defined by a mAb which inhibited 40–70% of the IgE binding from five individual sera (Le Mao et al., 1992). Mourad et al. (1989) have described two mAbs, defining *Lol p I* epitopes, which inhibited a large proportion of the IgE binding from twelve sera.

For other allergens, epidemiological dominance of specific epitopes is supported by studies using pools of sera. These epitopes include a linear epitope of *Bet v I*, defined by a 10 amino acid residue peptide which absorbed 76% of the human IgE specific for the allergen (Elsayed et al., 1989), and several mAbs defined epitopes. Two mAbs, defining conformation-dependent epitopes of *Amb a I*, inhibited each 50% of the IgE antibody binding to the allergen (Olson and Klapper 1986; Smith et al., 1988). Two Mabs defining two epitopes of grass Group I allergens inhibited cumulatively 80% of the IgE binding to *Fes e I*, and a Mab defining an epitope of *Sin a I*, a 14-kDa major allergen of *Sinapsis alba*, inhibited 50% of the IgE binding to this allergen (Esh and Klapper, 1989; Mourad et al., 1989; Menendez-Arias et al., 1990). Finally, Marc-Series et al. 1990 have described a mAb, defining an epitope of *Bet v I,* which inhibits to 49% the human IgE binding to this allergen.

These observations should be confirmed by analyses of individual antibody responses. It is important to consider that it has been repeatedly pointed out that the highly inhibitory activity displayed by mAbs in antibody competition assays could reflect the presence in human sera of antiidiotypic antibodies cross-reacting with epitopes of the variable region of murine immunoglobulins (Mourad et al., 1988; Castracane and Rocklin 1988).

The data presented in the preceding paragraphs strongly suggest that the antibody response to the allergens of *Parietaria* has a restricted specificity and is dominated by IgE and IgG antibodies directed towards the 6a epitope. This conclusion, suggested by the experiments with pooled sera, showing that rabbit anti-6a extensively competed with human IgE and IgG antibodies for the binding to purified radiolabeled *Par o I* (DARIA: Table 13-4) and the 6a fusion protein absorbed 90% of IgE antibodies binding to *P. officinalis* pollen allergens (ELISA: Fig. 13-8), was supported by the analysis of individual sera. Absorption with 6a fusion protein inhibited 67% of the binding of IgGs of an individual serum to purified *Par o I* (DARIA: Fig. 13-9); in antibody competition ELISA, rabbit anti-6a inhibited from 53 to 67% of the binding of human IgE antibodies to *P. officinalis* pollen allergens from 13 sera (Fig. 13-10).

Although observations obtained using different assays and distinct experimental designs consistently supports the notion that a major fraction of the antibodies specific for the allergens of *Parietaria* were directed toward the 6a epitope, it is important to stress that possibly the repertoire of specificities monitored in these assays is only a portion of that of the native allergen. For example, there is evidence that solid phase binding may alter the conformation of proteins: thus it has been shown that the repertoire of epitopes of *Amb a I* available for antibody binding in ELISA is different according to whether this allergen is bound to polyvinylchloride or to solid phase mAbs (Olson and Klapper, 1986; Le Mao et al., 1992; Mourad et al., 1988, 1989).

That the fine specificity of the human antibodies to allergens is restricted has been suggested by other authors. A study by Mourad et al. (1989) described two mAbs, defining *Lol p I* epitopes, each of which inhibited a large proportion of the IgE binding from 12 sera.

MAbs specific for epitopes of *Lol p I* (Esh and Klapper, 1989), *Poa p I* (Lin et al., 1988) and *Sin a I* (Menéndez-Arias et al., 1990) inhibiting a large proportion of the binding of IgE of human pools to the respective allergens have been described. Elsayed et al. (1980, 1981, 1989) have shown that short peptides defining linear epitopes of *Gad c I* and *Bet v I,* (Table 13-1) were able to absorb a large proportion of the IgE from human pools specific for the respective allergens. The restriction of the fine specificity repertoire of antibodies displayed by serum pools does not prove that a similar restriction exists in individual responses as it could reflect the presence of epidemiologically dominant epitopes.

The delineation of individual patterns of reactivity to epitopes of *Der p I* and *Bet v I* has failed to reveal dominance of specific epitopes. The analysis of epitope-specific antibody response of eight patients to *Der p I* (Greene et al., 1991) indicated that individual sera possessed antibodies toward several of the epitopes defined by a panel of recombinant peptides. A study of 38 tree allergic patients showed that IgE antibodies of individual sera identified several different epitopes of tree Group I allergens (Ipsen et al., 1992).

Evidence suggesting restriction of the specific repertoire of IgE antibodies and oligoclonality of IgE producing B cells come from different areas of study. One line of evidence is provided by the occurrence of idiotypes on allergen-specific antibodies, discussed previously. More recently, studies on the mechanisms of IgG H chain switch recombination have shown that the frequency of B lympho-cytes switching to IgE expression is relatively low. Oligoclonality may be impor-tant to avoid the triggering of mast cells induced by cross-linking between receptor-bound IgE and antigen which is hindered by the presence of IgE of too many distinct specificities per mast cell (Siebenkotten et al., 1992).

Species Cross-Reactivity

Not unexpectedly, considering the high cross-reactivity between the pollen ex-tracts of the two species (Corbi et al. 1985b), the 6a epitope was conserved among the major allergens of *P. judaica* and *P. officinalis*.

Earlier studies have shown that pollen extracts of distinct species of ragweed and grasses, when assayed for their immunochemical reactivity with IgE antibod-ies of allergic patients, are cross-reactive (Leiferman and Gleich, 1976; Leiferman et al., 1976). More recently, it has been shown that purified major pollen allergens of distinct species and genera of trees (Ispen et al., 1992; Rohac et al., 1991) and grass (Esh and Klapper, 1989; Mourad et al., 1988; Lin et al., 1990) are cross-reactive. So extensive is the cross-reactivity that for a large group of patients three recombinant timothy grass pollen allergens (*Phl p I, Phl p V,* and timothy grass profilin) and two recombinant birch pollen allergens have been shown to be sufficient to diagnose allergy to grass and trees (Valenta et al. 1991b; 1992b). Cross-reactivity is not limited to the human immune response, but has been observed also in the antibody response of rabbit immunized with tree (Ipsen et al., 1985) and grass (Standring et al., 1987) allergens.

Conclusions

The data presented are consistent with the conclusion that *Par o I* and *Par j I*, the major allergens of *P. judaica* and *P. officinalis* share a linear B-cell epitope (6a epitope), encoded by a 26 bp cDNA sequence, epidemiologically dominant in the IgE and IgG antibodies responses of *Parietaria*-allergic patients. Although the relative contribution of antibodies specific for the 6a epitope to the repertoire of antibodies specific for *Par o I* and *Par j I* cannot be ascertained, our data suggest that it is possible that for a considerable proportion of *Parietaria*-allergic patients, manifestations of clinical allergy are mediated primarily by antibodies with the 6a specificity. Considering that approximately one-third of *Parietaria*-allergic patients are monosensitized (Table 13-2) and that a further third produce an antibody response directed primarily toward the major allergens (Ford et al., 1986; Coscia et al., 1989), this proportion may be calculated as approaching 10% of all *Parietaria*-allergic individuals.

Recent advances have delineated IgE-specific T-helper cell function in the immune response to allergens (Parronchi et al., 1991; Yssel et al., 1992). The question of whether the ability to elicit IgE antibodies response is based on some intrinsic, although not obvious quality, of allergens or depends on specific ways and/or routes of immunization has been reformulated and appears to be amenable to experimental testing through structure/function studies. The dissection of allergens into epitopes is one relevant way of providing tools to this aim.

The wealth of data accumulated during the last few years on the structure of allergenic proteins and of the encoding genes and the dissection of the immune response into epitope-specific components prompt some considerations. At least three qualities common to allergens differentiate them from environmental antigens such as antigens of viral and bacterial parasites infecting man.

First, the differences among these categories of antigens consist in the relative restriction of the number of components of the sensitizing mixture that induce IgE antibody response. Following simultaneous immunization with a large number of distinct potentially antigenic components, man and experimental animals have been shown to produce an antibody response preferentially directed toward some of the components of the mixture. Thus, in the human response to cytomegalovirus, a ubiquitous virus that infects the majority of the population, of the 20–25 structural proteins detectable, only two induce detectable antibodies when the whole virus is used as an antigen (Urban et al., 1992). The cellular and molecular bases of this phenomenon, often referred to as intermolecular competition, are not well understood. A further restriction is imposed on the number of components eliciting IgE antibodies. For example, of the 52 components that induce IgG response in ragweed allergic patients, only 22 were shown to be allergenic (Lowenstein and Marsh, 1983). It has been observed that the range of antigenic specificity recognized may be further limited by the occurrence of cross-reactivity among distinct components of the pollen of the same species (Olsen et al., 1991; Mucci et al., 1992). This phenomenon suggests the existence of unknown mechanisms limiting the array of specificities of IgE antibodies.

Second, allergens are distinct from parasite antigens in exhibiting extensive cross-reactivity. Major allergens are cross-reactive among genera and species; profilin, a minor birch allergen has been shown to be cross-reactive from plants to man (Valenta et al., 1992a). In contrast, bacterial and viral antigens, are usually characterized by polymorphism of the antibody binding sites. Mechanisms of coevolution, in the battle of reciprocal adaptation between microbe and host, are thought to be responsible for creating diversity in domains of parasite proteins that have more inherent antigenicity. For example, the repeat domains, which are major antigenic regions for several protozoan parasites, exhibit intra- and interspecies species variability (McCuthan et al., 1988); in the human antibody response to the HA1 subunit of influenza virus hemagglutinin, the five dominant conformational B-cell epitopes correspond to sites of intrastrain polymorphism (Gerhard et al., 1981); the unique linear dominant epitope of gp 86, a 700 aa structural protein antigen of the human cytomegalovirus, a ubiquitous virus infecting the majority of the world population, is highly strain-specific (Urban et al., 1992).

Finally, recent reports suggest that genomic organization in multigene families could be a common feature of allergens. Allergens that have been shown to be encoded by members of multigene families include *Amb a I, Amb a II* (Rafnar et al., 1991; Bond et al., 1991; Griffith et al., 1991a). Poa p IX (Silvanovitch et al., 1991), *Lol p Ia* (Griffith et al., 1991b; Perez et al., 1990). The probability of any protein to belong to a multigene family is high, as it has been calculated that about one-quarter of the protein sequences present in current databanks belong to one of 120 protein families (Chothia, 1992). During evolution, proteins have diversified by point mutations, insertion, and deletion, as well as by processes that duplicate genes, creating new gene families and scrambling along the genome, segment of genes (exon shuffling). Proteins that have diversified in relatively recent times or have evolutionary conserved domains because of functional constraints, conserve a trace of these processes. Thus, for example, the phylogenetic analysis of catalytic domains conserved among 65 protein kinases of different species, ranging from yeast to man, has shown that domains conserved among species were also conserved among dispersed members of the family (Hansk et al., 1988). These genomic mechanisms may explain cross-reactivity among distinct allergens of the same species (Olsen et al., 1991).

Acknowledgment

This study was supported by the CNR Target Project: Biotechnology and Bioinstrumentation and by grant B102-CT92-0363 of the European Economic Community.

References

Atassi H., and M. Z. Atassi. 1986. Antibody recognition of ragweed allergen Ra3: localization of the full profile of the continuous antigenic sites by synthetic overlapping peptides representing the entire protein chain. *J. Immunol.* 16:229–235.

Ayuso, R., F. Polo, and J. Carreira. 1988. Purification of *Par j I* the major allergen of *P. judaica* pollen. *Mol. Immunol.* 25:49–56.

Benjamin, D. C., J. A. Berzofsky, I. J. East, E.R.N. Gurd, C. Hannum, S. J. Leach, E. Margoliash, J. G. Michael, A. Miller, E. M. Prager, M. Reichlin, E. E. Sercarz, S. J. Smith-Gill, P. E. Todd, A. C. Wilson. 1984. The antigenic structure of proteins: a reappraisal. *Ann. Rev. Immunol.* 2:67–101.

Berzofsky, J. A. 1988. Ir genes: Antigen-specific genetic regulation of immune response, in *The Antigens,* vol. VII, ed. M. Sela, pp. 1–113.

Bond, J. F., R. G. Garman, K. M. Keating, T. J. Briner, T. Rafnar, D. G. Klapper, and B. L. Rogers. 1991. Multiple *Amb a I* allergens demonstrate specific reactivity with IgE and T cells from ragweed allergic patients. *J. Immunol.* 146:3830–3853.

Bose, R., D. G. Marsh, J. Duchateau, A. H. Sehon, and G. Delespesse. 1984. Demonstration of autoantiidiotypic antibody crossreacting with public idiotypic determinants in the serum of rye sensitive allergic patients. *J. Immunol. 136:* 2474–2478.

Breiteneder, H., K. Pettenburger, A. Bito, R. Valenta, D. Kraft, H. Rumpold, O. Scheiner, and M. Breitenbach. 1989. The gene coding for the major birch pollen allergen *Bet v I* is highly homologous to a pea disease resistance response gene. *EMBO J.* 8:1935–1938.

Breiteneder, H., F. D. Ferreira, K. Hoffmann Sommergruber, C. Ebner, M. Breitenbach, D. Kraft, and O. Schneider. 1992. Cloning and expression in *E. coli* of four isoforms of *Cor a I* the major allergen of Hazel pollen. *Abstracts of Symposium on Molecular Biology and Immunology of Allergens,* Vienna, p. 20.

Brown, W. R., P. J. MacKinnon, A. Villasanté, N. Spurr, V. J. Buckle, and M. J. Dobson. 1990. Structure and polymorphism of human telomere-associated DNA. *Cell* 63:119–132.

Castracane, J. M. and R. E. Rocklin. 1988. Detection of human antiidiotypic antibodies (ab2): isolation and characterization of ab 2 in the serum of ragweed immunotherapy treated patient. *Int. Arch. Allergy Appl. Immunol.* 86:288–294.

Chothia, C. 1992. One thousand families for the molecular biologist. *Nature* 357:543–544.

Chua, K. Y., R. J. Dilworth, and W. R. Thomas. 1990. Expression of *D. pteronyssinus* allergens *Der p II* in *E. coli* and the binding studies with human IgE. *Int. Arch. Allergy Appl. Immunol.* 91:124–129.

Chua, K. T., W. K. Greene, P. Kehal, and W. R. Thomas. 1991. IgE binding studies with large peptides expressed from *Der p II* cDNA constructs. *Clin. Exp. Allergy* 21:161–166.

Corbi, A. L., V. Ley, F. Sanchez Madrid, and J. Carreira. 1985. Isolation of the major IgE binding protein from *P. judaica* pollen using monoclonal antibodies. *Mol. Immunol.* 22:1081–1089.

Corbi, A. L., Z. Pelaez, E. Errigo, and J. Carreira. 1985. Crossreactivity between *P. judaica and officinalis. Ann. Allergy* 54:142–147.

Coscia, M. R., O. Tarentino, U. Oreste, V. Santonastaso, A. Scotto d'Abusco, and A. Ruffilli. 1989. Immunoblot analysis of IgE binding components of *P. judaica* pollen and inflorescence extract. In *Pollinosis in the Mediterranean Area,* eds. G. D. Amato, A. Ruffilli, G. Sacerdoti, p. 202. Rocco Curto, Napoli.

Coscia, M. R., P. G. de Berardinis, C. de Santo, T. Menna, A. Ruffilli, A. Scotto d'Abusco, M. Tortora, and U. Oreste. 1992. Purification of a basic allergen from *P. officinalis* pollen. *Allergie et Immunologie* 24:55.

D'Amato, G., A. Ruffilli, and C. Ortolani. 1991. Allergenic significance of *Parietaria* pollen in *Allergenic Pollens and Pollinosis in Europe*, eds. G. D'Amato, E.T.M. Spieskma, and S. Bonini, pp. 113–118. Blackwell Scientific Publications, Oxford.

Dessain, A., S. T. Ju, M.E. Dorf, B. Benacerraf, and R. N. Germain. 1980. IgE response to synthetic polypeptide antigens. *J. Immunol.* 124:71–76.

Elsayed, S., K. Titlestad, J. Apold, and K. Aas. 1980. A synthetic exadecapeptide derived from allergen M. imposing allergenic and antigenic reactivity. *Scand. J. Immunol.* 12:171–175.

Elsayed, S., U. Ragnarsson, J. Apold, and E. Florvaag. 1981. Allergenic synthetic peptides corresponding to the second calcium-binding loop of cod allergen M. *Scand. J. Immunol.* 14:207–211.

Elsayed, S., L. Holen, and T. Dybendal. 1989. Synthetic allergenic epitopes from the amino-terminal regions of the major allergen of hazel and birch pollen. *Int. Arch. Allergy Appl. Immunol.* 89:410–419.

Esh, R. E. and D. G. Klapper. 1989. Isolation and characterization of a major crossreactive grass group I allergenic determinant. *J. Mol. Immunol.* 26:557–561.

Ford, S. A., B. A. Baldo, D. Geraci, and D. Bass. 1986. Identification of *P. judaica* pollen allergens. *Int. Arch. Allergy Appl. Immunol.* 79:120–126.

Geha, R. S. 1982. Elicitation of the Prausnitz-Kustner reaction by antiidiotypic antibodies. *J. Clin. Invest.* 69:735–741.

Gerhard, W., J. Yewdell, M. E. Fraenkel, and R. Webster. 1981. Antigenic structure of influenza virus hemagglutinin. *Nature* 290:713–717.

Giuliani, A., C. Pini, S. Bonini, N. Mucci, L. Ferroni, and G. Vicari. 1987. Isolation and purification of a major allergen from *P. officinalis* pollen. Allergy 42:434–440.

Gosh, B., M. P. Perry, and D. G. Marsh. 1991. Cloning the cDNA encoding the *Amb t V* allergen from giant ragweed (*Ambrosia trifida*) pollen. *Gene* 101:231–238.

Greene, W. K., K. Y. Chua, G. A. Stewart, and W. R. Thomas. 1990. Antigenic analysis of *gp I* house dust mite allergens using random fragments of *Der p I* expressed by recombinant DNA libraries. *Int. Arch. Allergy Appl. Immunol.* 92:30–38.

Greene, W. K., J. C. Cyster, K. Y. Chua, R. M. O'Brien, and W. R. Thomas. 1991. IgE and IgG binding of peptides expressed from fragments of cDNA encoding the major house dust mite allergen *Derp p I J. Immunol.* 147:3768–3773.

Greene, W. K. and W. R. Thomas. 1992. IgE binding structures of the major house dust allergen. *Der p I. Mol. Immunol.* 29:257–262.

Griffith, I. J., J. Pollock, B. L. Rogers, and A. K. Nault. 1991a. Sequence polymorphism of *Amb a I* and *Amb a II*, the major allergens in *A. artemisiifolia* (short ragweed). *Int. Arch. Allergy Apl. Immunol.* 96:296–304.

Griffith, I. J., P. M. Smith, J. Pollock, P. Theerakulpisut, A. Avjioglu, S. Davies, T. Hough, M. B. Singh, R. J. Simpson, L. D. Ward, and B. Knox. 1991b. b) Cloning and sequencing of *Lol p I*, the major allergenic protein of rye grass pollen. *FEBS Letters* 279:210–215.

Hanks, S. K., A. M. Quinn, and T. Hunter. 1988. The protein kinase family; conserved features and deduced phylogeny of the catalytic domain. *Science* 24142–24152.

Holley, L. H., J. Goudsmit, and M. Karplus. 1991. Prediction of optimal peptide mixtures to induce broadly neutralizing antibodies to human immunodeficiency virus type I. *Proc. Natl. Acad. Sci. USA* 88:6800–6804.

Ipsen, H., H. Bowadt, H. Janniche, B. N. Petersen, E. P. Munch, J. A. Wihl, H. Lowenstein. 1985. Immunochemical characterization of reference alder (*Alnus glutinosa*) and hazel (*Corylus avellana*) pollen extracts and the partial immunochemical identity between the major allergens of alder, birch and hazel pollens. *Allergy* 40:510–518.

Ispen, H., J. A. Wihl, B. Nuchel Petersen, and H. Lowenstein. 1992. Specificity mapping of patients IgE response towards the tree pollen major allergens *Aln g I, Bet v I* and *Cor a I. Clin. Exp. Allergy* 22:191–199.

Johnsen, G. and S. Elsayed. 1990. Antigenic and allergenic determinants of ovalbumin III MHC Ia binding peptide (OA 323-33) interacts with human and rabbit specific antibodies. *Mol. Immunol.* 27:821–827.

Leiferman, K. M. and G. J. Gleich. 1976. The crossreactivity of IgE antibodies with pollen antigens. I. Analysis of various species of grass pollens. *J. Allergy Clin. Immunol.* 58:128–139.

Leiferman, K. M., G. J. Gleich, and R. T. Jones. 1976. The crossreactivity of IgE antibodies with pollen antigens. II. Analysis of various species of ragweed and other fall weed pollens. *J. Allergy Clin. Immunol.* 58:140–147.

Le Mao, J., A. Meyer, J. C. Mazie, S. Rouyre, F. Marchand, A. Le Gall, and B. David. 1992. Identification of allergenic epitopes on *Der f I* a major allergens of *D. farinae*, using monoclonal antibodies *Mol. Immunol.* 29:205–211.

Lerner, R. A. 1984. Antibodies of predetermined specificity in biology and medicine. *Adv. Immunol.* 36:1–44.

Lin, Z., A.K.M. Ekramoddouullah, F. T. Kisil, J. Hebert, and W. Mourad. 1988. Isolation and characterization of *Poa p I* allergens with a murine monoclonal anti *Lol p I* antibody. *Int. Arch. Allergy Appl. Immunol.* 87:294–202.

Lin, Z., A.K.M. Ekramoddoullah, K. S. Jaggi, and J. Dzuba Fisher. 1990. Mapping of epitopes on *Poa p I* and *Lol p I* allergens with monoclonal antibodies. *Int. Arch. Allergy Appl. Immunol.*, 91:217–223.

Lombardero, M., P. W. Heymann, T.A.E. Platts-Mills, J. F. Fox, and M. Chapman. 1990. Conformational stability of B cell epitopes on group I and group II. *D. pteronyssinus* allergens. *J. Immunol.* 14:1353–1360.

Lowenstein, H. and D. G. Marsh. 1983. Antigens of *Ambrosia elatior* (short ragweed) pollen. III. Crossed radioimmunoelectrophoresis of ragweed allergic patients sera with special attention to quantification of IgE response. *J. Immunol.* 130:727–731.

Lowy, I., A. Prouvost-Danon, Abadie., A. Théze. 1980. Fine specificity and idiotypic analysis of the IgE response to the synthetic terpolmyer L-glutamic acid -L alanine -L tyrosine (GAT) and its dinitrophenyl conjugate (DNP-GAT). *J. Mol. Immunol.* 17:1033–1038.

Manfioletti, G. and C. Schneider. 1988. A new and fast method for preparing high quality lambda DNA suitable for sequencing. *Nucl. Acid. Res.* 16:2873–2884.

Marc-Series, I., Y. Boutin, E. R. Vrancken, and J. Hébert. 1990. Mapping of Bet *v* I epitopes using monoclonal antibodies. *Int. Arch. Allergy Appl. Immunol.* 92:226–232.

Marsh, D. G. 1990. Immunogenetic and immunochemical factors determining immune responsiveness to allergens: studies in unrelated subjects. In *Genetic and Environmental Factors in Clinical Allergy,* eds. D. G. Marsh and M. N. Blumenthal, pp. 97–123. University of Minnesota Press, Minneapolis, MN.

Mazur, G., X. Baur, and V. Liebers. 1990. Hypersensitivity to hemoglobins of the diptera family *Chironomidae:* structural and functional studies of their immunogenic/allergenic sites. In *Molecular Approaches to the Study of Allergens,* Monogr. Allergy, ed. B. A. Baldo, pp. 121–137. *28,* Karger, Basel.

McCuthan, T. F., F. V. de La Cruz, M. F. Good, and T. E. Wellems. 1988. Antigenic diversity in Plasmodium falciparum. *Prog. Allergy.* 41:173–192.

Menendez-Arias, L., J. Dominguez, I. Moneo, and R. Rodriguez. 1990. Epitope mapping of the major allergen from yellow mustard seeds. *Sin al. Mol. Immunol.* 27:143–150.

Mourad, W., S. Mecheri, G. Peltre, B. David, and J. Hebert. 1988. Study of the epitope structure of purified *Dac g I* and *Lol p I* the major allergens of *Dactilis glomerata* and *Lolium perenne* pollens, using monoclonal antibodies. *J. Immunol.* 141:3486–3491.

Mourad, D., M. Bernier, J. Jobin, and J. Heberet. 1989. Mapping of *Lol p I* allergenic epitopes by using murine monoclonal antibodies. *Mol. Immunol.* 26:1051–1057.

Mucci, N., P. Liberatore, R. Federico, F. Forlani, G. Di Felice, C. Afferni, R. Tinghino, F. De Cesare, and C. Pini. 1992. Role of carbohydrate moieties in cross-reactivity between different components of *P. judaica* pollen extract. *Allergy* 47:424–430.

Nagpal, S., K. N. Shanthi, R. Kori, H. Schroeder, D. D. Metcalfe, and P. V. Subba Rao. 1989. Induction of allergen specific IgE and IgG responses by antiidiotypic antibodies. *J. Immunol.* 142:3411–3415.

Olsen, E., L. Zhang, R. D. Hill, F. T. Kisil, A. H. Sehon and S. S. Mohapatra. 1991. Identification and characterization of *Poa p* IX Group of basic allergens of Kentucky bluegrass pollen. *J. Immunol.* 147:205.

Olson, J. R. and D. G. Klapper. 1986. Two major allergenic sites on ragweed pollen allergen antigen E identified by using monoclonal antibodies. *J. Immunol.* 136:2109–2115.

Oreste, U., M. R. Coscia, A. Scotto d'Abusco, V. Santonastaso, and A. Ruffilli. 1991. Purification and characterization of *Par o I,* major allergen of *P. officinalis* pollen. *Int. Arch. Allergy Appl. Immunol.* 96:19–27.

Parronchi, P., D. Macchia, M. P. Piccinni, P. Biswas, C, Simonelli, E. Maggi, M. Ricci, A. A. Ansari, and S. A. Romagnani. 1991. Allergen and bacterial antigen-specific T-cell clones established from atopic donors show a different profile of cytochine production. *Proc. Natl. Acad. Sci.* USA 88:4538–4542.

Perez, M., G. Y. Ishioka, L. E. Walker, and R. W. Chestnut. 1990. cDNA cloning and immunological characterization of the rye grass allergen *Lol p 1*. *J. Biol. Chem.* 265:16210–16215.

Platts-Mills, T.A.E., M. J. Snajdr, K. Ishizaka, and A. W. Frankland, 1978. Measurement of IgE antibody by antigen binding assay; correlation with PK activity and IgG and IgA antibodies to allergens. *J. Immunol.* 120:1201–1210.

Polo, F., R. Ayuso, M. Lombardeiro, O. Duffort, and J. Carreira. 1990. Separation of two polypeptide components of *Par j 1* allergen. *Clin. Exp. Allergy* 20:72.

Raffaelli, M. 1977. Note corologiche sulle specie Italiane del genere *Parietaria*. *Webbia* 31:49–68.

Rafnar, T., J. Griffith, M. Kuo, J. F. Bond, B. L. Rogers, and D. G. Klapper. 1991. Cloning of *Amb a I* (antigen E), the major allergen family of short ragweed pollen. *J. Biol. Chem.* 266:1229–1236.

Rogers, B. L., J. P. Morgenstern, I. J. Griffith, X. B. Yu, C. M. Counsell, A. W. Brauer, T. P. King, R. D. Garman, and M. C. Kuo. 1991. Complete sequence of the allergen *Amb a II*. Recombinant expression and reactivity with T cells from allergic patients. *J. Immunol.* 147:2547–2552.

Rohac, M., T. Birkner, I. Reimitzer, B. Bohle, R. Steiner, M. Breitenbach, D. Kraft, O. Scheiner, F. Gabl, and H. Rumpold. 1991. The immunological relationship of epitopes on major tree pollen allergens. *Mol. Immunol.* 28:897–906.

Saint-Remy, J.M.R., O. M. Lebrun, S. J. Lebecque, and O. L. Masson. 1986. The human immune response against major allergens from house dust mite, *D. pteronyssinus*. II. Idiotypic cross reactions of allergen specific antibodies. *Eur. J. Immunol.* 16:575–580.

Saint Remy, J.M.R., S. J. Lebecque, and P. M. Lebrun. 1988. Human immune response to the allergens of house dust mite *D. pteronyssimus*. III. Crossreactivity of bystander idiotypes on allergen specific IgE antibodies. *Eur. J. Immunol.* 18:77–81.

Sanger, F., S. Nicklen, and A. R. Coulson. 1977. DNA sequencing with chain terminating inhibitors. *Proc. Nat. Acad. Sci. USA* 74:5463–5469.

Scotto d'Abusco, A., U. Oreste, M. R. Coscia, G. Sacerdoti, O. Tarantino, D. Centis, R. Tosi, V. Santonastaso, and A. Ruffilli. 1989. Association between HLA DR specificity and IgE antibody response to allergens of the pollen of *Parietaria*. In *Pollinosis in the Mediterranean Area* eds. G. D'Amato, G. Sacerdoti, and A. Ruffilli, p. 70, Rocco Curto, Napoli.

Scotto, d'Abusco, A., U. Oreste, M. R. Coscia, V. Santonastaso, G. Sacerdoti, F. Lo Schiavo, and A. Ruffilli. 1990. Isolation of Poly(A)+ RNA from inflorescences of *P. judaica* and characterization of *in vitro* translated proteins. *Int. Arch. Allergy Appl. Immunol.* 91:411–416.

Scotto, d'Abusco, A., T. Menna, C. De Santo, M. R. Coscia, U. Oreste, and A. Ruffilli. 1992. Characterization of a cDNA encoding a major epitope of *Par o L*. In *Molecular Biology and Immunology of Allergens*, eds. D. Kraft and A. Sehon CRC Press, Boca Raton, FL, in press.

Siebenkotten, G., C. Esser, M. Wahl, and A. Radbruch. 1992. The murine IgG1/IgE class switch program. *Eur. J. Immunol.*, in press.

Silvanovich, A., J. Astwood, L. Zhang, E. Olsen, F. Kisil, A. Sehon, S. Mohapatra, and R. Hill. 1991. Nucleotide sequence analysis of three cDNA coding for *Poa p IX* isoallergens of Kentucky bluegrass pollen. *J. Biol. Chem.* 266:1204–1210.

Singh, M. M., T. Hough, P. Theerakulpisut, A. Avijoglu, S. Davies, P. M. Smith, P. Taylor, R. J. Simpson, L. D. Ward, J. Mc Cluskey, R. Puy, and B. Knox. 1991. Isolation and cDNA encoding newly identified major allergenic protein of rye-grass pollen: intracellular targeting to the amyloplast. *Proc. Natl. Acad. Sci. USA* 88:1384–1388.

Smith, J. J., J. R. Olson, and D. G. Klapper. 1988. Monoclonal antibodies to denatured ragweed pollen allergen. *Amb a I:* characterization specifically for the denatured allergen and utilization for the isolation of immunogenic peptides of *Amb a I. Mol. Immunol.* 25:355–.

Standring, R., V. Spackman, and S. J. Porter. 1987. Distribution of a major allergen of ryegrass (*Lolium perenne*) pollen among other species. *Int. Arch. Allergy Appl. Immunol.* 83:96–103.

Urban, M., W. Britt, and M. Mach. 1992. The dominant linear neutralizing antibody-binding site of glycoprotein gp 86 of human cytomegalovirus is strain specific. *J. Vir.* 66:1303–1311.

Valenta, R., M. Duchene, K. Pettenburger, C. Sillaber, P. Valent, P. Bettelheim, M. Breitenbach, H. Rumpold, D. Kraft, O. Scheiner, 1991a. Identification of profilin as a novel pollen allergen: IgE autoreactivity in sensitized individuals. *Science* 253:557–559.

Valenta, R., M. Duchene, S. Vitala, T. Birkner, C. Ebner, R. Hirschwehr, M. Breitenbach, H. Rumpold, O. Scheiner, and D. Kraft. 1991b. Recombinant allergens for immunoblot diagnosis of tree pollen allergy *J. Allergy Clin. Immunol.* 88:889–894.

Valenta, R., M. Duchene, C. Ebner, P. Valent, C. Sillaber, P. Deviller, F. Ferreira, M. Tejkl, H. Edelman, D. Kraft, and R. Scheiner. 1992a. Profilin constitutes a novel family of functional plant pan-allergens. *J. Exp. Med.* 175:377–385.

Valenta, R., S. Vrtala, C. Ebner, D. Kraft, and O. Scheiner. 1992b. Diagnosis of grass pollen allergy with recombinant Timothy grass (*Phleum pratense*) pollen allergens. *Int. Arch. Allergy Immunol.* 97:287–294.

van't Hof, W., P. C. Drieduk, M. van den Berg, A. G. Beck-Sickinger, G. Jung and R. C. Alberse. 1991. Epitope mapping of the *D. pteronyssinus* house dust mite major allergen *Der o I* using overlapping synthetic peptides. *Mol. Immunol.* 28:1225–1232.

Weber, E. A., C. H. Heusser, and K. Blaser. 1986. Fine specificity and idiotype expression of antiphosphorylcholine IgE and IgG antibodies. *Eur. J. Immunol.* 16:919–923.

Yssel, H., K. E. Johnson, P. V. Schneider, J. Wideman, A. Terr, R. Kastelein, and J. E. De Vries. 1992. T cell activation inducing epitopes of the house dust mite allergen *Der p I.* Proliferation and lymphokine patterns by *Der p I* specific CD4+ T cell clones. *J. Immunol.* 148:738–745.

14

Profilin: A Novel Pan-Allergen and Actin-Binding Protein in Plants

Rudolf Valenta, Ines Swoboda, Monika Grote,
Susanne Vrtala, Fatima Ferreira, Michael Duchêne,
Erwin Heberle-Bors, Dietrich Kraft, and Otto Scheiner

Introduction

Using serum-IgE from a pollen-allergic patient we isolated a cDNA from a birch pollen expression library (Valenta et al., 1991a). The deduced amino acid sequence of this cDNA clone showed significant homology with sequences from other eukaryotic profilins. Profilins function as actin-binding proteins (Isenberg et al., 1980) and interact with phosphatidylinositol 4,5 bisphosphate (PIP_2) (Goldschmidt-Clermont et al., 1990). In addition a role of profilins in the signal transduction cascade via CAP was proposed (Vojtek et al., 1991). These data indicate that profilins are essential components of eukaryotic cells. This chapter concerning the biological functions of plant profilins illustrates that these proteins are ubiquitously distributed in plants and are structurally similar. For this reason plant profilins can function as targets for cross-reactive IgE antibodies of allergic patients and represent a novel class of pan-allergens.

Purification of Plant Profilins

During the purification of proline hydroxylase by affinity chromatography on poly-(L-proline)-Sepharose it was found that profilin and the profilactin complex also bound to this matrix. Poly-(L-proline)-Sepharose can therefore be used to purify profilins via a single step affinity procedure. Under physiological conditions profilin binds to Sepharose coupled poly-(L-proline) and can be eluted using buffers containing urea or dimethylsulphoxide (Tanaka and Shibata, 1985; Lindberg et al., 1988). Figure 14-1 shows Coomassie blue stained gels containing poly-(L-proline) affinity purified profilins from human platelets, *Entamoeba histolytica* trophozoites, timothy grass (*Phleum pratense*) pollen, and *Escherichia coli* expressing recombinant birch profilin. Using this procedure we were able

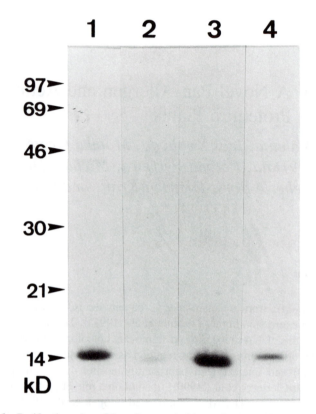

Figure 14-1 Purification of profilins. Coomassie blue stained SDS-polyacrylamide gels of poly-(L-Proline) affinity purified profilins from man (lane 1), *Entamoeba histolytica* trophozoites (lane 2), timothy grass (*Phleum pratense*) pollen (lane 3), *E. coli* expressing recombinant birch profilin (lane 4).

to purify profilins from pollens of distantly related plants such as trees, grasses, and weeds. All these profilins share IgE epitopes and therefore represent a novel class of pan-allergens (Valenta et al., 1992a). Buffers that maintain the profilactin complex during the protein extraction allow isolation of profilin and actin via poly-(L-proline) affinity indicating that profilin is an actin-binding protein also in plants (Valenta et al., 1993).

Expression of Profilin in Different Plant Tissues and Under Different Stimuli

Using profilin specific antibodies we investigated the distribution of profilin in several plant tissues. Figure 14-2 shows that profilin can be detected in pollen, leaves, and female inflorescences of white birch. In pollen two bands at 14 kD

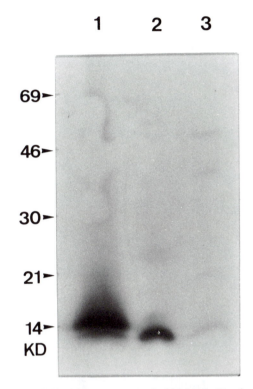

Figure 14-2 Expression of profilin in tissues of white birch (*Betula verrucosa*). Immunoblot containing protein extracts from birch tissues probed with a rabbit anticelery profilin antibody. Lane 1, pollen proteins; lane 2, leaf proteins; and lane 3, proteins from female inflorescences.

and 13 kD are detected whereas in leaves and female inflorescences only the 13 kD band was found. A differential expression of profilin isoforms in the tested tissues would be consistent with the observation of an expression of profilin isoforms in *Physarum polycephalum* (Binette et al., 1990).

Figure 14-3 shows that in birch callus, which was grown on plates either in light or in darkness, only the 13 kD profilin could be detected. Suspension cultures of birch callus treated with a series of agonists such as dichlorophenoxacetic acid (2,4-D), light, $CaCl_2$, cytochalasin B, phosphatidylinositol 4,5 bisphosphate (PIP_2), epidermal growth factor, or hydroxyurea had no effect on the protein expression pattern. Figure 14-4 shows that in all protein extracts a similar 13-kD band was detected as in the solid calli.

These data suggest a regulation of profilin function by redistribution and binding to different partners such as actin or PIP_2 rather than by different transcription or translation. An occurrence of profilin isoforms in different tissues that have different affinities to actin or PIP_2 would be consistent with data describing

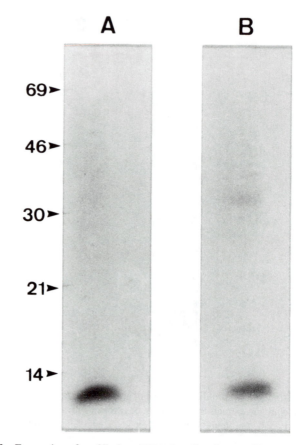

Figure 14-3 Expression of profilin in solid birch callus. Immunoblot containing proteins extracts from solid birch calli was probed with profilin-specific serum-IgE. Lane 1, proteins from green callus grown in light and lane 2, proteins from white callus grown in the dark.

the expression of actin isoforms in plant tissues (Meagher, 1991) and the observation that profilins bind with different affinity to actin isoforms in chicken (Ohshima et al., 1989). Moreover, recently a developmental upregulation of profilin expression during tobacco pollen maturation was reported (Mittermann et al. 1995).

Immunohistochemical Localization of Profilin in Pollen

Reports on the possible role of the dissociation of the profilactin complex in the initiation of the acrosomal reaction in *Thyone sperm* (Tilney and Inoue, 1982; Tilney et al., 1983) led us to study profilin-actin interactions in pollens during germination. In vitro germinated tobacco pollen (Benito-Moreno et al., 1988a, 1988b) was chosen as a model and the distribution of profilin and actin in resting

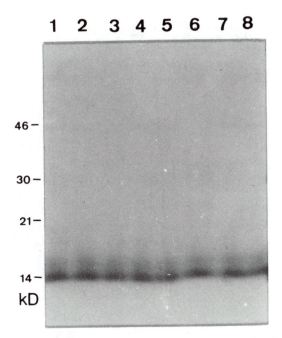

Figure 14-4 Expression of profilin in birch callus suspension cultures. Protein extracts from stimulated birch suspension callus cultures were separated by SDS-PAGE (Laemmli, 1970), blotted to nitrocellulose and probed with a rabbit anticelery profilin antibody. Lane 1, callus without addition of agonist; lane 2, dichlorphenoxyacetic acid (2,4-D); lane 3, light; lane 4, $CaCl_2$; lane 5, cytochalasin B; lane 6, phosphatidylinositol 4,5 bisphosphate (PIP_2); lane 7, epidermal growth factor and lane 8, hydroxyurea.

and germinating pollen was studied by immune-electronmicroscopy. In particular we observed actin staining in the middle pectic layer of the protruded intine, which had already been described by Heslop-Harrison as containing "protein-tubules" (Heslop-Harrison, 1987). Pollen tube sections showed continuous actin staining in the inner layer of the pollen tube wall and in the cytoplasm of the pollen tube (Grote et al., 1995). Profilin was mainly detected in the cytoplasm of the pollen tube. Figure 14-5A shows a cross-section of a tobacco pollen tube probed with a rabbit antiprofilin antibody and Figure 14-5B shows a longitudinal control section incubated with normal rabbit serum.

One possible interpretation of the immunolocalization of profilin and actin during pollen germination would suggest that the dissociation of the profilactin complex initiates pollen tube growth by actin polymerization.

Sequence and Structure Similarities of Profilins

Figure 14-6 illustrates the amino acid sequence similarity of different profilins. The sequence data were used to calculate a tree reflecting the similarities of

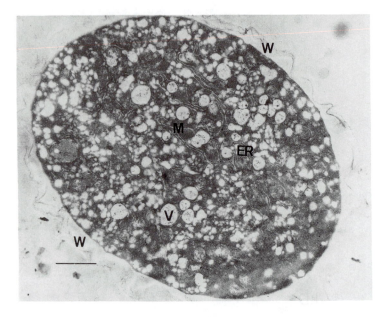

5A

Figure 14-5 Immune-electronmicroscopical localization of profilin in tobacco pollen tubes. *A*, Cross-section of a tobacco pollen tube probed with affinity purified rabbit antibirch profilin antibody.

different profilin amino acid sequences (Felsenstein, 1988). Profilins from birch and timothy grass share 80% sequence identity whereas a lower sequence identity of plant profilins with other eukaryotic profilins in the range of 35% was found. Despite the different degree of similarity of profilins, we were able to demonstrate that plant profilins not only bind to plant actins but also to animal actins (Valenta et al., 1993). In first experiments to identify regions of plant profilins involved in actin binding we could identify the profilin C-terminus as an actin-binding domain. The importance of this region was already suggested by chemical cross-linking of profilin and actin in *Acanthamoeba* (Vandekerckhove et al., 1989).

Profilins Represent a Novel Class of Plant Pan-allergens

Because of the ubiquitous occurrence of profilins in eukaryotic cells and their structural similarity, profilins from different sources are capable of eliciting allergic reactions in sensitized patients. The clinically most relevant allergic cross-reactions are observed with profilins from plant origin (pollen, plant derived food) (Valenta et al. 1991b, 1992a, 1992b, Vallier et al. 1992, van Ree et al.,

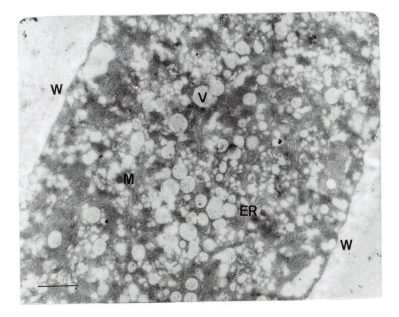

5B

Figure 14-5 B, Longitudinal control section of a pollen tube probed with a normal rabbit serum purified as the antibody. Fixation was done in aqueous paraformaldehyde. Abbreviations used are: V, vacuole; W, pollen tube wall; ER, endoplasmatic reticulum; M, mitochondrium. The bar represents 1μm. Detection of bound antibodies was done with the method using protein A coupled to gold particles as described (As described in Grote, 1991).

1992, Hirschwehr et al., 1992). About 20% of all pollen allergic patients show reactivity to plant profilins, which we therefore designated as plant pan-allergens. In addition profilins specifically bind to patients' IgE and IgG and release histamine from basophils of sensitized patients via cross-link of high affinity Fcε-receptors (Valenta et al., 1992c).

In contrast to other allergens, profilins can be found in monocotyledonic and dicotyledonic plants and are therefore responsible for allergic cross-reactivities towards taxonomically distantly related plant species. The phenomenon of allergic cross-reactivities toward profilins may be one example of how the biological functions of an allergen are linked to clinical manifestations observed in allergic patients.

Acknowledgments

This study was supported by grant 3/8167 of the "Fonds zur Förderung der gewerblichen Wirtschaft" and by grants S06703MED, S6002-Bio and S6003-Bio of the "Fonds zur

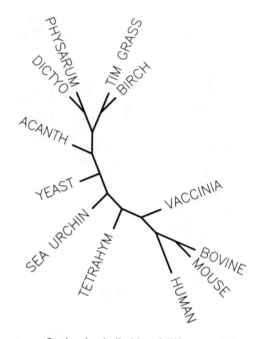

Figure 14-6 Gene tree reflecting the similarities of different profilin sequences calculated according to Felsenstein (1988). A profilin sequence derived from vaccinia virus (Blasco et al., 1991), which is closely related to mammalian profilins, was included. The plant profilin sequences [birch, timothy grass (TIM GRASS)] show greatest sequence homology to the profilin sequences derived from *Physarum* (PHYSARUM), *Dictyostelium* (DIC-TYO) and *Acanthamoeba* (ACANTH), less homology to the profilins of *Saccharomyces cerevisiae* (YEAST), *Tetrahymena piriformis* (TETRAHYM) and sea urchin (SEA UR-CHIN). The mammalian profilins are most distantly related.

Förderung der wissenschaftlichen Forschung" Vienna, Austria. We thank P. Deviller, Immuno-Virologie Moleculaire et Cellulaire, Lyon for providing us a rabbit anticelery profilin serum.

References

Benito Moreno, R. M., F. Macke, M. T. Hauser, A. Alwen, and Heberle-Bors, E. 1988a. Sporophytes and male gametophytes from in vitro cultured immature pollen. In *Sexual Reproduction in Higher Plants*, eds. M. Cresti, P. Gori, and E. Pacini, pp. 137–142. Springer Verlag, New York, Berlin, Heidelberg.

Benito Moreno, R. M., F. Macke, A. Alwen, and E. Heberle-Bors. 1988b. *In-situ* seed production after pollination with *in-vitro*-matured, isolated pollen. *Planta* 176:145–148.

Binette, F., M. Bénard, A. LaRoche, G. Pierron, G. Lemieux, and D. Pallota. 1990. Cell-specific expression of a profilin gene family. *DNA and Cell Biol.* 9:323–334.

Blasco, R., N. B. Cole, and B. Moss. 1991. Sequence analysis, expression, and deletion of a vaccinia virus gene encoding a homolog of profilin, a eukaryotic actin-binding protein. *J. Virol* 65:4598–4608.

Felsenstein, J. 1988. Phylogenies from molecular sequences: inference and reliability. *Ann. Rev. Genetics* 22:521–565.

Goldschmidt-Clermont, P. J., L. M. Machesky, J. J. Baldassare, and T. D. Pollard. 1990. The actin-binding protein profilin binds to PIP₂ and inhibits its hydrolysis by phospholipase C. *Science* 247:1575–1578.

Grote, M. 1991. Immunogold electro microscopy of soluble proteins: Localization of *Bet v* I major allergen in ultra-thin sections of birch pollen after anhydrous fixation techniques. *J. Histochem. Cytochem.* 39:1395–1401.

Grote, M., I. Swoboda, R. B. Meagher, and R. Valenta. 1995. Localization of profilin and actin-like immunoreactivity in *in vitro* germinated tobacco pollen tubes by electron microscopy after special water-free fixation techniques. *Sex. Plan Reprod.* 8:180–186.

Heslop-Harrison, J. 1987. Pollen germination and pollen-tube growth. In Pollen Cytology and Development, eds. K. L. Giles and J. Prakash, pp. 1–78. Academic Press, Orlando, FL.

Hirschwehr, R., R. Valenta, C. Ebner, F. Ferreira, M. Rohac, H. Rumpold, O. Scheiner, and D. Kraft. 1992. Identification of common allergenic structures in hazel pollen and hazelnuts: A possible explanation for sensitivity to hazelnuts in tree pollen allergic patients. *J. Allergy Clin. Immunol.*, 90:927–936.

Isenberg, G., U. Aebi, and T. D. Pollard. 1980. An actin-binding protein from *Acanthamoeba* regulates actin filament polymerization and interactions. *Nature* 288:455–459.

Laemmli, U. K. 1970. Cleavage of structural proteins during the assembly of the head of bacteriophage T4. *Nature* 227:680–685.

Lindberg, U., C. E. Schutt, E. Hellsten, A.-C. Tjäder, and T. Hult. 1988. The use of poly(L-proline)-Sepharose in the isolation of profilin and profilactin complexes. *Biochim. Biophys. Acta.* 967:391–400.

Meagher, R. B. 1991. Divergence and differential expression of actin gene families in higher plants. *Int. Rev. Cytol.* 125:139–163.

Mittermann, I., I. Swoboda, E. Pierson, N. Eller, D. Kraft, R. Valenta, and E. Heberle-Bors. 1995. Molecular cloning and characterization of profilin from tobacco (*Nicotiana tabacum*): increased prolifin expression during pollen maturation. *Plant Mol. Biol.* 27:137–146.

Ohshima, S., H. Abe, and T. Obinata. 1989. Isolation of profilin from embryonic chicken skeletal muscle and evaluation of its interaction with different actin isoforms. *J. Biochem.* 105:855–857.

Tanaka, M., and H. Shibata. 1985. Poly(L-proline)-binding proteins from chick embryos are a profilin and a profilactin. *Eur. J. Biochem.* 151:291–297.

Tilney, L. G., and S. Inoue. 1982. Acrosomal reaction of *Thyone* sperm. II. The kinetics and possible mechanism of acrosomal process elongation. *J. Cell Biol.* 93:820–827.

Tilney, L. G., E. M. Bonder, L. M. Coluccio, and M. S. Mooseker. 1983. Actin from *Thyone* sperm assembles on only one end of an actin filament: a behavior regulated by profilin. *J. Cell Biol.* 97:112–124.

Valenta, R., M. Duchêne, K. Pettenburger, C. Sillaber, P. Valent, P. Bettelheim, M. Breitenbach, H. Rumpold, D. Kraft, and O. Scheiner. 1991a. Identification of profilin as a novel pollen allergen; IgE autoreactivity in sensitized individuals. *Science* 253:557–560.

Valenta, R., M. Duchêne, S. Vrtala, T. Birkner, C. Ebner, R. Hirschwehr, M. Breitenbach, H. Rumpold, O. Scheiner, and D. Kraft. 1991b. Recombinant allergens for immunoblot diagnosis of tree-pollen allergy. *J. Allergy Clin. Immunol.* 88:889–894.

Valenta, R., M. Duchêne, C. Ebner, P. Valent, C. Sillaber, P. Deviller, F. Ferreira, M. Tejkl, H. Edelmann, D. Kraft, and O. Scheiner. 1992a. Profilins constitute a novel family of functional plant pan-allergens. *J. Exp. Med.* 175:377–385.

Valenta, R., S. Vrtala, C. Ebner, D. Kraft, and O. Scheiner. 1992b. Diagnosis of grass pollen allergy with recombinant timothy grass (*Phleum pratense*) pollen allergens. *Int. Arch. Allergy Immunol.* 97:287–294.

Valenta, R., W. R. Sperr, F. Ferreira, P. Valent, C. Sillaber, M. Tejkl, M. Duchêne, C. Ebner, K. Lechner, D. Kraft, and O. Scheiner. 1993. Induction of specific histamine release from basophils with purified natural and recombinant birch pollen allergens. *J. Allergy Clin. Immunol.*, 91:88–97.

Valenta, R., F. Ferreira, M. Grote, I. Swoboda, S. Vrtala, M. Duchêne, P. Deviller, R. B. Meagher, E. McKinney, E. Heberle-Bors, D. Kraft, and O. Scheiner. 1993. Identification of profilin as an actin-binding protein in higher plants. *J. Biol. Chem.* 268:22777–22781.

Vallier, P., C. Dechamp, R. Valenta, O. Vial, and P. Deviller. 1992. Purification and characterization of an allergen from celery immunochemically related to an allergen present in several other plant species. Identification as a profilin. *Clin. Exp. Allergy* 22:774–782.

Vandekerckhove, J. S., D. A. Kaiser, and T. D. Pollard. 1989. Acanthamoeba actin and profilin can be cross-linked between glutamic acid 364 of actin and lysine 115 of profilin. *J. Cell Biol.* 109:619–626.

Van Ree, R., V. Voitenko, W. A. van Leeuwen, and R. C. Aalberse. 1992. Profilin is a cross-reactive allergen in pollen and vegetable foods. *Int. Arch. Allergy Immunol.* 98:97–104.

Vojtek, A., B. Haarer, J. Field, J. Gerst, T. D. Pollard, S. Brown, and M. Wigler. 1991. Evidence for a functional link between profilin and CAP in the yeast S. cerevisiae. *Cell* 66:497–505.

Index